FRACTIVISM

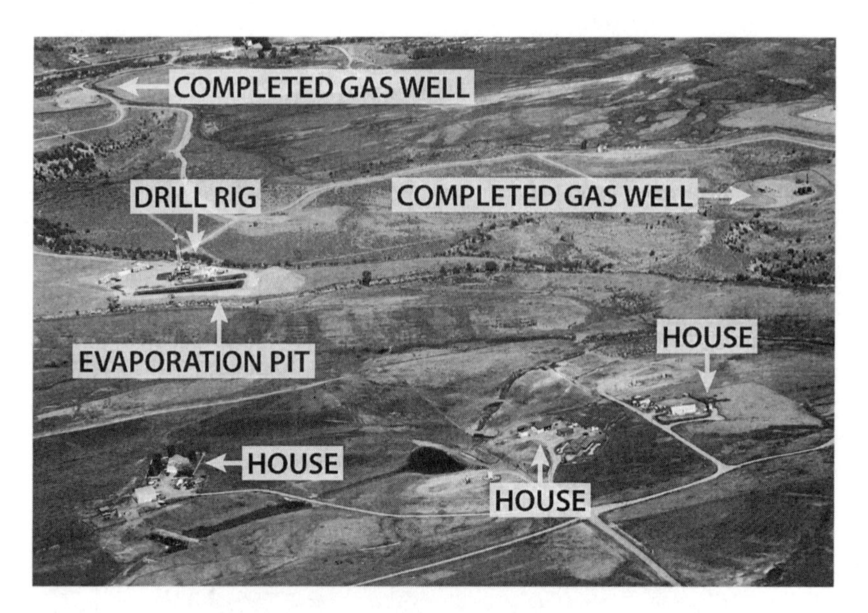

Oil and gas wells next to homes in Colorado, 2006.

Experimental Futures

TECHNOLOGICAL LIVES, SCIENTIFIC ARTS, ANTHROPOLOGICAL VOICES
A series edited by Michael M. J. Fischer and Joseph Dumit

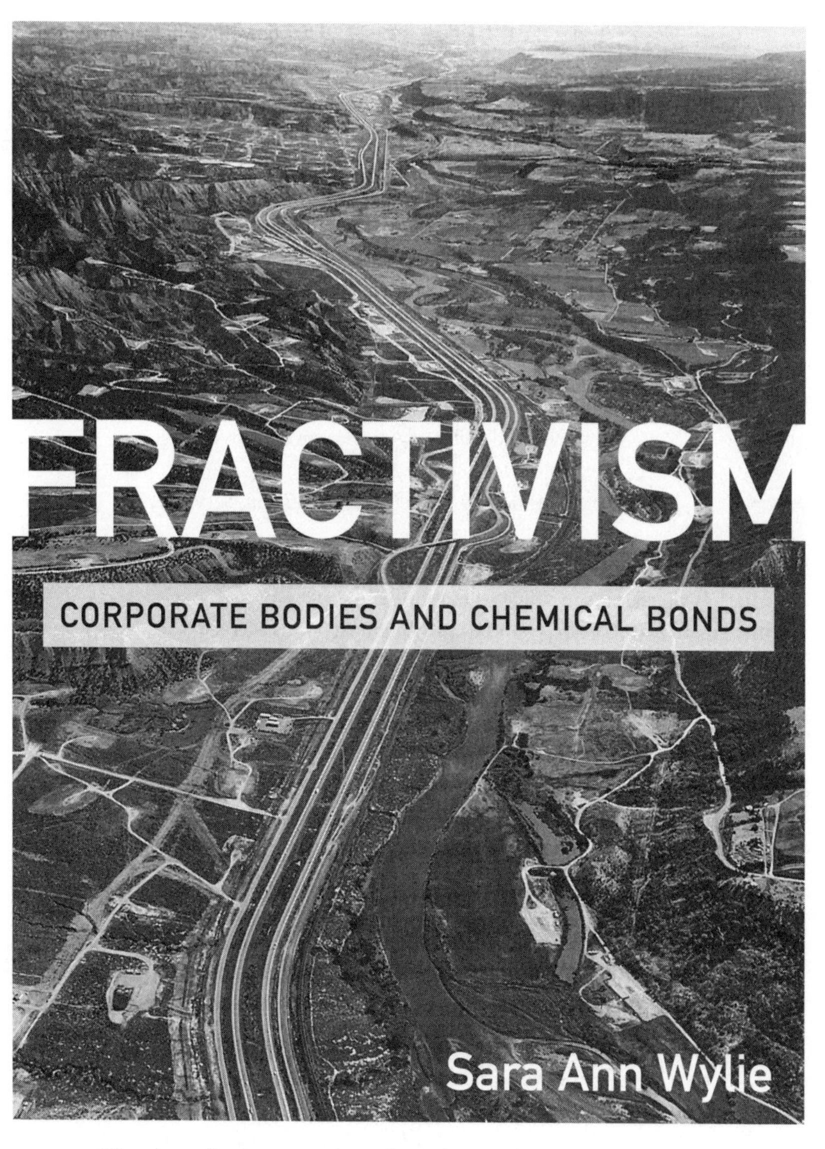

Oil and gas development along the Colorado River and I-70 in 2006.

Duke University Press

DURHAM AND LONDON 2018

Printed in the United States of America on acid-free paper ∞
Designed by Matt Tauch
Typeset in Minion by Westchester Publishing Services

Library of Congress Cataloging-in-Publication Data
Names: Wylie, Sara Ann, [date—] author.
Title: Fractivism : corporate bodies and chemical bonds / Sara Wylie.
Description: Durham : Duke University Press, 2018. | Series: Experimental futures | Other title, changed in ECIP : "Shale Gas" | Includes bibliographical references and index.
Identifiers: LCCN 2017036329 (print) | LCCN 2017056103 (ebook)
ISBN 9780822372981 (ebook)
ISBN 9780822363828 (hardcover : alk. paper)
ISBN 9780822369028 (pbk. : alk. paper)
Subjects: LCSH: Hydraulic fracturing—Health aspects. | Hydraulic fracturing—Environmental aspects.
Classification: LCC TN871.255 (ebook) | LCC TN871.255 .W955 2018 (print) | DDC 338.2/7283—dc23
LC record available at https://lccn.loc.gov/2017036329

Cover art: Oil and gas development along the Colorado River and I-70 in 2006.
Photo courtesy of TEDX.

Fractivism is a neologism combining "fracking" and "activism."

As WellWatch and Landman Report Card are no longer online,
materials from these websites can be found on Sara Wylie's website:
https://sarawylie.com/publications/fractivism-corporate-bodies-and-chemical-bonds/.

This book is dedicated to those who made it possible:

THEO COLBORN—never have I encountered a more powerful personality, a being so bent on transforming the world around herself. She was a mentor and an inspiration. I feel honored to have known her. I hope this book does justice to her work and spirit.

DR. JOAN RUDERMAN—without the hours spent in her lab, without the grudging trip I took to the library to dig up *Our Stolen Future*, without our conversations muddling through endocrine disruption and assays, this project would never have begun.

CHRIS CSIKSZENTMIHÁLYI—without Chris's invitation to collaborate on ExtrAct, I would not have made the move to "making" as well as writing critique. He fundamentally changed my approach to scholarship; he gave me the concrete tools, confidence, and opportunity to begin putting STS into practice.

This project has been years in the making; I had my first interview with Theo in 2005. Along the way, countless people have played pivotal roles in shaping its and my own evolution. My hope is that the book itself does justice to their contributions because I can't write them all out here without writing a second book. This work is dedicated to all of them, to the community organizers, landowners, academics, artists, scientists, engineers, programmers, friends, and family who daily work in micro and macro ways to build more equitable, just, and responsible futures. May this work contribute one more ripple toward changing how we collectively inhabit this amazing world.

CONTENTS

PREFACE

Unconventional oil and natural gas extraction in the early twenty-first century has transformed social, physical, economic, legal, and biological landscapes in the United States. From 2000 to 2014, shale gas production increased from near zero to approximately 40 billion cubic feet per day, making the United States the world's largest natural gas and oil producer (Smith 2014; EIA 2015). The dramatic growth of U.S. oil and gas prices produced record low prices for oil and gas in 2015, leading to declines in production, job losses, and bankruptcies (Frazier 2016; Hunn 2016; Scheyder 2016). Hydraulic fracturing (fracking) that entails high-pressure injection of synthetic chemical mixtures into subsurface formations made drawing oil and natural gas from previously unreachable reserves possible. Thirty-two states now produce unconventional fossil fuels (EIA 2013b) at an unprecedented scale (EIA 2013a). Extraction from the Brakken Shale in North Dakota produced new human settlements visible from space (J. Amos 2012; Swanson 2014).

Industry and regulators promote unconventional gas and oil production as the key to U.S. energy independence and as a bridge to a low-carbon economy (MITEI 2011; Graves 2012; G. Zuckerman 2013). But there is another side to the story. In 2014, New York State became the first state rich in unconventional natural gas to ban fracking due to human health and environmental concerns (Lustgarten 2014). Debates over fracking are being fought state by state everywhere that the practice spreads; the 8,696 unconventional shale wells drilled in Pennsylvania from 2000 to 2014 incurred 5,983 violations from the Department of Environmental Protection (Kelso 2014). Earthquakes in Alabama, Ohio, Oklahoma, and Texas are linked to the injection of fracking wastewater—3.5 billion barrels in Texas alone in 2011 (up from 46 million in 2005) (Hargrove 2014). Research in Pennsylvania, Texas, and

Colorado shows fracking-contaminated ground and surface water is destroying the lives and livelihoods of landowners (Jackson et al. 2013; Warner et al. 2013; Darrah et al. 2014; Kassotis et al. 2014). Furthermore, natural gas may be no more "green" than coal. Life-cycle analyses of its production reveal that tons of methane are released into the atmosphere when natural gas is processed and transported (Howarth, Santoro, and Ingraffea 2011; Karion et al. 2013; S. Miller et al. 2013; Pétron et al. 2014). Methane is a far more potent greenhouse gas than carbon dioxide (Myhre et al. 2013).

This book explores the emergence of both the gas boom and its controversies, offering innovative scientific approaches to studying gas extraction's harmful impacts on human health and the environment. Via participant observation within a small scientific advocacy organization, The Endocrine Disruption Exchange (TEDX), I follow the development of the first database of chemicals used in natural gas extraction, a database that documents not only the (often proprietary) constituents of fracking chemicals but also their bodily and ecological effects. My ethnographic analyses of TEDX's database demonstrate how it transformed an information vacuum around fracturing into fierce regional and national debates about the public health effects of this activity.

Expanding on TEDX's databasing methodology, the book describes the research, development, and impacts of a set of online, user-generated databasing and mapping tools designed to interconnect communities encountering the corporate forces and chemical processes animating gas development. Fracking is an intensive, technological practice that requires the delicate calibration of corporate, governmental, and legal apparatuses in order to proceed. The industry operates at county, state, federal, and international levels, and it has successfully organized regulatory environments suited to rapid and lucrative gas extraction. Amid such multiscalar forces, communities have little legal or technical recourse if they have been subjected to chemical and corporate influences that undermine their financial, bodily, and social security. ExtrAct, a research group I cofounded and directed with the artist and technologist Chris Csikszentmihályi, sought to empower isolated local communities by developing a suite of online mapping and databasing tools through which gas-patch communities exposed to fracking could share information, network, study, and respond to industry activity across states. Using ExtrAct as an example, I explore how social sciences and the academy at large can invest in developing research tools, methods, and programs designed for noncorporate ends to help redress the informational and technical imbalances faced by communities dealing with large-scale multinational industries.

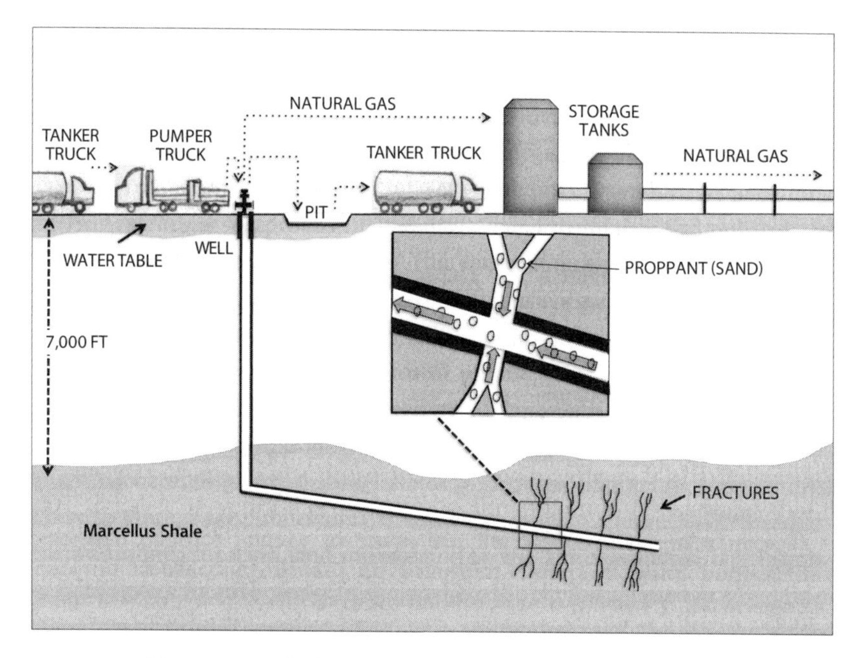

FIGURE P.1. The process of hydraulic fracking.

In unconventional gas reserves, gas is distributed throughout a porous matrix like coal or sandstone. Fracking creates a route through which this gas can be brought to the surface (EPA 2004) by injecting a mixture of synthetic chemicals under extremely high pressure into the porous matrix. A high-volume hydraulic frack is a large-scale industrial operation. A cavalcade of 18-wheeler trucks bearing containers filled with the thousands of gallons of fluid and associated machinery necessary for the procedure draws up to a frack site (see figure P.1). These containers are arranged around the "head" of the well to be fracked and then connect to the wellhead to form what looks like an octopus of piping.

Within the containers, blending machinery mixes fracturing fluids composed of dry or nondiluted stores of chemicals and other materials such as proppants, sand, or other grainy materials used to prop the fractures open (Montgomery and Smith 2010: 30).[1] This mixture is forced into the well by powerful diesel engines capable of producing 15,000 hhp (hydraulic horsepower), which is roughly equivalent to about twenty-three 650 hp 18-wheel truck engines roaring to life as a frack operation begins (Montgomery and Smith 2010: 30). A frack operation can continue for many hours as the mixture is pumped underground at a high-enough pressure to break pathways in the subsurface gas-bearing layer thousands of meters below the surface. The force of these fluids generates a mini-seismic event. A single frack can require one million gallons of fluid, and a well might be fracked from three to 40 times during its life cycle. On average, each horizontally drilled well in the United States undergoes 10 fracking cycles (Montgomery and Smith 2010: 27, 28, 35). Most horizontally drilled wells in the Marcellus Shale region, the largest natural gas-producing region in the United States in 2013, extend two kilometers below the surface and over a kilometer into shale beds (Kemp 2014).

The combination of horizontal drilling and fracking has transformed oil and gas resource extraction. Rather than sinking individual wells into pockets of gas or oil, the oil and gas industry can now collect fossil-fuel resources from across a formation, drawing gas and oil from kilometers of pipe drilled laterally through a shale bed.

ACKNOWLEDGMENTS

Now to the practical and less poetic acknowledgments: first, I sincerely thank the academics who inspired this project methodologically, theoretically, and materially; chief among them are (in alphabetical order) Joseph Dumit, Mike Fischer, Kim Fortun, Stephen Helmreich, Susan Silbey, and Chris Walley.

At each stage of this work, I was also fortunate to work with research groups whose conversation and collaborations inspired the project's development:

- BioGroup at MIT: Sophia Roosth, Natasha Myers, and Etienne Benson, with leadership from Stephan Helmreich. We delved into questions of sensing and making sense that permeate this whole narrative.
- ExtrAct: Of course, this group is central to this project. The ExtrAct Team, particularly Chris Csikszentmihályi, Dan Ring, Christina Xu, Matt Hockenberry, Lisa Sumi, Jennifer Goldman, and Tara Meixsell, as well as our great research assistants and interns—I know I don't do justice to your characters and efforts here; would that there could be more volumes. Also thanks to the rest of the Computing Culture group for your inspirational work performing critique, and thanks to the larger community of the Center for Future Civic Media for not just asking but trying to answer the question of how to build infrastructures for informed communities and democratic accountability.
- The Rhode Island School of Design (RISD) Environmental Justice Research Cluster: Special thanks to Kelly Dobson for creating the opportunity for me to teach at RISD and lead a research group as well as inspiring me with her thought-provoking work and lasting friendship. Also to Jeff Warren who co-led the group and Megan

McLaughlin who worked tirelessly on the hydrogen sulfide monitoring project.

- Public Lab: Nothing makes me prouder than my part in cofounding Public Lab, an online community for developing open-source hardware and software for community-based environmental monitoring. Many of the ideas behind Public Lab animate the ending of this book; thanks to fellow cofounders Liz Barry, Shannon Dosemagen, Adam Griffith, Stewart Long, Matt Lippincott, and Jeff Warren. It is amazing to see this seed bearing fruit.
- The Social Science Environmental Health Research Institute (SSEHRI) at Northeastern University: SSEHRI has been a welcome academic home, allowing for drafts of multiple chapters to be reviewed. I'd particularly like to thank Phil Brown for welcoming me to SSEHRI. Special thanks also to Len Albright, who coauthored chapter 10 of this book, and to the other members of our hydraulic fracturing research group, particularly Elisabeth Wilder, Jacob Matz, and Lauren Richter. Deep thanks to Lourdes Vera for her amazing support in figuring out the book's figures and her fantastic research work on the hydrogen sulfide monitoring project.

Beyond participating in these research collectives, contributing to three edited volumes helped connect me and this book to the larger scholarly efforts in social sciences to comprehend, track, and reform unconventional energy extraction and to develop engaged forms of scholarship. I thank all of the authors and editors of *Subterranean Estates: Life Worlds of Oil and Gas*, particularly Michael Watts, Hannah Appel, and Susanna Sawyer whose work shaped this book. Working with Matt Ratto and Kirk Jalbert on the special issue of *Information Society* about critical making helped develop my thinking on civic technoscience. Finally, coediting a special issue of *Journal of Political Ecology* on hydraulic fracking with Anna Willow brought together a powerful group of anthropologists who are still actively connected through the Society of Applied Anthropology's extraction working group.

Adding to this wealth of scholarly support and inspiration are my collaborators in making and doing STS, namely, Max Liboiron, Nick Shapiro, and Dvera Saxton, who advised on the final chapters of this book.

I am deeply grateful to be part of such a thriving community of academics and part of an era in history where cultural critique is valued, so that research such as this is funded. The work described here has been made possible with funding from the National Science Foundation IGERT graduate research training program, the Knight Foundation's News Challenge, the American Anthropological Association Environmental Section's small grants program, numerous public contributions to Public Lab through Kickstarter.org, Northeastern University's College of Social Sciences and Humanities, and the JPB Foundation's Environmental Health Fellowship program, organized through the Harvard School of Public Health. It is my fervent hope that we can together prove the social, ethical, and scientific worth of increasing our collective investment in making cultural critique vital to education, policy, and industry.

Connected to, mixed with, and yet still distinct from these academics are the remarkable community members and nonprofit organizations that collaborated on many parts of this story. First, The Endocrine Disruption Exchange (TEDX), founded by Theo Colborn to continue her database of research on endocrine disruption and carried on by Carol Kwiatkowski. TEDX is a unique and vital organization to the larger movement of endocrine disruption research and reform. My thanks goes to the entire staff, particularly Lynn, Mary, and Kim.

Through TEDX, I had the great pleasure of coming to collaborate with Earthworks/Oil and Gas Accountability Project (OGAP) on ExtrAct. Particular thanks to Lisa Sumi and Jennifer Goldman who were essential parts of the ExtrAct team and to Bruce Baizel, Gwen Lachelt, and Sharon Wilson who supported this and our current collaborative projects.

OGAP led ExtrAct to work with many remarkable landowners and citizens' organizations from San Juan Citizens Alliance to North East Ohio Gas Accountability Project. I'll forever remember traveling the gas patch with Rick Roles, whose story begins and ends this book, as well as time spent with Sug and Jack in New Mexico and Calvin Tillman and Sharon Wilson in Texas, as well as hearing the stories of Deb Thomas, Kari Matsko, Mark and Carol, Dee, and the Fitzgeralds. The measure of success in reforming this industry must be material improvement in the lives of those it negatively impacts. Let's hope that working together we can meet that mark.

Then there are the editors. It was no small task to reduce a manuscript from 700 pages to its finely honed present-day form. Luckily I benefited from great support in the effort. Many, many thanks to Julia Ravell for editing and fact-checking the entire work. I also had the great fortune to work with Dr. Jo-Linda Butterfield, whose practical advice on time management and thought untangling not only created beautiful post-it art all over my walls but also literally helped order my thoughts into the linear format that the written word demands. This linear, sequential format is utterly insufficient to the task of thanking my mother for the hours she spent working through drafts of each and every chapter with me. There are too few women out there benefiting from having an academic mum, who has already charted the academy's rough terrain. I could not hope for a better field guide and more constant companion than Janet (Heasman) Wylie. A sea of cross-current thoughts and emotions could not express my gratitude.

Which brings me to the heart of the matter: the friends and family who reminded me that not all life is work and that the deepest joy is in the smallest moments of intersection during our chaotic, unique, and fleeting courses through this material world. Friends, Father, Brothers, Sisters, Husband, Son, you make life worth living and a lot of fun.

INTRODUCTION

An STS Analysis of Natural Gas Development in the United States

An Ethnographic Tour of the Gas Patch

I was talking to one of the neighbors night before last, at the last forum. They were asking about the condensate [a term for fluid produced during natural gas and oil development], and he [a representative from a gas company] said, "Ah, it's just water, nothing to worry about." Oh, I told him, I wish I'd have been there! I'd go and get me a quart jar and carry it around in the truck from now on, of condensate,[1] and when he says that, I'm going to break his jaw, and then make him drink it. Then I'll tell him, "Can you still use your cell phone? Because you'd better get a paramedic over here quick. It ain't going to do much good though, because they don't know what the chemicals are over at the hospitals or nothing. They can't treat you."—RICK ROLES (summer 2006)

Rick Roles has 19 natural gas wells on his property. It was a sweltering August morning in 2006 when the rancher and landowner from the western slope of Colorado told me the story above.[2] He had just arrived in his beaten-up, faded blue pickup truck, fuming. Rick's eyes were partially shaded by his weatherworn white cowboy hat, and his tangle of long hair was in a ponytail. He was tired, and angry, but ready to talk, ready to give a tour of life in the Garfield County gas patch. "Gas patch" is a term used by residents and the oil and gas industry to describe an area devoted to drilling for natural gas reserves. Roles, like many others, was becoming convinced that fluids produced during gas extraction were extremely dangerous.

Theo Colborn, one of the world's most prominent environmental health scientists;[3] Lisa Sumi, a veteran activist; and I, then a graduate student from MIT studying Theo's research process as part of my PhD in anthropology of science, listened to Rick under a corrugated iron shelter that gave some respite from the heat. The shelter was intended for picnickers who might stop for a rest from Interstate I-70, just outside the small rural ranching community of Rifle, Colorado. It was less than scenic: a bit of asphalt and scrub stuck between the Colorado River and I-70, but it counted as a town park, and people went there to fish and picnic, "not too far from the sewage treatment ponds," as Theo observed. With our spread of fresh and homemade food on the table, we probably looked like picnickers. Theo felt it important to feed Rick up. His body had become temperamental and ill since gas drilling had begun on his ranch.

The story of natural gas drilling brought us together. Lisa Sumi, of Japanese descent, self-described as "little, yellow, different," grew up in Wawa, Canada, a small mining town on the Great Lakes.[4] Her mining town's story, where no one wanted to talk about arsenic pollution, drew her into studying the health hazards of extractive industries. She was then the research director of a small advocacy organization—the Oil and Gas Accountability Project (OGAP)—based in another small town, Durango, in another large gas patch, the San Juan Basin.[5] She had come to catch up on the database that Theo was amassing of potentially hazardous chemicals used in gas development. But mainly she was here to tour the Rifle gas patch and to meet Rick Roles.

Theo had organized the tour as well as the food. The fried chicken she had prepared sat thawing in the sun in Tupperware containers. She had frozen it to keep it fresh during the drive over the Grand Mesa—a great table, indeed, America's largest flat-topped mountain. Now she was worrying that it was too frozen. It was. Luckily we had peaches, fresh from the local organic farmstand on the road leading out of Paonia, the town where Theo lived and worked. Eating organic foods had become important to Theo after she helped found the field of endocrine disruption. Endocrine disruption refers to chemicals that can interfere with hormonal signaling in the body, particularly during fetal development, and thereby disturb reproductive, neurological, and immunological functions (Colborn, Dumanoski, and Myers 1997; Krimsky 2000). Many pesticides have been recognized as endocrine disruptors (Colborn et al. 1997). Colborn's master's and PhD research showed how low levels of toxic chemicals in high-altitude streams in Colorado had induced lasting biological effects (Colborn et al. 1997). During the 1990s she helped transform public, regulatory, and scientific evaluation of synthetic chemicals. This book describes how, between 2004 and 2011, Colborn and her research organization, The Endocrine Disruption Exchange (TEDX), based in Paonia, Colorado, raised national awareness about the public health problems associated with natural gas extraction. Paonia, a coal-mining and ranching community, is in a valley perfect for growing peaches and cherries. In the shadow of high mountains, its altitude and the mixture of snowmelt and agricultural runoff moderate Colorado's baking summer heat. It is an intriguing combination of sagebrush aridity, fruit trees, and coal mines.

Fresh fruit and home-cooked food, that is what Rick's body needed, according to Theo. He lives in a changing biological and physical landscape. He also carries it with him as he later explained, showing me the results of his blood work, revealing trace levels of benzene, toluene, ethylbenzene, and xylene (BTEX) in his body, in his blood.

Rick was the reason for our visit. We had come to hear how his body and his physical landscape had been changed by natural gas development. He told us about a school meeting, where concerned parents and faculty had gathered to learn more about a gas well that was to be located within a few hundred feet of the school grounds. "What about the toxic chemicals they are going to use?" someone had asked. "Toxic chemicals?" the company representative replied. "There are no toxic chemicals, just guar gum, sand, and water. That's it." "That's it!?" An outraged Rick reported, "They just lied to them."

Rick no longer believes what is said by gas industry reps. He lost faith when his property's value plummeted after the 19th well went in on his ranch and when his goats developed growths on their necks and gave birth to sacs of water. But he became really angry when his body stopped listening to him, when his limbs went numb randomly in the mornings and his hands cramped, so that he could not eat his breakfast because he could no longer pick up a spoon; when painkillers failed to assuage the wandering pains that traveled through his body and his mind slowed down, so that he could not remember things. A case of "classical chemical sensitivity," according to Theo. "Poisoning," says his holistic healer. Rick has a piece of paper, the evidence on his dashboard that he is not crazy, that there is something else in his body and in his land. Is this evidence enough, and who will listen?

After hearing Rick's stories, I joined Rick in his truck, so I could ask him questions, while Lisa and Theo followed behind on the dusty backroads of Rifle, a small town that does not really invite you to stop over. Most tourists, including myself before fieldwork brought me here, zip past on I-70, admiring the steep valleys of banded ochre yellows, blushing reds, and clinging sage green, experiencing the cinematic recall of old Westerns—except for the rigs beside roads and heavy truck traffic. The road follows the Colorado River, through the wide valley it has cut between the Grand Mesa and the Roan Plateau, of which both are preserved wild lands protected as national forest and also heavily contested sites for natural gas development. Drill rigs flank us to the left and right of the road, up hills, in valleys, on mesas, by the river. In addition to the rigs there is a network of supporting infrastructure; a maze of unsigned, rutted roads, with 18-wheelers kicking up dusty clouds; large cylindrical condensate tanks; and supply yards and migratory labor camps. A lattice of pipelines and pressurizing compressor stations keeps the gas moving. Harder to detect from the ground are the waste pits, where fluids produced from drilling and fracking are stored. They are difficult to see from a car because they are flat. You can find them by looking from above, or, as Rick pointed out, by looking for changes in the soil around berms: ridges of built-up earth around the ground on which a well or pit stands. You can spot them because they make unusually even lines on the horizon. Rick had been tracking the use of a particular chemical called Soli-Bond, and sometimes its ashy gray color can be distinguished along the rims of the berms.

Soli-Bond is used to bind hazardous chemicals and to prevent them from being washed away. However, it contains toxic materials such as n-hexane and methylcyclohexane; is harmful to aquatic organisms; and is dangerous to inhale. European warning labels state that it is "highly flammable, harmful and dangerous to the environment."[6] Rick had noticed clean-up crews half-draining these pits, mixing up the remaining sludge with Soli-Bond, and then simply plowing the ground over, pit liner and all. These pits were left unmarked; only a discerning, well-trained eye could spot them. "They are creating hundreds of individual, unmarked acres of toxic land where nothing will grow," Rick vented. One of Theo's goals was to determine exactly what sorts of chemicals might be in these pits and to understand what hazards might be left behind in the ground.

Theo had become concerned about fracking after a 2004 Environmental Protection Agency (EPA) study found it to be safe, leading to its exemption from the Safe Drinking Water Act in 2005 (Energy Policy Act 2005).[7] As a result there remains no federal requirement to make public the chemicals used in fracking or to monitor the process for its impact on drinking-water quality. Theo began making a database to document the chemicals used in the oil and gas industry in 2004 when she met Weston Wilson, an EPA whistle-blower. Wilson criticized the EPA study upon which fracking's regulatory exemptions are based because its authors had significant conflicts of interest. Moreover, he said, the report's analysis relied entirely on industry-supplied data and had involved no field studies (EPA 2004; Wilson 2004).

Theo's interest increased dramatically when she was contacted in 2004 by a Garfield County, Colorado, resident and outfitter, Laura Amos, who had developed a rare adrenal cancer after her domestic water well blew up during a fracking operation. Laura's home was within 900 feet of a well pad. She had come across a document Theo submitted in 2002 to the regional directors of the U.S. Forest Service and the Bureau of Land Management contesting permits issued to drill and frack on the Grand Mesa, in the watershed for the valley where Theo had raised her family.[8] In the letter, Theo noted links between adrenal cancers and a commonly used fracking chemical, 2-BE (2-Butoxyethanol). At the time she submitted the letter, Theo thought it was a local issue and that no one would start reporting health problems from exposure to such chemicals as 2-BE for many years, if ever. She submitted her letter for future generations, rather than for people here and now. But Rick, Laura, and their neighbors' emerging health problems suggested that effects might manifest much sooner than she had thought. With this new and alarming impetus, Theo's research group, TEDX, started its database in 2005, documenting the potential health effects of chemicals used in natural gas extraction. I worked on developing this database with TEDX, and our research had led to this gas-patch tour.

We drove across I-70 to the lush side of Rifle, where irrigation from the Colorado River has allowed fields and farms to grow. Horses graze in fields and there are numerous small ranch houses, a mixture of subdivisions and rural life. Every place we passed had some kind of story for Rick. We headed past the former home of a woman, Chris Mobaldi, who was sick with degenerative neurological problems. She and her doctor linked her illnesses (rashes, blisters, nosebleeds, pituitary tumors, and a very rare neurological condition known as foreign-language syndrome) to exposure to oil and gas chemicals. Paint, exposed to the breezes from nearby wells and pits, had peeled from the side of her house. Her dogs also developed tumors (Mobaldi 2007). She was diagnosed with chemical exposure, so she installed industrial air scrubbers in her home and wore a respirator while outside. Still, she fell ill, forcing the couple to abandon their home without selling it. Chris died in 2010 due to complications from her third surgery for pituitary tumors (Colson 2010b).[9]

We visited one pit that had recently been covered over. Sludge oozed from the corners where the mix had not yet dried. From the pit we could see across the highway where irrigation and fields had clearly been given up for the production of other resources. Three drill rigs were crowded onto one small, jutting mesa.

A few weeks earlier, Theo and I had taken a flight over this landscape to take a look at the gas patch from above. The pits, roads, and pads are easy to see from the air. They are clear as day, white-yellow lines creeping up the sides of the mesas, terminating in what look like "gopher holes" (figure I.1). The pits could be swimming pools (figure I.2), if they were not out in the wilderness and colored with reds, yellows, and white scum. The relative size of a tanker truck to these three waste pits shows the sheer scale of produced waste fluids (figure I.3).

Further examination of these aerial photos showed one peculiar facility that we could not identify. It included two very large waste pits, but also what looked like an area of straw and rust red about twice as big as the pits (figure I.4). So now we were looking for this site by road, hoping to view it from the ground.

"Let's go take a look over there," I suggested.

"I can take you. It's private property, though," said Rick. "I can't promise we won't be asked to leave."

"Let's give it a shot anyway," Theo chimed in.

We drove under the highway and over to the piece of property we had been viewing from a distance. Rick explained that the owner of this relatively large ranch was lucky to own his mineral rights, potentially easily making him hundreds of thousands of dollars a year from the royalties on his wells.[10] Throughout Colorado, many landowners are in what are known as "split estate" situations in which they own the surface of their land but not the minerals beneath. If an oil and gas company manages to lease the minerals from the mineral rights owner, the company has the right to develop wells on the surface owner's property. Mineral owners are in a much stronger position to negotiate how natural gas is produced from their land. Although this rancher had turned his entire ranch into a gas patch, he was amazingly, according to Rick, still herding his cattle through the land covered by wells, pits, and compressors.

This was an active area. Gleaming white pickup trucks, specialized oilfield services company vehicles, and 18-wheel "water" trucks picked up and removed waste fluids and condensate. Many of these workers would have traveled not from their local homes but from "man camps," migratory labor camps. Despite their name, the man camps also housed some female workers. A local friend of mine, Sandra, a slight, tough-as-nails blonde with a

Stetson hat, had recently taken a job driving a water truck for $80,000 a year, far above what she had been making selling industrial refrigeration units. She took her pit bull and her handgun with her to the man camp and lived there, working 12-hour shifts in the gas patch. Although one of very few women, she told me how she felt right at home in the man camps, having grown up in a gruff, male-dominated mining community. Her stories of people enjoying communal meals painted a very different picture from those in the local newspapers filled with stories of meth-using, sexually and physically violent "roughneckers," a colloquial term for gas and oilfield workers (Farrell 2005; Siegler 2007).

Sandra spent her weeks off back home in her house in the nearest big city, Grand Junction. Grand Junction is where oilfield services companies like Halliburton and Schlumberger house central warehouses for fracking chemical and equipment storage. These companies are contracted by oil and gas companies to perform specialized technical work, particularly fracking operations. There is constant traffic between these central storage areas, the man camps, and the individual wells in development. Red Halliburton and blue Schlumberger trucks loaded with complex machinery are common sights. On that day with Rick, we kicked up dust right along with them, though probably Theo's Prius stood out.

We stopped at each passing facility, trying to build up an on-the-ground anatomy of the gas infrastructure through photos. As we took pictures, Rick explained what things were. Black pipelines, carrying waste fluids, ran all over the arid properties, emptying into pits and separator tanks. We stopped to look at a "flex rig" drill rig, the latest innovation in drilling technology (figure I.5) that can drill at an angle, or even horizontally, from the well pad. This means that one well pad can have many wells that reach out all around into the subsurface. The rig itself does not require many hands because much of the process is computerized. However, human bodies are required to double check that the lengths of pipe fit well together and descend straight. We watched one man attached to a safety wire scale to the top of the gantry (figure I.6). It is precarious and risky work.

We also inspected the pallets of chemicals waiting to be added to the drilling muds: viscous chemical mixtures that keep drill bits lubricated and cool as they cut through rock. Drilling muds also keep the appropriate pressure on a drill bit as it shifts through layer upon layer of rock, each layer requiring higher and higher pressures to cut and stabilize the pressure within the well shaft. They are vital for keeping the tunnel pressurized and keeping force behind the bit.

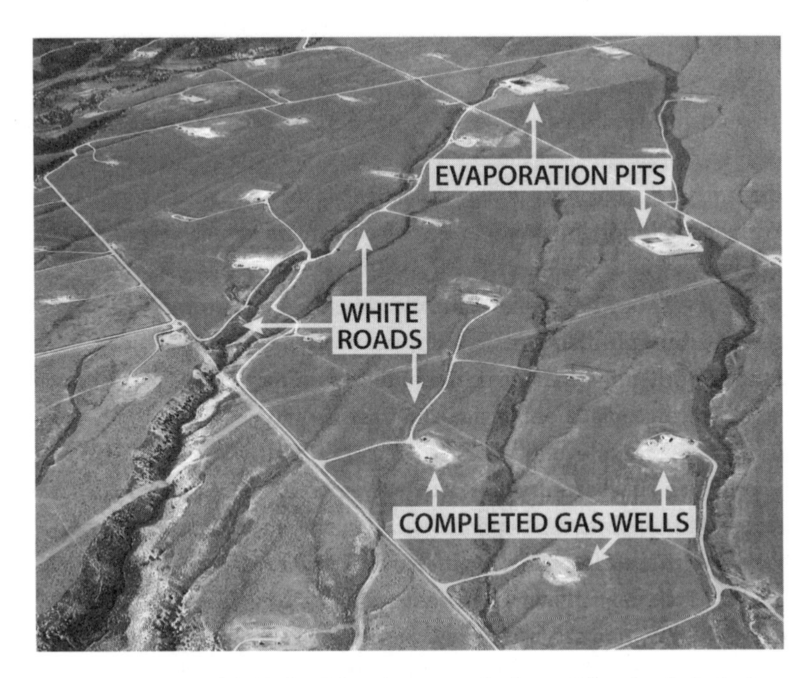

FIGURE I.1. View of the Rifle, Colorado, gas patch, showing "gopher holes," taken in the summer of 2006. Photo by the author.

FIGURE I.2. Swimming pools? Various waste pits around Rifle, taken in the summer of 2006. Photo by the author.

FIGURE I.3. An 18-wheeler tanker truck (shown for scale) along three waste pits. Photo by the author.

FIGURE I.4. Photo of a facility we could not identify, summer 2006. Courtesy of TEDX.

FIGURE I.5. Flex rig for directional drilling. Photo by the author.

FIGURE I.6. A worker scaling a flex rig. Photo by the author.

We followed lines of black piping from this rig to a large waste pit. Theo recalled that on a previous visit, one of the women with her had suddenly gone blind and developed a crushing headache after breathing fumes from a pit. Had I not known about the potential contents of gas-patch waste, I would have had a hard time seeing it as anything but water reflecting the blue Colorado sky. I strained to smell something incriminating, then concentrated to see if I had a headache. Nothing. "It affects us each differently," Theo said.

Just up the hill, along County Road 246 and through industrial chicken-wire fence, we finally saw the area we had viewed from the airplane. We risked trespassing to take a closer look. Clean-cut guys approached us in big trucks from the bottom of the site as we entered. We explained we were just curious and asked permission to look around. "Sure," they agreed, much to our surprise. As the social controversy in the area about the safety of gas extraction has increased, security around such sites has become much stricter.

A gray-graveled and well-maintained industrial zone stretched downhill into twin waste pits, bigger than two Olympic swimming pools. The views of both the waste disposal site and the valley below were amazing. There were four fountains bubbling away in one pool. Stretching up from the pools, a massive black tarpaulin encased the hillside. A sprinkler system sprouted, row after row, up the hill along the length of the enormous tarp. The tarpaulin formed the backdrop to an industrial irrigation system that, rather that distributing water, sprayed gas-patch wastes up into the air, serving like the fountains to speed the evaporation of the condensate, wastewater, and volatile organic chemicals. The "straw" effect we had seen from the air was actually the crusted orange, red, and yellow residues left by the evaporating wastes. Lisa took pictures of the scene (figure I.7). This industrial fountain on the hillside sprayed waste returned from deep below the earth up into the air. This waste traveled with the wind, over the Colorado River and the I-70 and into homes and bodies, changing the Colorado landscape and its people.

Historical Background and Theoretical Significance

What is happening in this landscape? Why does Rick think gas drilling is making him ill? How is Colorado's social, physical, and legal landscape being transformed in the early 21st century by drilling for natural gas? How do the dynamics of this natural gas boom relate to those of other extractive industries that have shaped the social and environmental history of this region (BBC Research and Consulting 2008; Casselman 2008; Colorado School of

FIGURE I.7. A waste pit seen up close, summer 2006. Courtesy of Earthworks.

Mines 2009)? What might studying this natural gas boom tell us about the contemporary fossil-fuel industry in the age of what scholars have variously characterized as "post-modernity," "globalization," "late" or "millennial capitalism," and the "Anthropocene," especially given that the boom-bust cycles of fossil-fuel extractive industries are classically "modern" problems (Yergin 1991; Black 2000; Comaroff and Comaroff 2000; Crutzen and Stoermer 2000)?

This book follows the efforts by people across the United States (scientists, journalists, lawyers, gas-patch residents) to make visible and actionable the damaging changes produced by the recent U.S. boom in natural gas extraction. Making such impacts legally, scientifically, and socially apparent is no easy business, just as it is no easy business to locate and bring to the surface a flammable, odorless gas from miles below ground, gather it in pipes, and circulate it thousands of miles across the country to power factories, fuel buses, and light stoves, as well as generate electricity and petrochemicals. This book examines how the processes of visualizing, extracting, transporting, marketing, and valuing natural gas strategically make the problems of people like Rick invisible. It also examines and develops novel scientific and social scientific tools for combating that invisibility.

The gas-patch trip described here can now be taken all over the United States in one form or another: in suburban Ohio and rural Pennsylvania, in New Mexico, California, Texas, Louisiana, and Michigan. While fracking's wide-scale application to extracting gas from shale is necessary for the present gas boom, it is not the only activity that opened up this new energy frontier (Colorado School of Mines 2009). To track the variety of activities—scientific, geological, chemical, social, political, national, regional, economic, cultural, corporate—that made this new frontier, this book draws on the history and theory of science and technology development, employing ethnographic fieldwork as its primary method. I am particularly interested in the role that contemporary science and technology played in producing—as well as resisting—the creation of this frontier. I analyze how relationships between corporations and the academy, together with state, local, and federal agencies, aligned to speed the extraction of natural gas. I show that this alignment has a series of social and organic consequences. To understand what I term the "chemical bonds" that tie Rick's illnesses to his environmental conditions, this book examines new ways to map and unsettle "corporate bodies"—those networked and peopled, physical and environmental assemblages that are presently creating the global oil and gas industry.

This story is not unique. The dynamics of the gas boom are similar to those that have occurred and are occurring in extractive industries the world over. They include environmental transformation and destruction, social destabilization, and structural and physical violence (Appadurai 1990; Nash 1993; Coronil 1997; Peluso and Watts 2001; Watts 2003; Sawyer 2004; Tsing 2004; Ferguson 2005, 2006; Ong and Collier 2005; Santiago 2006; Zalik 2004, 2008, 2009).[11] Historians, anthropologists, and activists have endeavored to describe such dynamics since the first use of oil and gas for illumination, and then as fuel, replacing coal and transforming humanity's capacity to produce mechanical work, speed, heat, and light (Yergin 1991; Black 2000). It is hard to overstate the importance of fossil fuels to contemporary human life; their very necessity and integration into every aspect make their scale and importance nearly impossible to see unless some crisis interferes with supply. Any really serious restriction of supply could rapidly cause basic services such as air, car, and train travel, as well as the defense industry, electricity, and petrochemical production to collapse. Since World War I, much of international foreign policymaking has aimed at preventing any such event. From the perspective of fossil fuel–dependent industries, the resulting policies have been quite successful (Yergin 1991).[12]

This book draws from four fields in order to situate the ethnography of natural gas extraction in this massive industry: environmental history, postcolonial anthropology, science and technology studies (STS), and digital-media studies.[13] Each field offers theoretical tools, case studies, and methods that, when combined, offer novel ways to analyze and intervene in the complex social, physical, and academic terrain of the oil and gas industry. Since the current natural gas boom depends on the use of fracking by oilfield services companies, and the health effects of chemicals used in this process are controversial, this book also speaks to the larger issues of toxic chemicals and environmental health. It combines theory and case histories from environmental history, anthropology, and STS to argue for new digital-media approaches to studying both chemicals and corporations in order to improve regulatory, social, and scientific methods for identifying and addressing emerging environmental health issues.

Chapter Previews

I begin by examining how the chemical bonds forged in the natural gas extraction process are being revealed by emerging forms of "civic science." Anthropologists Kim and Mike Fortun coined this term to analyze the emergence of forms of epidemiology and toxicology that seek to counter the influence of corporate science and, frequently, the corporate production of manifestly ideological knowledge about matters of scientific and social concern (Brown and Mikkelsen 1997; Fortun 2001; Allen 2003, 2004; M. Fischer 2003, 2009; Fortun and Fortun 2005). I trace the construction and influence of a database collaboratively built and distributed by scientists, landowners, whistle-blowers, and advocacy organizations to illustrate the potential human and environmental health effects of chemicals used in natural gas extraction. I follow how this database helped to identify natural gas extraction as a potentially massive public health threat, leading to a congressional inquiry into the practice, Oscar and Golden Globe–nominated documentaries, and moratoriums and bans on natural gas drilling (Waxman 2007; Fox 2010, 2013; Lustgarten 2014). Mapping and understanding connections between chemical exposures and illnesses is further advanced, this book argues, when we use participatory interactive digital tools designed to help communities struggling to understand extractive industries.

Rarely are these communities able to collectively shape the extractive process. We lack a process for studying and responding to the impacts of multinational corporations at the level of individual and community health. Here, I expand on Fortuns' notion of civic science by experimentally developing new media for public-interest science: for example, digital databases that enable collaborative community and academic research about infrastructural industries such as the oil and gas industry with the aim of improving communities' ability to shape extractive industries. Such participatory and recursive databases could act as a vital medium for civic science by cultivating corporate accountability, regulatory efficiency, and effective public engagement in shaping energy extraction and academic relevance (Kelty 2008).

Chapter 1 looks at the history of how fracking came to be exempted from the Safe Drinking Water Act in 2005. It illuminates how specific information, spaces, and people were sequestered in order to enable a boom in natural gas extraction using fracking. The United States' natural gas development frontier emerges from and builds upon this sort of enclaving tactic (Bowker 1994; Ferguson 2005, 2006; Watts 2005; Santiago 2006). My discussion of landowner stories and events experienced in western Colorado illustrates how these tactics initially made it impossible to link emerging illnesses to the chemicals used in natural gas extraction.

Chapters 2 and 3 explore the longer history of petrochemical industries, which actively underdevelop and hinder the environmental health sciences from connecting environmental exposures to illnesses through their use of shared corporate strategies, investment in particular modes of science, public relations, and political influence. I analyze how the environmental illnesses and damage produced by these industries have been systematically rendered hard to study. Chapter 2 examines Colborn's work historically to analyze how she developed research methodologies and social strategies that unsettled the "regimes of imperceptibility" around the ability of industrial chemicals to disturb hormonal signaling, producing the field of endocrine disruption research. I take the term "regimes of imperceptibility" from Michelle Murphy's study (2006) of Sick Building Syndrome, and I explore scientific and later social scientific methods to undo them.

Chapter 3 examines how Colborn applied her effective research methodologies for databasing and social strategies to form her novel science-advocacy organization, TEDX. TEDX maintains a database of all publications about endocrine disruption, and it develops research reports for policymakers, scientists, and advocacy organizations to help reduce exposure to, and impacts of, endocrine disruptors. Ethnographic analysis of the database provides material with which to examine the relationship between corporations and social movements in the context of contentious scientific policy debates in the United States over energy resources. Building on Kim Fortun's notion of "informating" environmentalism, I identify the development of a mode of civic scientific research that I call health environmental impact response science, or HEIR science (Fortun 2004; Fortun and Fortun 2005), influenced by Colborn's research model for studying endocrine disruptors. HEIR science (or HEIRship) recognizes how chemicals form and transform relationships across species and across generations within species by making use of familial resemblances across species, such as shared biochemical signaling pathways (e.g., the estrogen receptor) and shared ecological positions in food chains.[14] Recognizing chemicals as a biocultural inheritance that ties together human and environmental health by potentially transforming the biology of future generations requires public and regulatory attention to the unpredicted hazards produced by synthetic chemistry. HEIRship is made more challenging due to the academic, regulatory, and economic capital of chemical production and consumption industries. In response, new research infrastructures such as TEDX and their database of research on endocrine disruption are growing.

Chapter 4 examines how TEDX and Colborn's HEIRship led to their examination of the emerging local health problems related to the chemicals used in natural gas extraction. I analyze how Colborn's database of the potential health effects of chemicals used in gas extraction is being employed as a map in order to transform people's perspectives on the effects of natural gas development. TEDX's database allows landowners to relate their diverse illnesses and experiences to natural gas development. It provides predictive maps of health problems related to chemicals, and it also articulates how holes in scientific data and regulatory oversight make the environmental human health impacts of fracking hard to study. I illustrate how TEDX and its collaborators successfully sparked a scientific and social controversy about the chemicals used in fracking.

Connecting chemical exposure and illnesses needs detailed validation and is challenging scientifically and socially. Interactive online tools for popular epidemiology, mapping incidents, local circumstances, and the spread of industry tactics (Brown and Mikkelson 1997) can further address these challenges. Chapter 5 begins to analyze whether online participatory mapping and databasing tools can provide new ways to transform community relations with extractive industries beyond the specific practice of fracking. Given the historical interrelationship between these industries and the academy, I examine the need for an engaged or "activist" social science that actively works to develop civic infrastructures. Among other things, these infrastructures can help communities potentially impacted by extractive industries to collect and keep in the public domain information about illnesses, toxicities, water quality, spills and leaks, aquifer protection, and experiences elsewhere. In general, they can provide the information necessary for an informed citizenry. I name this novel branch of STS, enabled by digital media, "STS in practice": engaged social scientific research that actively involves communities, scientists, and engineers in transforming the processes of scientific research and technology development. It aims to reduce power asymmetries that persist and pattern scientific inquiry, and that are embedded in technical infrastructures. Tools developed by STS in practice, I argue, can increase the industry's social, environmental, and legal accountability.

Chapter 6 methodologically grounds the development of STS in practice by introducing the work of a research group called ExtrAct. ExtrAct's primary project was to develop web tools for the oil and gas patch communities. Digital media tools mapping pipelines, spills, and wells have become commonplace since ExtrAct.[15] ExtrAct was in many ways ahead of its time, providing the first interactive, online map, for instance, of oil and gas wells in five different states (WellWatch), a tool for evaluating and sharing interactions with industry employees (Landman Report Card), and an online platform for sharing news about oil and gas development (News Positioning System). The tools, unlike many that have followed, were specifically designed to both facilitate information sharing and build online community in accordance with the larger goal of this Knight Journalism Foundation–funded center which, under Chris Csikszentmihályi's directorship, was to develop new forms of "civic media" by using digital media.[16] This chapter describes how by combining participatory design practices with ethnographic fieldwork involving community organizations, NGOs, environmental scientists, and lawyers, ExtrAct co-conceptualized and developed a set of online databasing and mapping tools in order to link communities managing issues related to the oil and gas industry.

Chapter 7 studies the technical issue of how to develop open-source, open-access online tools for community organization in a contentious and polarized field like that of natural gas development. Separating the front-end website design and development from back-end or code development, this chapter looks at the difficulties of developing tools that can adequately protect users of ExtrAct websites and also fulfill the websites' goals of increasing industry accountability and community awareness.

Chapter 8 examines ExtrAct's back-end database development ethnographically. It explores how databases can be designed so that nonprogrammers can become involved in supporting and extending online databases they use, without requiring them to become proficient in coding. Designing for participatory development of database infrastructure extends models of community formation developed in open-source software communities to generate participatory databasing and mapping tools for grassroots monitoring of the oil and gas industry.

Chapter 9 evaluates the use of the ExtrAct tools, particularly WellWatch, a wiki-based tool for community monitoring of the oil and gas industry. This chapter analyzes WellWatch's efficacy as a platform for multisited ethnography and discusses its potential benefits for both communities and academics.

Chapter 10 memorializes Theo Colborn by theorizing the "fossil-fuel connection" between natural gas and petrochemicals (Colborn 2014). It examines how these two industries need to be studied together in order address the twin challenges of endocrine disruption and climate change.

Throughout this book I analyze the interrelationships among the oil, gas, and chemical industries, whose vast social, political, and environmental influences come together within the bodies of landowners like Rick. The book's conclusion revisits Rick's story to systematically draw together the book's theoretical and methodological interventions. Rick and the other landowners I describe in this book struggle to connect their illnesses to these industries in part due to scientific, regulatory, technical, and social structures developed in the service of industry imperatives. These industries remain unaccountable. They are also inseparable in our fossil fuel–dependent economy. Oil, gas, and petrochemicals produce a vast range of products from fuel to fertilizer and pesticides that are central to our industrial economies (Kamalick 2009; IHS 2011). To understand and account for our corporate bodies and chemical bonds, we need to develop new forms of science and social science that can keep up with the legal, social, and technical changes that enable them and that create blindness to many trade-offs and costs erased from corporate accounting sheets and business plans. TEDX and ExtrAct are situational responses to the impact of the contemporary boom in natural gas extraction. ExtrAct's STS in practice and TEDX enable new forms of social science involving civic science, interactive media, and informed social movements to build and strengthen empirical databases and explanatory models that stand up to scientific and legal challenges, offering new modes of monitoring the environmental health of our communities as well as our bodies.

Securing the Natural Gas Boom

Oilfield Service Companies and Hydraulic Fracturing's Regulatory Exemptions

On a busy Thursday morning in April 2008, a gasfield worker walked into Mercy Regional Medical Center's emergency room, complaining of feeling sick and lightheaded. From 20 feet away, the nurse on duty, Cathy Behr, could smell "a sweet kind of alcohol-hydrocarbon" odor. It would be the last thing she smelled for three weeks. Behr put on rubber gloves and a thick paper mask, and went to the worker. She now realizes that she should have told him to wait in the parking lot while she put on more protective gear. "I took him straight to the shower. Mistake. This is the embarrassing part of the story," Cathy said. Another nurse took the materials safety data sheets from the man's supervisor and looked up the non-proprietary chemicals in a computer database. Meanwhile, Cathy noticed the patient's boots were damp. She removed his clothing and boots and double-bagged them in plastic sacks. The other nurse almost vomited while taking the bags outside. But Cathy didn't notice, because she had already lost her sense of smell. At the time, though, she felt fine—just a little headache, which she put down to not eating lunch (Hanel 2008b).

A few days later, after coming down with what she thought was the flu, Cathy experienced liver, heart, and respiratory failure. Her intensive care doctor decided to treat her for chemical exposure but when he called the company they refused to release the information on the proprietary chemicals to which she had been exposed, because the information was a trade secret. The same privacy rules kept Cathy from telling a reporter for the *Durango Herald* which chemical had made her ill. The rules also prevented Mercy officials from revealing the gasfield worker's employer (Hanel 2008b). The site and scale of the spill on a BP well site on the Southern Ute reservation were not reported to either tribal authorities or regulatory agencies and were unknown until August (Hanel 2008c; Moscou 2008). Cathy spent over 30 hours in intensive care, and after a long recovery she returned to work in July 2008 (Moscou 2008). The oilfield worker reportedly had no further symptoms but was fired (Frankowski 2008).

Why was the chemical information Behr and her doctors desperately needed unavailable? Why does fracking involve chemicals that pose human health hazards? Why was the operator able to refuse to provide information about the proprietary chemicals? Answering these questions requires looking deeper into the role of oilfield services companies in the contemporary U.S. gas boom and particularly asking why sequestrating information on their proprietary chemicals is vital to maintaining their market edge. This chapter describes how a novel public health threat from chemicals used in fracking is related to the corporate structure of the oilfield services industry and its close association with both science-based regulatory agencies and the nation's premier science and technology centers. This health threat is directly related to how the oilfield services companies have developed in order to retain control over their intellectual property and as a result of the technical challenges of extracting gas from unconventional gas reserves. The world's largest oilfield services companies, Schlumberger and Halliburton, made record profits throughout the early twenty-first century's U.S. gas boom (Casselman 2008) in part because their services have been required to stimulate unconventional gas reserves using proprietary fracking techniques (Manama 2010a, 2010b). The process of protecting oilfield services companies' intellectual property around fracking is creating a novel form of petro-violence—widespread public health threats from chemicals that are made structurally impossible to monitor (Peluso and Watts 2001; Watts 2005). Highlighting the role of science-based agencies and the academy in this process generates new opportunities for resistance and transformation and new responsibilities on the part of academics to intervene.

Oil and gas extraction is very much a guessing game, because it is hard to know what lies beneath our feet. Mapping the subsurface has required the technical development of alternatives to and expansion of the human senses, including the development of electromagnetic and acoustic seismic imaging. Except perhaps for the U.S. military or NASA, no industry is more invested in developing alternative means of sensing and mapping than the oil and gas industry, since success in the business depends on identifying and accessing subsurface reserves of oil and gas (Bowker 1987, 1994; D. MacKenzie 1990; Masco 2006). Oilfield services companies are heavily invested in the cultivation of alternative modes of human sensing required for this activity (Bowker 1987, 1994). In his history of the company's early development, Bowker describes how Schlumberger made its market niche by black boxing one mode of sensing, the differential electrical conductivity of geological layers, and linking that with potential fossil-fuel reservoirs (Bowker 1987, 1994). Schlumberger has maintained its position by dramatically expanding its technical sensing ability.

J. Robinson West, the chairman of the energy-consulting firm PFC Energy, remarked in a 2008 news article that characterized Schlumberger as a "stealth oil giant," that Schlumberger is "the indispensable company": "They are involved in every major project in every important producing country" (Reed 2008). Schlumberger has come a long way from being a struggling company founded by two brothers in 1919. It is a colossus that in 2014 made $48.6 billion (Bowker 1994; Schlumberger 2015). Schlumberger works in more than 85 countries and has roughly 120,000 employees of 140 different nationalities.[1] It operates 125 different research and development units worldwide.[2] The company expanded from its initial market niche of downhole oil and gas reservoir analysis to be involved in practically every aspect of fossil-fuel extraction.

Bowker's history of Schlumberger, *Science on the Run* (1994), analyzes how the company forged its market niche by developing elaborate technical and social protections for its intellectual property, because that property was its most valuable resource.[3] Schlumberger began by using electrodes to map differences in conductivity in order to locate subsurface oil and gas. When it realized that this technique was relatively easily copied, it made its machines overly complex in order to appear hard to follow and removed data analysis from sites of detection (Bowker 1987, 1994). Schlumberger successfully maintained its market niche by protecting these costly investments.

The term *costly investments* comes from a sentence in Michael Watts's work that is crucial to understanding the relationship between natural gas extraction environments in the United States and those of oil and gas production worldwide. When we examine how federal politics facilitated the present natural gas boom, the United States begins to look like Watts's description of a "petro-state": a state whose economy, society, and governance structure are entangled with the extraction of oil. In petro-states Watts notes, "The security apparatuses of the state (often working in a complementary fashion with the private security forces of the companies) ensure that costly investments are secured" (2005: 9.7–9.8). Watts's use of "security" refers to the physical use of the state's military force. What if the definition of the security apparatus of the state is expanded to account for other means of imposing a state's power? In his lectures titled *Territory, Security and Population* (2007), Foucault develops an alternative definition of security. For Foucault, security apparatuses seek to produce particular environmental and social structures conducive to sustaining a territory and developing a population. Producing a healthy nation means measuring and optimizing social and environmental conditions that increase the rates of perceived social goods over those of social ills. Foucault argues that security mechanisms rely on the sciences, particularly the science of bureaucracies: statistics, mapping, and databasing. States use these tools to count their populations and measure the rates of particular social issues such as crime or trade. Taking Foucault's equation of security and science into account, the petro-state's work of protecting oilfield services investments could be rewritten: "The [science] apparatuses of the state (often working in a complementary fashion with the private sector [scientific] forces of the companies) ensure that costly investments are secured" (Watts 2005: 9.7–9.8).

The next three sections analyze how the contemporary shale gas boom hinges on the complementary operations of both state and private-industry science apparatuses. I examine fracking's exemption from the Safe Drinking Water Act (SDWA) in 2005, first through industry lobbying and second through influence over the science-based regulatory process. Next I analyze the close relationship between oilfield services companies and the academy. I argue that the practice of securing intellectual property by using the state's scientific and technical apparatus is producing a mode of violence proper to fracking—that is, chemical contamination that is impossible to track.

A Brief History of Hydraulic Fracturing

Fracking was unregulated until the mid-1990s. This method of "well stimulation" was pioneered in the 1940s and patented in 1949 by Halliburton Oil Well Cementing Company (Howco) (Montgomery and Smith 2010: 27). In response to the energy crisis of the 1970s, the federal government supported the development of fracking by collaborating on demonstration projects, by offering tax credits on unconventional energy production, and by funding research (Shellenberger et al. 2012). The practice gradually developed in the field through a process of trial and error; explosions during tests in the 1970s "blew the pipe out of the well about 600 feet high" (Begos 2012). Machinery, such as surplus World War II airplane engines, was adapted to pump the fluids (Montgomery and Smith 2010: 28). Fracking has since become technically and chemically complex, requiring the services companies to "furnish several million dollars' worth of equipment" (Montgomery and Smith 2010: 30).[4] In 1977 the Department of Energy first successfully demonstrated massive-scale fracking and the combination of fracking and horizontal drilling in 1986 (Shellenberger et al. 2012). Three-dimensional micro-seismic imaging developed for coal mining via Sandia National Laboratory proved to be vital for identifying shale gas reserves.[5] Fracking for shale gas was not considered financially viable until 1998 when George Mitchell, a Texas oilman, brought the cost of a fracking operation from $300,000 or $250,000 down to $100,000 on his 36th attempt (Begos 2012; Shellenberger et al. 2012).[6] Between 2000 and 2010 the industry spread rapidly with the advent of successful gas shale, coal-bed methane (CBM), and tight sands drilling methods. It facilitated a boom in unconventional natural gas that tripled service company fracking revenues between 1999 and 2007 from $2.8 billion to $13 billion (EPA 2004; Wagman 2006; Montgomery and Smith 2010: 35–36). Oilfield services companies fought vigorously to preserve sole control over the processes and chemicals involved in fracking.

Frack fluids were developed to suit the unique challenges of surface to subsurface operations, where temperatures in deep wells can reach over 2,500°F and sheer pressures are intense (Montgomery and Smith 2010). Engineering a substance that remains fluid in such conditions is a challenge (LaGrone, Baumgarther, and Woodroof 1985). Moreover, fracking fluids must be mixed and combined within the chemical and physical environment of the surface, and transferred into very different subsurface conditions, after which they are returned again to the surface. This mixing creates various problems. Surface

FIGURE 1.1. High-volume hydraulic fracking in progress. Courtesy of TEDX.

water is used in very large quantities (see figure 1.1). According to a 2009 report by the Ground Water Protection Council (GWPC), between two to four million gallons of water is used during each frack.

Oxygenated surface water has a bacterial load, much of which will not survive the heat of deep wells. However, a small proportion of the bacterial populations found in this water will thrive in the low-oxygen, high-heat environment "down hole." These surviving bacteria may destabilize the fracking fluids and reduce gas yields by populating the fractures in the form of biofilms that limit gas flow. Biocides are added to kill the bacteria in surface water and prevent the formation of such blockages. Fracking fluid returning to the surface is therefore laced with biocides that are toxic to surface life and hazardous to dispose (Rimassa et al. 2011).

Other chemicals used in fracking create similar engineering, environmental, and public health quandaries. For example, acids are used to clean the well of drilling muds and other debris, as well as serve as a form of well stimulation (Kalfayan 2008; GWPC 2009).[7] After well completion, approximately 5,000 gallons of diluted acetic or hydrochloric acid are pumped down the well at a flow rate of about 500 gallons a minute, enough to simultaneously fill 50 bathtubs with an average volume of 100 gallons within 10 minutes (GWPC 2009: 59).[8] Flowback water containing acids is frequently stored in open-air pits where hydrochloric acid can volatilize and form the precursors to acid rain. Surface storage of acid also poses risks, as seen in Leroy Township, Pennsylvania, where 4,700 gallons of hydrochloric acid leaked out of its container (Hrin 2012).

Chemical additives like acids also create hazards for well infrastructure and require the addition of counteracting chemicals, such as corrosion inhibitors, to neutralize excess acids. Oxygen scavengers such as ammonium or sodium bisulfite are used to remove oxygen from fracking fluids as oxygen can react with chemicals in the fluid and destabilize the composition of the gel (Walker et al. 1995). Oxygen also rusts the pipes' infrastructure, potentially threatening the wells' integrity under the high pressures used in fracking (GWPC 2009). Oxygen scavengers and corrosion inhibitors are both very reactive and may act as sterilizing agents. Thus the challenges of creating and preserving linkages between the surface and the subsurface create hazards to surface life. Each fracking operation involves 18 different cycles, each cycle requiring changes in the chemistry and composition of fracking fluids (GWPC 2009; McKenzie et al. 2012). Exempting fracking fluids from federal reporting requirements and monitoring makes it extremely difficult to evaluate the risks posed by the chemicals.

High-volume fracking started in the United States in the late 1980s to extract gas from unconventional reserves in Alabama. In 1994, a Florida-based nongovernmental organization (NGO), the Legal Environmental Assistance Fund (LEAF), petitioned the Environmental Protection Agency (EPA), arguing that fracking ought to be regulated by the state under the SDWA after numerous families in Alabama experienced contamination of their water wells and strange health effects coincident with fracturing operations. As one report noted, "In June of 1989, the Hocutt family's water well became contaminated with brown, slimy, petroleum smelling fluid that was similar to the discharged hydraulic fracturing fluid that traveled downhill from the USX-Amoco methane well near their house (reportedly killing all plant and animal life in its path). . . . Ms. Hocutt and her husband have both experienced a variety of diseases including cancers of unknown etiology. At least 8 more neighbors also have some form of cancer of unknown etiology" (NRDC 2002b: 3).

The SDWA requires states to regulate any threats from underground injection to drinking water. Fracking, LEAF argued, is plainly underground injection and should be regulated by the states (LEAF 1997). In its 1995 response to a LEAF petition, the EPA argued that fracturing was not underground injection because the primary goal of fracturing is not to leave chemicals underground but rather to promote natural gas rising to the surface. LEAF did not believe this argument held up under scrutiny (because it contradicts the SDWA) and so it appealed to the 11th Circuit Court of Appeals, which in 1997 concluded that fracturing indeed constituted underground injection according to the plain language of the SDWA (LEAF 1997).

The potential impact of this ruling was not lost on the oil and gas industry. It potentially meant that every state would have to stop the practice until it had reviewed and developed an underground injection control (UIC) plan that ensured that there was no contamination of water sources. The SDWA statute requires that the UIC plan be "an effective program (including adequate record keeping and reporting) to prevent underground injection which endangers drinking water sources" (LEAF 2001). The *Drilling Contractor* journal reported that the "industry is working feverishly to develop legislative remedies specifically exempting hydraulic fracturing from the underground injection program" (2000: 43). In 2001, an unprecedented opportunity to develop such legislative remedies emerged.

A former CEO of Halliburton, Dick Cheney, was elected as vice president along with President George W. Bush in 2000, and in 2001 Cheney formed the National Energy Policy Development Group, more commonly known as the Energy Task Force (GAO 2003). Details about the Energy Task Force are hard to gather because Cheney, in violation of the Federal Advisory Committee Act, which requires the disclosure of activities and details about federal advisory committees, refused to release documentation on the Energy Task Force's members, meetings, or discussions. Civil liberties groups filed four lawsuits, which eventually resolved into two lawsuits on file: one by the Natural Resources Defense Council (NRDC) against the Department of Energy under the Freedom of Information Act (FOIA) and the other by the nonprofit governmental accountability organization Judicial Watch against the Energy Task Force itself (GAO 2003: 4). As these lawsuits and FOIA requests filed by the NRDC proceeded, approximately 40,000 pages of documents filtered into the public domain, although many of them had redacted text (Milbank and Blum 2005; Abramowitz and Mufson 2007).

Republished by the NRDC online, these documents show that the Energy Task Force was directly lobbied on regulatory issues related to fracking. In

May 2001, members of the Energy Task Force received documentation about the possible impacts that the regulation of fracking could have on the development of CBM production. Emails exchanged about this documentation imply that this information was shown in a presentation to the Energy Task Force on March 27, 2001. An industry lobbying firm, Advance Resources International (ARI), produced parts of the presentation (NRDC 2002a: 6681). Making direct reference to the LEAF ruling, the presentation by ARI, reproduced here, argues that regulation on fracking would cause a more than 50% decline in CBM production in the United States by 2010 (NRDC 2002a: 6667):

> Hydraulic Fracturing. In the case of Leaf vs. EPA, the court ruled that the injection of fluids for the purpose of hydraulic fracturing constitutes underground injection as defined in the Safe Water Drinking Act (SWDA) [*sic*], that all underground injection must be regulated, and that the hydraulic fracturing of CBM wells in Alabama was not regulated under Alabama's UIC program. . . . It is our understanding based on a review of EPA's proposed study methodology, that the regulations would govern the injection of all fluids. Given the scope of this ruling, this would affect virtually every CBM well drilled in the U.S. Under the "worst case" scenario (i.e., all CBM wells affected), only a limited development would take place. Therefore, current production would gradually decline over the next 10 years to 700 Bcf [billion cubic feet] with little or no replacement drilling (Exhibit 2). (NRDC 2002a: 6667) (see figure 1.2)

The methods used to produce these projections are not described in the supplied materials. Regulating fracking was identified as one of three pending issues that could affect the goal of producing 1,684 bcf of CBM by 2010 (see figure 1.2). Based on such industry lobbying, the first draft of the Energy Policy Bill that was debated and defeated in Congress in 2003 contained specific provisions exempting fracking from the SDWA.

The federal exemption from the SDWA (later criticized as cronyism) was arguably less important to the future of fracking than the subtle work of generating scientific evidence that fracking did not threaten drinking water. In 2004, the EPA released a white paper with this crucial statement: "EPA has determined that the injection of hydraulic fracturing fluids into CBM wells poses little or no threat to USDWs [underground sources of drinking water]. Continued investigation under a Phase II study is not warranted at this time" (EPA 2004: ES-16).

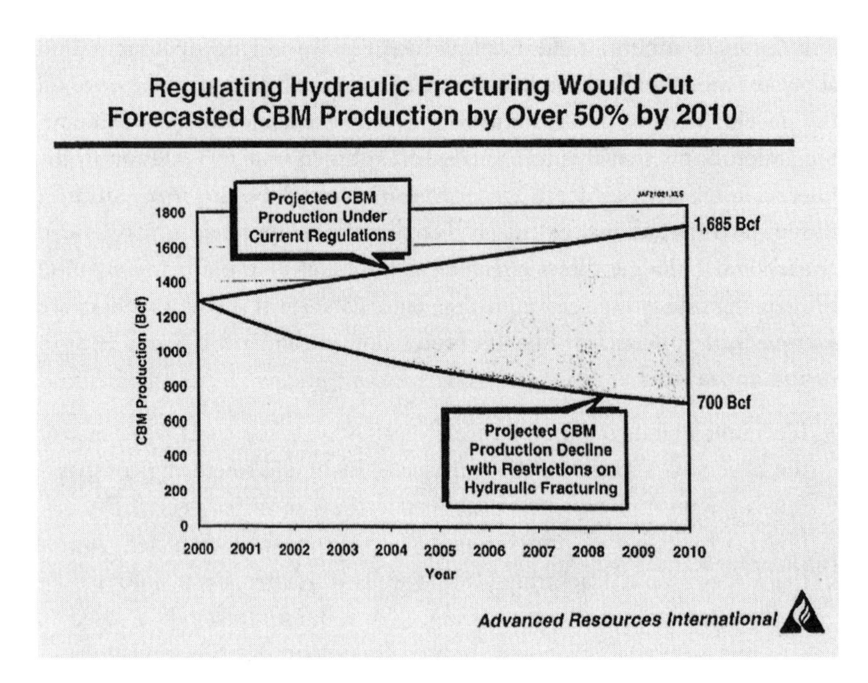

FIGURE 1.2. Diagram illustrating a more than 50% decline in natural gas production by 2010 if hydraulic fracturing were to be regulated. NRDC (2002a: 6680). Reprinted with permission from the Natural Resources Defense Council.

After the ruling against LEAF's petition to protect drinking water in relation to fracking, the EPA commissioned a study on fracking. Initially this was to be a two-stage study, the first being a literature review and the second field studies. Following the LEAF ruling, in July 2000, the EPA called for stakeholders to participate in designing a study of fracking's potential to impact underground sources of drinking water (EPA 2004: 2–1). The EPA study was limited from the outset to only USDW, based on its perceived mandate under the SDWA. It did not consider fracking holistically. It did not include hazards created by the practice's extended infrastructure such as waste pits and chemical storage sites, or other routes of environmental and human health hazards such as through surface-water, air, or soil contamination. The study was built around available peer-reviewed literature, further winnowing the possibility of identifying and analyzing such risks. No information was subpoenaed and no proprietary information was passed over (EPA 2004: 2–3): "In Phase I, EPA did not incorporate new, scientific fact-finding, but instead used existing sources of information, and consolidated pertinent data in a summary report to serve as the basis for the study" (EPA 2004: App. B-7).

This approach was limited by the paucity of studies on the health and environmental impacts of fracking at the time. The majority of the literature reviewed focused on improving the engineering efficacy of fracking. Furthermore, the threshold for performing follow-up field research to fill in gaps around environmental health questions was set very high from the outset of the study: "Specifically, EPA determined that it would not continue into Phase II of the study if the investigation found that no hazardous constituents were used in fracturing fluids, hydraulic fracturing did not increase the hydraulic connection between previously isolated formations, *and* reported incidents of water quality degradation could be attributed to other, more plausible causes" (EPA 2004: 2–2).

The methods section of the EPA report italicizes "and" in this statement as if to emphasize that all of these conditions must be met to trigger a Phase II review. A whistle-blower from the Denver EPA office, Dr. Weston Wilson, argued in a letter to Congress in 2004 that the study was fundamentally flawed. Wilson pointed out that the EPA did find that the first condition was met. Fracking used hazardous constituents. Further review was neglected because the EPA only considered impacts if the water and methane-bearing formations were previously structurally separated from each other and if there were not other "more plausible" causes for water-quality degradation. The EPA was not required to prove these other potential causes were the actual cause or to materially rule out fracking as the cause of contamination through field research.

The EPA study is a confusing document. The agency found that "10 of the 11 basins may lie, at least in part, within USDWs" (EPA 2004: ES-13) and that "in many coalbed methane-producing regions, the target coalbeds occur within USDWs, and the fracturing process injects stimulation fluids directly into the USDW" (EPA 2004: 1–6). Additionally the study reports that

> in any fracturing job, some fracturing fluids cannot be recovered and are said to be "lost" to the formation. Palmer [citation omitted] observed that for fracture stimulations in multi-layered coal formations, 61 percent of stimulation fluids were recovered during a 19-day production sampling of a coalbed methane well in the Black Warrior Basin. He further estimated that from 68 percent to possibly as much as 82 percent would eventually be recovered. (EPA 2004: 3–23)

Considering the volumes of fluid used in fracking, a loss rate of 39% is extraordinarily high, particularly since the fluid may well be moving into drinking-water aquifers. The study also notes that multiple chemicals used in fracking fluids pose health hazards and that there is a lack of data on this point (EPA 2004: 4). However, since these regions of underground drinking water were structurally interconnected, dilution should be sufficient to resolve any concerns, the report argues, without performing any of its own field studies to validate this assumption. On the basis of this first report, which involved short site visits to view four fracking operations, plans were cancelled for the second study, which could have included long-term field analysis (EPA 2004: App. B). The report's one achievement was a voluntary, though unmonitored, agreement between oil and gas companies to no longer use diesel as the primary fracking fluid. Wilson pointed out that this ban itself created internal contradictions. Why ban diesel if there was no concern about water-supply contamination?

Wilson's letter stated, "EPA decisions are not consistent with the findings of its study nor have EPA decisions complied with the purpose of the SDWA." This conclusion was based on the fact that the EPA established that fracking fluids contain "toxic and carcinogenic" substances, yet it decided there was no need for further study (Wilson 2004). Wilson told Congress that the EPA reported that there was a lack of field data on whether or not such fluids could contaminate the water supply, even in the face of evidence from the San Juan Basin that CBM production had indeed contaminated water wells with natural gas. Wilson concluded as follows:

> In June of this year [2004], EPA produced a final report pursuant to the Safe Drinking Water Act that I believe is scientifically unsound and contrary to the purposes of the law. In this report EPA was to have studied the environmental effects that might result from the injection of toxic fluids used to hydraulically fracture coal beds to produce natural gas. In Colorado, coal beds that produce natural gas occur within aquifers that are used for drinking water supplies. While EPA's report concludes this practice poses little or no threat to underground sources of drinking water, based on the available science and literature, EPA's conclusions are unsupportable. EPA has conducted limited research researching the unsupportable conclusion that this industry practice needs no further study time. (Wilson 2004)

The composition of the EPA-appointed peer review panel and the report's findings demonstrate the structural alignment of the state's scientific apparatus and the scientific forces of industry in the process of developing this report. Wilson discusses how the review committee did not comprise EPA experts in toxicology and hydrogeology, but rather external individuals drawn heavily from industry. He states that by the EPA's own standards, five of seven members of the review panel had conflicts of interest. The panel was made up of one petroleum engineer from BP Amoco, a technical advisor from Halliburton, an engineer from the Gas Technology Institute, a former employee of BP Amoco, an engineer from the Colorado Oil and Gas Conservation Commission (COGCC), and a former employee of Mobil Exploration. No one on the panel was qualified to analyze the human and environmental effects of chemicals used in fracturing, and the panel did not have any groundwater experts able to speak to the dynamics of aquifers (Wilson 2004: 14).

Despite the outcry from environmental organizations (H. Anderson 2005) when the matter was reconsidered in 2005, the Energy Policy Act passed. Section 322 contains provisions exempting fracking from regulation under the SDWA (Energy Policy Act 2005). This exemption meant that chemicals used in fracking would not be publicly disclosed, and state UIC plans would not be developed to monitor and ensure against harms to drinking water. There would be no public record about fracking operations. With passage of this exemption, the scientific research branch of the state, the EPA, coordinated with industry science to protect and secure the costly investments made by the oilfield services industry.[9]

Sustaining Hydraulic Fracturing's Exemption: Industry and Academic Relationships

The close relationship between industrial scientific infrastructure and the state, which in this case helped to secure the investments of the oilfield services industry, extends beyond the direct policy process and regulatory agencies and into the structure of academic science and technology development. The network's impact is demonstrated in the present policy and scientific debate about the safety of fracking and the necessity of shale gas extraction for national security. Partnering with universities to develop technologies and scientific knowledge that promote and enable gas extraction provides two benefits to oilfield services companies: it increases the technical feasibility of unconventional gas extraction, and it gives credence to industry policy recommendations.

This cyclical relationship between industry and academia is exemplified by MIT's relationships with Schlumberger and the larger energy industry through its energy initiative, MITEI (pronounced "mighty"). As a research and development-driven company, Schlumberger is interested in participating in and developing useful research programs and technologies at MIT that might assist company development. In 2005, Schlumberger moved its primary research and development headquarters to a position across the street from MIT (Tuz 2004). The company reports that the move has been a great success:

> "We made the move after 50 years in Connecticut and already it's paying off for us tenfold," said Peyret [Olivier Peyret, a former vice president for university collaborations and recruitment at Schlumberger]. "The last time I visited, I was amazed to see how many MIT faculty were in the Schlumberger cafeteria as a matter of routine. It demonstrates the value to us of proximity, as well as helping people studying at MIT learn about Schlumberger." (*Innovation Quarterly* 2008: 5)

Schlumberger has an executive management-training program within MIT's Sloan School.[10] The company has given proprietary mapping and data visualization tools to the geology and geophysics departments (MIT News 1999). This software was key to establishing a new seismic research laboratory in 1999, which "expands its research activities in petroleum reservoir imaging and monitoring, borehole seismology and acoustics, environmental geophysics, geologic mapping and remote sensing," all tools that instrumentally support oil and gas extraction (Halber 1999).[11]

Continuing these industry partnerships is a celebrated part of MITEI according to MIT's president, Susan Hockfield: "This exciting partnership between MIT and BP epitomizes what the MIT Energy Initiative is designed to accomplish: the pairing of innovative MIT researchers across the entire campus with results-oriented scientists, engineers and planners in industry, working together to transform the world's energy marketplace" (MIT News 2007).[12] MITEI's stated goal was to "help transform the global energy system to meet the needs of the future and to help build a bridge to that future by improving today's energy systems."[13] MITEI brought together researchers from all five of MIT's schools in a cross-institutional research effort that oil and gas companies supported. The founding corporate members of the MITEI, such as BP technologies and Eni SpA, "operate in the oil and gas, electricity generation and sale, petrochemicals, oilfield services construction and engineering industries" (MITEI 2012, 2014a).[14] Founding members have a seat on the MITEI Executive Committee. They may place a researcher in a participating MIT faculty member's lab and direct 75% of their contribution to MITEI to targeted research. Founding members also have "the option to obtain a worldwide, royalty-bearing commercial license for patented technology, with the right to sublicense." Founding members gave MIT $5 million per year for five years to support MITEI (MITEI 2009a, 2009b).

As part of the partnership, MITEI focused research efforts in numerous areas vital to natural gas development such as "Multiscale simulation of gasification; Synthesis gas cleanup and upgrade; Gasification technology development; New processes for converting synthesis gas to liquid fuels; Process integration and design for operability; and Fuels market and policy analysis."[15]

These research foci bore fruit and became foundational to MITEI studies on "hydrocarbons products and processing" (i.e., oil and gas) (MITEI 2014a). In January 2007, MITEI reported that "MIT engineers have developed a mathematical model that could help energy companies produce natural gas more efficiently and ensure a more reliable supply of this valuable fuel," and, in 2006, it reported on novel acoustic technologies produced to help identify "sweet spots" for natural gas extraction, particularly in tight sands gas reserves (Halber 2006; Stauffer 2006).[16] Thus, while MITEI was pushing research forward on energy sources not based on fossil fuels, it was also actively promoting and producing technologies that encouraged and enabled the expansion of natural gas extraction. MITEI researchers not only developed technologies that enable natural gas extraction in collaboration with oil and gas companies and oilfield service companies, but also authored policy papers that advocate that natural gas is the bridge fuel:

> The two-year study, managed by the MIT Energy Initiative . . . , examined the scale of U.S. natural gas reserves and the potential of this fuel to reduce greenhouse-gas emissions. . . . "Much has been said about natural gas as a bridge to a low-carbon future, with little underlying analysis to back up this contention. The analysis in this study provides the confirmation—natural gas truly is a bridge to a low-carbon future," said MITEI Director Ernest J. Moniz in introducing the report.[17]

This draft (released in 2010) and the finalized report (released in 2011) focused on the potential of unconventional natural gas formations, such as shale or tight sands, finding that "the United States has a significant natural gas resource base, enough to equal about 92 years' worth at present domestic consumption rates. Much of this is from unconventional sources, including gas shales. While there is substantial uncertainty surrounding the producibility of this gas, there is a significant amount of shale gas that can be affordably produced."[18]

The report found that unconventional reserves were "rapidly overtaking conventional resources as the primary source of gas production" and that the "U.S. currently consumes around 22 Tcf [trillion cubic feet] per year and has a gas resource base now thought to exceed 2,000 Tcf."[19] It concluded that such intensive resource extraction requires fracking as well as other technological developments: "In order to ensure the optimum development of these important national assets, it is necessary to build a comprehensive understanding of geochemistry, geological history, multiphase low characteristics, fracture properties and production behavior across a variety of shale plays" (MITEI 2010: 14).

MITEI recommended that the Department of Energy sponsor research and development "in collaboration with industry and academia, to address some of the fundamental challenges of shale gas science and technology, with the goal of ensuring that this national resource is exploited in the optimum manner" (2010: 14). Further, MITEI advocated continued work by the United States Geological Survey (USGS) to improve methodologies for assessing unconventional reserves. The report cited the following environmental impacts associated with natural gas development:

> 1. Risk of shallow freshwater aquifer contamination, with fracture fluids; 2. Risk of surface water contamination, from inadequate disposal of fluids returned to the surface from fracturing operations; 3. Risk of excessive demand on local water supply, from high-volume fracturing operations; 4. Risk of surface and local community disturbance, due to drilling and fracturing activities. (MITEI 2010: 15)

Gas-patch communities would likely recognize this list of problems as generally sound, but there is no discussion of the research on which it is based. The report neglected to acknowledge that many of these risks are understudied and impossible to study, given present regulations. It added that "with over 20,000 shale wells drilled in the last 10 years, the environmental record of shale gas development is for the most part a good one," providing no evidence whatsoever to support this statement (MITEI 2010: 15). Further, it asserted that the "protection of freshwater aquifers from fracture fluids has been a primary objective of oil and gas field regulation for many years" (MITEI 2010: 15). This claim is hard to support, given the regulatory exemptions from the SDWA accorded to the industry. Rather than investigating whether there have been cases of aquifer contamination during gas extraction, it offered a table that indicates that "there is substantial vertical separation between the freshwater aquifers and the fracture zones in the major shale plays. The shallow layers are protected from injected fluid by a number of layers of casing and cement, and as a practical matter fracturing operations cannot proceed if these layers of protection are not fully functional" (MITEI 2010: 15).

EPA research in Pavillion, Wyoming; in Alabama; and in the New Mexico and Colorado's San Juan Basin and West Virginia has contradicted this finding, but passed unmentioned (EPA 2004; EPA Region 8 2009; Horwitt 2011). Finally, while calling for disclosure of fracturing chemicals, it concluded that "good oilfield practice and existing legislation should be sufficient to manage this risk" (MITEI 2010: 15). No evidence supported this statement and it is not clear that any research was performed to make such a determination.

Given the contemporary state of scientific knowledge about the environmental and human health hazards of natural gas extraction, definitive statements on its environmental and human impacts could not yet be made. But evidence from those living in the gas patch contradicts the view from the oil and gas extraction industry (Amos and Amos 2005; Clarren 2006b; Nijhuis 2006; Lustgarten 2008a, 2008b; EPA Region 8 2009; Fox 2010, 2013; Urbina 2011a, 2011e, 2011f, 2011g). MITEI's presentation lacks a gas-patch perspective in part because communities affected by oil and gas extraction lack the close connections to research institutions that companies are capable of forming (Noble 1977; Hightower 1978; Downey 2007). Natural gas extraction is a sociotechnical discourse, and the role of an academic report should be considered holistically. One outcome of the 2010 MITEI draft report was the ardent pursuit of natural gas development, thereby fulfilling the document's predictions. The production of "white papers," none of them peer-reviewed academic studies that make policy recommendations, deserves further academic investigation, particularly the production and publication of "draft" white papers like the 2010 MITEI report.

The MITEI white paper perpetuates and occupies a gray zone of academic production that is certified by both the author's expertise and the institution's reputation, and yet it was framed as a provisional "draft" that was not formally peer reviewed as a scientific publication. A discursive function of this draft's publication in 2010 amid growing public controversy about gas extraction from the Marcellus Shale was arguably to alleviate public concern. Such documents are important to analyze because they circulate with high credibility in both industry and policy circles, despite their provisional nature.[20] Indeed, the "draft" status further functions to convey academic transparency, as if the document might be modified by public comment, though there is no structure for such feedback. It protects the document, authors, and institution by creating deniability for errors in analysis because of its "draft" status. Academic study is not removed from this discourse. It has a fundamental role in producing gas economies because the choice of which technical and scientific questions are worthy of investigation in the long term shapes what is known and what can be known (Noble 1977; Hightower 1978; Winner 1980; Downey and Rogers 1995; Downey 1998; Proctor and Schiebinger 2008). Exemplifying the need for agnotological analysis (the study of structured forgetting; see Proctor and Schiebinger [2008]) of this industry, a regime of imperceptibility, taking the form of regulatory exemptions, and institutional, technical, and social biases have developed around natural gas extraction and play an integral role in allowing the unconventional gas boom to flourish (Murphy 2006).[21]

A regime of imperceptibility is produced when the very tools intended to investigate a problem actually work to render the problem less visible. Here, the EPA's 2004 report helped make fracking hazards imperceptible by assisting the gas industry's exemption from the 2005 Energy Policy Act. This exemption meant that Cathy, the nurse whose story began this chapter, and her fellow medical practitioners were unaware of the chemical risks posed by exposure to fracking fluids. The regime's logic is self-supporting. It is easy to argue that the extraction industry's environmental record is a good one when there have been no studies. In the regime of imperceptibility around fracking, we see a pattern proper to the corporate form of oilfield services companies—the alignment between corporate and state scientific and technical apparatuses—which, like the security alliance between the state security and oil and gas industry, must be unwound if we are to truly assess and address the environmental and human health costs of this industry.

Living In and Out of Exemptions: The Case of Laura Amos

The impact of this regime of imperceptibility is best appreciated by considering an individual case. Laura Amos's story of how she lost her health and her wilderness home from which she and her husband ran game-hunting expeditions received national media coverage (McKibbin 2004; Clarren 2006a, 2006b; Frey 2006; Nijhuis 2006; Lustgarten 2008a), before she was prevented from sharing it by a "gag order" on behalf of the well operator, the Canadian company Encana.[22] I first learned of Laura Amos from Colborn in

January 2006. Over dinner in Paonia, she told me of how Laura developed a rare form of adrenal cancer after her water well exploded, coincident with a fracking operation on a gas well just 1,000 feet from her front door. Like millions of people globally who own only the surface of their land and not the rights to the minerals beneath, Laura was powerless to refuse the natural gas industry's resource extraction. "Split estates" (*Grand Junction Daily Sentinel* 2006; Spaulding 2006a, 2006b, 2006c; D. Anderson 2009) are pivotal to the dynamics of western resource extraction because surface owners are legally required to allow reasonable access to holders of rights to the development of minerals under their property. Mineral rights are valueless without the necessary access through the surface (Feriancek 2008). Before surface owner protections bills were passed in New Mexico and Colorado in 2007, companies were not required to sign contracts with surface owners, and surface owners were not able to direct where well pads, pipelines, and associated roads could be placed (OGAP 2007a, 2007b).[23] Ballard Energy held the mineral rights to 280 billion cubic feet of subsurface in northwest Colorado and began installing gas wells around Laura's property in 1998. Alberta Energy bought Ballard Energy in 2001 and thereby acquired Ballard's Colorado holdings. Alberta Energy subsequently merged in 2002 with Pan Canadian Petroleum to form Encana (*New York Times* 2001).

Before these various acquisitions and mergers, in 2001 Ballard subcontracted an oilfield service company, BJ Services, to frack four gas wells near Laura's property. During one fracking operation, her water well erupted like a geyser. Liquid emerged from her well, sludgy and gray, heavy with sediment and bubbling with methane, as much as 14 milligrams per liter, "almost as much methane that water will hold at [that] elevation" (Amos and Amos 2005: iv–23, 24). BJ Services noted that its equipment showed no problems during the frack. Alberta Energy (which had bought out Ballard and became Encana in 2002) denied responsibility for water contamination, but it began delivering drinking and washing water to the Amos household two weeks after the event.

In 2003, Laura developed a rare form of nonmalignant adrenal cancer, primary hyper-aldosteronism or Conn's Syndrome, and had to have her adrenal gland removed. Following her water-well contamination and struggles for reparations, she became active in opposing oil and gas drilling and in 2004 came across a memo written by Theo Colborn in 2002. The memo to the Bureau of Land Management (BLM) and the Forest Service in Delta County (Amos's neighboring county) described the potential health effects of a chemical that Gunnison Energy Corporation proposed to use in the development of wells on the Grand Mesa: 2-Butoxyethanol (2-BE), or ethylene glycol monobutyl ether, a virtually odorless, colorless, and tasteless chemical that is easily absorbed by the skin, highly soluble and volatile at room temperature.[24] 2-BE can cause bleeding in the urinary tract and immunological problems, but its potential as an adrenal carcinogen caught Laura's attention:

> At the end of a two year chronic bioassay, elevated numbers of combined *malignant and nonmalignant tumors of the adrenal gland* were reported in female rats and male and female mice. Low survival rates in the male mice in this study may have been the result of the high rate of liver cancers in the exposed animals. This study revealed that long-term exposure to 2-BE often led to liver toxicity before the hemolytic effects were discernible. No human epidemiological studies are available to assess the potential carcinogenicity of 2-BE. However, from the results of laboratory studies, using Guidelines for Carcinogenic Risk Assessment (1986), 2-BE has been classified by the USEPA as a possible human carcinogen.[25] (emphasis added)

Laura investigated the connection and was assured by Encana that 2-BE was not used on her well: "Encana's spokesman, Walt Lowrey, assured several of our neighbors, and my husband and me that 2-BE was NOT used. In addition, Lowrey told the Associated Press and many reporters in western Colorado, and Denver that '2-BE was not used on the pad, or anywhere in this area'" (Amos and Amos 2005: iv–25).

2-BE is a surfactant specifically used in fracturing to enable fracking fluids' charged chemical constituents to mix into a homogeneous solution. Due to their highly variable polarity, fracking chemicals present an oil-and-water problem because they are immiscible. Surfactants are charged at one end but not at the other, allowing them to act as bridges and mix differently charged molecules (Klein et al. 2013). The hazards of surfactants are multiple: the molecules are toxic and can pass through skin and other membranes. When surfactants penetrate through the skin, they can also bring other harmful substances along. This is particularly dangerous in the context of fracking because of the complex mixture of chemicals and by-products from drilling, including heavy metals that could potentially associate with the surfactant. Laura found out that a second frack job occurred, even after the linkage between her water well and the gas-bearing formation was already known through the COGCC's 2001 analysis of the gas in her water well. The goal of this shallower frack was the Wasatch formation, from which the Amos family drew their water. Documents about this second frack revealed that in fact

2-BE had been used (Sumi 2005: iv-25). Even with this evidence of 2-BE's use, the COGCC still failed to support Laura. Laura described how Commission Director Brian Macke "told a CBS News Bureau Chief in Washington D.C. that I am crazy, and that my exposure to 2-BE may have come from Windex!" (Amos and Amos 2005: iv–25).

SUSTAINING A REGIME OF IMPERCEPTIBILITY

Although it became clear that 2-BE had been used in fracking around Laura's property, there was no way, years later, to prove that it was actually the causative contaminant in her water or that the fracturing fluid had directly caused her tumor. Scientific knowledge about human 2-BE exposure and cancer was uncertain and no one tested Laura's water for it. Furthermore, the lag between her illness and potential exposure worked against establishing the chain of causation that courts require to prove liability. The COGCC was unaware of 2-BE's potential hazards and performed no tests for it because of the federal exemptions on fracking. Finally the complex relationships among Encana, Albert Energy, Ballard, and the subcontractor obfuscated chemical use. From federal regulation to local enforcement, these obstacles compounded one another and created a landscape in which proving causation was impossible. The processes of frontier development in this industry exacerbate the obstacles faced by individuals like Laura in making their stories scientifically and legally legible.[26]

Companies generate relationships between scales that favor gas extraction through information sequestration and the structural compartmentalization of resource development. Information sequestration makes it hard for private individuals to move between the scales of local, state, national, and international regulatory and policy arenas in ways that are available to industry. A company the size of Encana is built from many individual lease agreements and subcontracted oilfield services companies. At the scale of surface ownership, contracts to develop natural gas, such as the surface owner agreement with Laura and the mineral rights lease to her land, were made separately. Laura and the mineral rights owner did not know each other, but Ballard/Encana knew both. These contracts and legal precedent gave the mineral owner or lessee (in this case, Ballard/Encana) dominant estate and the right to extract minerals. Subcontracts and corporate mergers and acquisitions made it legally and socially difficult to aggregate all the information about what happened during the development of wells around Laura. The industry comprises such networks: loosely associated, compartmentalized actors, or modules, like oilfield services, migratory labor camps, and individual landowner contracts (Bowker 1987, 1994; Appel 2011). These units actively sequester information and people into compartments that limit their ability to travel outside of the structured network. Loose connections between such compartments enable the industry to move across scales in ways that are hard to follow (Tsing 2004). On a state level, the COGCC was mandated to develop oil and gas resources. In 2004, COGCC was primarily staffed by individuals with oil and gas industry histories (Neslin 2009). At a federal level, political and scientific wrangling by loosely connected industry actors, from lobbyists to researchers, enable scientific research and energy policies that hastened natural gas development by exempting fracking from federal regulations.

A result of this amorphous, compartmentalized, and hard-to-follow network is that landowners' lives are twisted out of their individual control (Bowker and Star 2000). People like Laura are placed in a space of exemption, without the tools to create connections that would enable them to gain recompense from a system that adversely affected their lives and lands (Agamben [1995] 1998).[27] The final irony of this situation is that Laura was forced to contribute to the industry's information sequestration by signing a nondisclosure agreement or gag order in return for receiving enough money from the company to move away from the contaminated site.[28] This gag order contributed to the persistence of the regime of imperceptibility around fracking's health hazards. However, by contacting Theo Colborn, Laura became part of building an alternative network of noninstitutional science, whistle-blowing regulators, and NGOs that, over the course of 10 years, has effectively identified fracking as a pressing public health problem.

Methods for Following Chemicals

Seeing a Disruptive System and Forming a Disruptive Science

Laura Amos's and Theo Colborn's calls can be viewed metonymically as a transformative convergence between two frontiers: the social and environmental health hazards of energy extraction, and concerns about synthetic chemicals. Amos and her family became a bodily intersection between these two issues: the structural impossibility of confronting the energy industry, and the scientific impossibility of linking her illness to chemical contamination. Through Colborn, Amos connected with a network of researchers, policymakers, and NGOs that had been working for the past 20 years to regulate potentially dangerous synthetic chemicals. Laura's lived experience was a vital link between those experiencing fracking and research into the unforeseen hazards of synthetic chemicals. Colborn formulated the endocrine disruption hypothesis, the idea that many synthetic chemicals in the environment can disturb hormonal signaling and fundamentally reshape reproductive, immunological, and neurological systems during human and animal development.[1]

This chapter tracks Colborn's development of a unique approach to researching the effects of synthetic chemicals. The interconnection of synthetic chemicals, endocrine disruption, fracking, and fossil fuels is not incidental. It is based on a growing recognition of the adverse impacts of modernity's historical, material, and cultural investment in controlling biological environments.[2] Synthetic chemicals have been wielded to generate and stabilize boundaries in agriculture, manufacture, households, and laboratories as well as to control and influence basic human biology (J. Meikle 1995; Fenichell 1996; Russell 2001; Casper 2003; Ester 2005). A Tufts University laboratory provided one of the first illustrations of endocrine disruption in 1989. A cancer and cell biologist, Professor Ana Soto, was working with a cell line that proliferated in the presence of the hormone estrogen (an estrogen-sensitive cell line) when she was surprised by cell growth *before* she added estrogen.[3] She and her collaborator, Professor Carlos Sonnenschein, tore their laboratory apart to find the unknown source of estrogen. After long investigation, they identified the culprit: Corning test tubes in which the growth medium (the solution used to culture cells) had been stored. Corning confirmed that it had indeed changed the composition of its test tubes, but the tubes' chemical composition was a trade secret. Soto and MIT chemists reverse-engineered the tubes' composition and discovered that a plasticizer, nonylphenol, was the estrogen mimic. This finding affected the storage of her cell lines and raised the questions of whether the biological activity of plastics could contaminate other experiments and indeed influence estrogen responsive breast cancers (Soto et al. 1991; Colborn, Dumanoski, and Myers 1997; Krimsky 2000).

This discovery of a plasticizer that acts as a hormone turned the background infrastructure of an experiment, the test tube, into the experiment's surprise-generating subject or what the historian of science Hans-Jörg Rheinberger (1997) calls an epistemic thing. Such accidents or unforeseen circumstances when scientists make receptors for signals they are not seeking to find can completely transform the laboratory environment and the topic of research (Wylie 2011b). Corning's desire to protect its trade secret by withholding information about the chemical composition of its test tubes both thwarted Soto's efforts to solve her contamination problem and introduced a corporate intellectual property protection into her investigation. Institutionally enabled information sequestration became an active part of solving a scientific question. Scientific infrastructure developed to study the "natural" world was redirected to investigate the test tubes' chemical composition. This book will argue that in order to study chemicals one must often also investigate the corporations that produce them.

Most laboratory biology work uses plastic pipette tips, test tubes, and Petri dishes because they are disposable, sterile, and cheap. Endocrine disruption research problematizes what had culturally evolved into a sacred space for knowledge production, the controlled laboratory environment.[4] Plastics had been vital in establishing that control, yet Soto's study revealed they could in fact undermine it.

Such complexities became clear to me during my first postgraduate position testing for endocrine disrupting chemicals in frog eggs with Dr. Joan Ruderman at Harvard's Cell Biology Department. Ruderman's lab investigated fundamental aspects of cell cycles (how and when a cell divides, rests, or grows) using the oocytes (unfertilized eggs) of the African clawed frog, *Xenopus laevis*, which could be stimulated to mature into eggs capable of being fertilized when exposed to hormones. Ruderman thought that *Xenopus* oocytes might make a good model to test for chemicals that disrupt hormonal signaling. We realized that Soto's findings meant that we could no longer use plastics in our experiments and that even meticulously cleaned glassware test tubes might be problematic for trace contaminants.

Endocrine disrupting chemicals (EDCs) challenge traditional toxicological models of dose response. Very low doses—parts per million and parts per trillion—can produce remarkably strong effects (Vogel 2008b; Vandenberg et al. 2012), unlike most chemicals where "the dose makes the poison": the higher the dose, the greater the effects (Vogel 2008b). Figure 2.1 shows this idealized model of linear dose-response relationship.

Based on this assumption, regulators establish "safe levels" of chemical exposure by testing progressively lower doses of a chemical on model organisms until a no-observable-adverse-effect level (NOAEL) is reached. This dose is generally divided by a factor of 3 to 1,000 to determine the threshold limit value (TLV) for allowable exposure to a chemical (Sellers 1997; Murphy 2004, 2006; Vogel 2008a; Vandenberg et al. 2012: 28). However, the endocrine system works differently. Hormones can work as feedback systems, producing U-shaped dose-response curves (nonmonotonic) rather than linear (monotonic), as shown in figure 2.2.[5]

Hormonal systems can be responsive at very low doses, represented by the rising arc of the lighter gray line. A peak of responsiveness can happen at a low dose then; as the dose increases, the feedback system shuts down. Adverse effects increase again at much higher doses (beyond the middle ebb and up the darker gray line), but these effects can be unrelated to the endogenous hormonal feedback system. The darker gray line graphs the dosages tested in standard toxicological practice where the effects of lower doses are ignored because it is assumed that they produce lower effects (Vogel 2008b).

Developing the *Xenopus* assay was a nightmare, not just because of its demands for tiny precise doses and spotless glassware but also because of the possible exposure to chemicals that might adversely affect my future

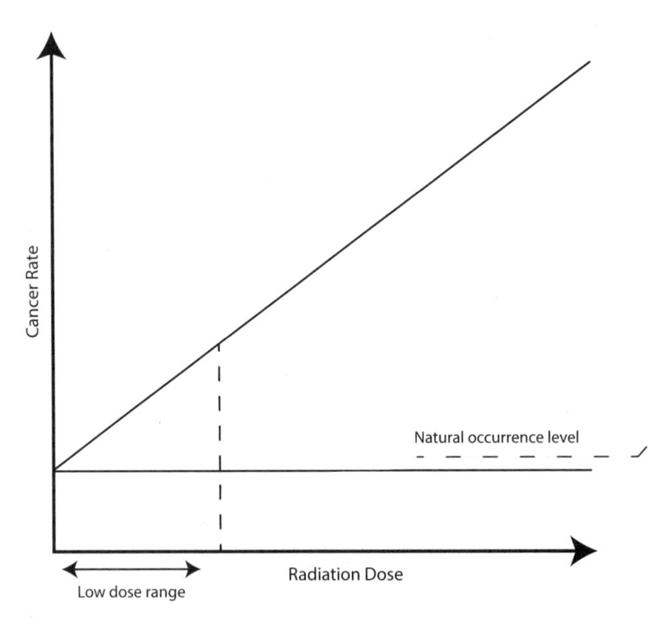

FIGURE 2.1. An idealized example of a linear dose-response curve.

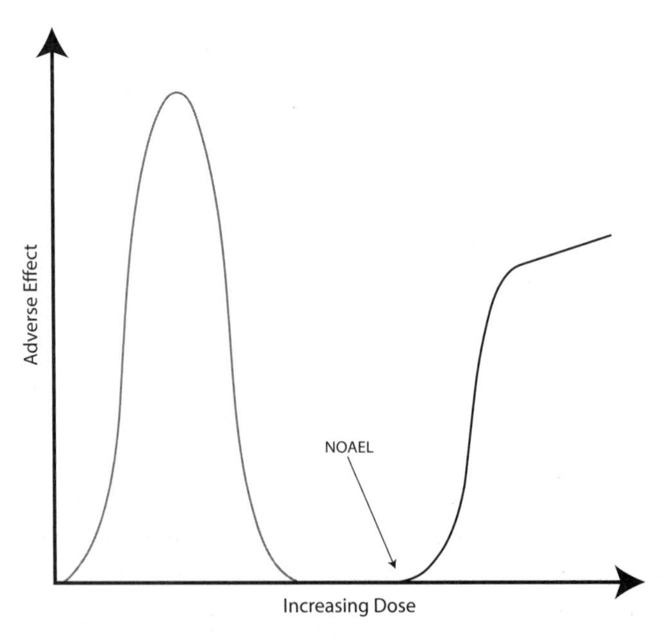

FIGURE 2.2. U-shaped dose-response relationship. NOAEL stands for "no observed adverse effect level."

children.[6] The frog oocytes' variability meant that eggs from different frogs matured at different rates, making it challenging to reliably assess whether the chemicals were changing the rates of oocyte maturation between experiments. Furthermore, the frogs' baseline health was questionable because, raised in plastic tanks as embryos and for their entire lifetimes, they were probably already exposed to EDCs.

The problem was fascinating for me, particularly as I'd grown up with stories of my parents' research on how nongenetic signals control early stages of embryo development.[7] Suddenly, the controlled space of the lab—the very instruments of that control, the disposable, sterile plastics—had become uncontrollable. How had this property of plastics been overlooked?

Sensing a Disruptive System

This section follows how Colborn formulated the endocrine hypothesis and first popularized the idea that many synthetic chemicals could be hormonally active. I pay particular attention to the metaphors she uses to describe discovery. Colborn describes metamorphosing into a beagle dog as she tracked scientific literature to evaluate the environmental health of the Great Lakes ecosystem. She "followed [her] nose" through secondary research material to describe the phenomenon that became endocrine disruption (Colborn et al. 1997: 25). This section examines Colborn's metaphors of smell for what they reveal about her research, data gathering, analysis, and synthesis.

TOXIC CHEMICALS = CANCER?

Colborn was an outlier. She received her PhD in zoology at 58 after careers in sheep farming and pharmacy. In 1987, she received an unexpected job offer from the Conservation Foundation to join a collaborative team with the Canadian Institute for Research on Public Policy in order to study the health of the Great Lakes straddling the U.S.-Canadian border. The United States and Canada signed the Great Lakes Water Quality Agreement in 1972, pledging to clean up the Great Lakes after decades of overfishing and industrial and agricultural pollution. The foundation aimed to develop an ecosystem approach in which the two countries would manage the Great Lakes as one system. Colborn took a pay cut for a job that she told me "felt like something [she] must do, [that] was just right" (phone interview with the author, August 18, 2005). She realized the project's magnitude and challenges early: six policy- and advocacy-oriented staff divided between Canada and the United States, and Colborn was the only biologist on the team.

Evaluating the "health of the Great Lakes" is an amorphous project. However, Colborn felt her path was well defined: toxic chemicals were synonymous with cancer. Colborn was a creature of her context. Rachel Carson's *Silent Spring* and the DES (diethylstilbestrol) scandal (the popular synthetic estrogen DES caused cancer in the children of mothers who were prescribed the drug) had embedded cancer as synthetic chemicals' primary insidious risk in both the public imagination and regulatory and safety testing regimes (Carson 1962; Dumit and Sensiper, in Dumit and Davis-Floyd 1998; Krimsky 2000: 2). From the Delaney Clause of the Federal Insecticide, Fungicide and Rodenticide Act (1958) to the Toxic Substances Control Act (1976) and the Occupational Safety and Health Act of 1970, carcinogens had been singled out as dangerous chemicals, unsafe at any dose. So Colborn went looking for cancer. She focused on public health records and epidemiological research for areas of heightened toxic chemical exposure as well as on reports of cancer in wildlife. She devoted months to looking for increased cancer rates in the Great Lakes Basin and surrounding areas, comparing rates across America and Canada. Colborn even attempted to narrow the data by looking for elevated rates of particular forms of cancer, such as clear cell adenocarcinoma, the hallmark cancer associated with DES exposure.

Colborn found no links. In fact, the Great Lakes' cancer rates were slightly lower overall than in the rest of the United States. She was "at a dead end" (Colborn et al. 1997: 19). She set out looking for cancer like a technician applying a prescribed template: toxic chemicals=cancer.[8] This template directed her on where to look, at areas of high exposures to toxic chemicals, assuming that the dose makes the poison. It also framed the kind of data available to her. The state cancer registries that she consulted collected and compiled data on regional cancer rates that excluded other health outcomes. Available scientific research was also geared toward cancer. For example, "cutting-edge" research at the 14th Annual Aquatic Toxicity Workshop dealt with polyaromatic hydrocarbon contaminants and fish tumors. Toxic chemicals=cancer was a highly robust and influential classificatory structure of research and regulation that motivated Colborn's searches.

This kind of template for identifying hazards has structured toxicology, producing bioassays that assumed cancer was the primary health problem from toxic chemicals and that dose-response curves were linear. The National Cancer Institute's 1970s bioassay program tested chemicals at three set dosages based on a linear dose-response model in standardized organisms

(F344/N inbred rats and the B6C3F1 hybrid mouse). This protocol required oral doses to be started at five to six weeks old (Bucher 2002: 199). Animals that had received these doses were then examined for predetermined outcomes such as cancerous or precancerous abnormalities (Bucher 2002: 201). The bioassay depended on and produced standardized biological tools—from the genetically standardized animals to the routine samples taken. These animals and samples became ideal types suited to finding cancer, and only cancer, as a disease outcome (Murphy 2006).

But no classificatory system is totalizing (Bowker and Star 2000: 304). *Our Stolen Future* describes how focusing on cancer became an impediment that had "blinded [Colborn] to the diversity of data she had collected" (Colborn et al. 1997: 19). Colborn's research shifted when she refocused to follow outliers, rather than discarding these cases. She followed the trail of scents that led her away from the cancer template and its ways of knowing to her eventual discovery of endocrine disruption.

Moving beyond the toxic chemicals=cancer equation proved to be the most important step in her journey, for as she looked at the material with new eyes, she gradually began to recognize important clues and follow them. Colborn describes "floundering around in a mass of undigested data," suggesting she felt that she was moving through the data rather than looking at it through a bird's-eye view (Colborn et al. 1997: 19). Diverse data and singularities set aside as outliers no longer blocked her inquiries. They became the very things to investigate. One example was "aberrant parental behavior." Colborn recalled conversations with the Canadian wildlife biologist Glen Fox, who researched behavioral changes in herring gulls. Fox's study of polluted areas of Lake Ontario recorded twice the number of eggs normally found in nests, suggesting that female birds were nesting together rather than within male-female couples. A wildlife toxicologist from University of California, Davis, Michael Fry, later found that DDT, DDE (a metabolite of DDT), and methoxychlor, when injected into previously uncontaminated eggs, "feminized" male birds, producing female cell types in testes and oviduct development in males. Fox sent gull samples to Fry, who found that five out of seven of the male offspring of female-female nesting-gull pairs showed cellular evidence of feminization.

Other "abnormalities" appeared when Colborn reexamined wildlife data, including reduced testicle size in fish exposed to organochlorine chemicals in the Baltic, sterility in mink populations fed on Great Lakes fish, and "abnormal" mating behavior of bald eagles whose aeries were "wasting" when healthy chicks began to languish "suddenly and unpredictably . . . and eventually die" (Colborn et al. 1997: 25). A disturbing study showed lagging neurological development in the children of mothers who consumed Great Lakes fish for two to three meals a month. Moreover, the higher the concentration of polychlorinated biphenyls (PCBs) found in the mothers' umbilical cord blood, the worse their children scored on neurological tests. Colborn "dug deeper" into the minutiae of tissue analysis in wildlife and humans and kept finding the same chemicals in "the troubled species": DDT, dieldrin, chlordane, lindane, and PCBs. She also looked at records of chemicals in human tissues and fats and was "shocked by the concentrations reported in the fat of human breast milk" (Colborn et al. 1997: 25).

LIKE A BEAGLE

Visual metaphors are hegemonic in the sciences: scientists *reveal* things, have great *insights*. They change how the world is *seen*.[9] Why diverge from this narrative tradition using phenomenological language for olfaction? Colborn's concept of endocrine disruption emerged when she caught a scent and "followed [her] nose" to track a collection of seeming accidents (Colborn et al. 1997: 25). She "discovered" endocrine disruption by following and organizing problems in a process called here *beagling* and defined as tracking problems across species and disciplinary boundaries to aggregate data previously dispersed across different scientific disciplines. Colborn reorganized these problems in relation to one another, tracked the flow of synthetic chemicals through the Great Lakes ecosystem, and was able to identify "endocrine disruption." Connecting this data also brought together diverse researchers, a disruptive pack of researchers. Ultimately, it founded a new field of inquiry. Colborn's research traversed many fields: human epidemiology, ecology, endocrinology, developmental biology, biochemistry, and toxicology. Collecting wildlife studies in 43 species of amphibians, fish, birds, and mammals, Colborn described herself as feeling "like a beagle . . . following its nose. She wasn't sure where she was headed, but propelled by her curiosity and intuition, she was hot on the trail" (Colborn et al. 1997: 25). Just as Colborn's language shifts perspective as she moves through the data, her metaphors change from the visual to the olfactory. Modern vision is bound with the ideology of the impartial observer with a God's-eye view from nowhere (see Haraway 1991). Smell has a long-standing association with the animal and physical. As Connor writes, "Smell has probably declined in importance because of the deodorization of the soul and the animalization of smell that has steadily taken place in the chilly, inodorous cultures of the

North" (2005: 3). Following your nose suggests a scent, the prey's biological product rather than its image, wafting unpredictably through air into your body. It is a metaphor of simultaneous bodily engagement with the object of pursuit and its environment rather than a metaphor of separation. Colborn was "on the ground," tracking, following her prey through an underbrush of paper and across streams dividing disciplines. She scented her path without the benefit of a bird's-eye view, mapping her quarry's course from a distance to identify what it was and where it was heading. She tracked something she knew neither how to see nor to directly detect. The data did not emerge from the "view from nowhere." Colborn's immersion in the data enabled her to follow partial imprints—the trace of scent on a leaf or broken branch to a print in the mud—that were left when the quarry breached the expected in what could be mistaken for an accident. With her deadline approaching and project money running dry, Colborn had gathered 2,000 scientific papers and 500 government documents with "so many tantalizing parallels, so many echoes among the studies, " that she had "pieces and patterns but no picture" (Colborn et al. 1997: 25).

A NEW MATRIX OF INTELLIGIBILITY FOR SYNTHETIC CHEMICALS

Switching metaphorical registers again from embodied following to the visual, Colborn thought she might get some "perspective if she laid it all out" (Colborn et al. 1997: 25). She generated a matrix by lining up the disorders she had encountered, such as population decline, reproductive effects, eggshell thinning, wasting, gross defects, tumors, immune suppression, behavioral changes, and generational effects along an x-axis (see figure 2.3). She listed species information, focusing on the 16 species showing the "greatest array of disorders" (Colborn et al. 1997: 25) along the y-axis.

This matrix is a visual log of the path that Colborn tracked. Arranging the singularities she had encountered made aspects of her quarries' shape, direction, and habits visible to her. Making the matrix was a way of thinking through where she had traveled and what she had encountered. Viewing and thinking with the data in this way made three realizations possible for Colborn. First, all of the animals charted with the most problems were top predators who fed on Great Lakes fish. Second, seeing this data arrangement confirmed Colborn's suspicion that although the adults in these populations appeared to be fine, their offspring expressed a broad range of health problems, including intersexuality, metabolic disorders, and gross birth defects.

Population, Organism, and Tissue Effects Found in Great Lakes Animals

Species	Population Decline	Reproduct. Effects	Eggshell Thinning	"Wasting"	Gross Defects	Tumors	Target Organ	Immune Suppress.	Behavioral Changes	Generational Effects
Bald eagle	x	x	x	x						x
Beluga whale	x		n/a		x	x	x	x		
Black-crowned night heron	x	x		x	x					
Caspian tern	x	x		x	x		x		x	x
Chinook/coho salmon	n/a	x	n/a			x	x			
Common tern	x			x			x	x	x	
Double-crested cormorant	x	x	x	x	x		x		x	x
Forster's tern	x	x	x	x	x		x		x	x
Herring gull	x	x	x	x	x		x	x	x	x
Lake trout	x	x	n/a	x					x	x
Mink	x	x	n/a	x			x			
Osprey	x	x	x							
Otter	x		n/a							
Ring-billed gull	x		x	x			x			
Snapping turtle	x	x		x	x		x			x

x = Observed effects that have been reported in the literature. Cells not marked do not necessarily mean there is no effect; only that no citation was found.

n/a = Not applicable.

Source: The Conservation Foundation.

Population, Organism, and Tissue Effects Found in Great Lakes Animals

Species	Population Decline	Reproduct. Effects	Eggshell Thinning	"Wasting"	Gross Defects	Tumors	Target Organ	Immune Suppress.	Behavioral Changes	Generational Effects
Bald eagle	x	x	x	x						x
Beluga whale	x		n/a		x	x	x	x		
Black-crowned night heron										
Caspian tern										x
Chinook/coho salmon										
Common tern	x			x			x	x	x	
Double-crested cormorant									x	x
Forster's tern									x	x
Herring gull									x	x
Lake trout	x	x	n/a	x					x	x
Mink	x	x	n/a	x			x			
Osprey										
Otter										
Ring-billed gull										
Snapping turtle										x

All top predators in Great Lakes Food Chain feeding on great Lakes fish.

Prominent Intergenerational Effects

All the effects can be linked to disruption of the Endocrine System during development

x = Observed effects that have been reported in the literature. Cells not marked do not necessarily mean there is no effect; only that no citation was found.

n/a = Not applicable.

Source: The Conservation Foundation.

FIGURE 2.3. Two versions of Colborn's matrix, published in Colborn et al. (1990). The second version is marked to show the three key realizations behind the formulation of the endocrine disruption hypothesis.

She then wondered if these disorders could be related to something being "derailed" during offspring development. Recalling Fry and Fox's work on DDT mimicking estrogen and intersex phenotypes as well as the recurring chemicals she had tracked in wildlife tissue analyses, Colborn had the third realization that all the "disorders" could be related to similar "disruptions" in hormonal signaling during development.[10]

Colborn went on another hunt, this time directed to the chemicals appearing in wildlife tissues from fish fat to breast milk. She gathered a list of 17 chemicals including DDT; DDE; methoxychlor (introduced to replace DDT); dioxin herbicides 2,4D and 2,4,5T; lindane, a commonly used pesticide; and PCBs (Krimsky 2000: 20). She also encountered reports by scientists about the estrogen-mimicking ability of the plasticizers bisphenol A (BPA) and p-Nonylphenol, common components of polyvinylchloride and polystyrene (Colborn et al. 1997: 129, 130). Each of these synthesized chemicals was developed and used as a technological means of controlling boundaries. As described in Russell's *War and Nature*, DDT was first used to kill mosquitoes during World War II in order to keep American soldiers alive long enough in foreign environments to win the war (2001: 112). DDT returned home to continue the war against pests (2001: 169). The herbicides 2,4D and 2,4,5T, though initially invented as plant-hormone mimics, were used due to their dramatic ability to kill broad-leafed plants. Of these, 2,4D was first sold domestically in 1945 as a lawn-care product (Doyle 2004: 131). It was also widely used as a defoliant by the Department of Transportation and the Forest Service when laying railway tracks and roads (Doyle 2004: 65). Both chemicals played significant roles in the Vietnam War as defoliants, where 2,4,5T became infamous as the source of dioxin in Agent Orange (Doyle 2004: 54).[11]

A myriad of synthetic chemicals made possible the production of standardized food and spaces such as lawns, highways, and railway systems that fed the emergence of North America's twentieth-century mass markets. Industrialized agriculture patterned the American landscape with monocrops, from midwestern fields of wheat and corn to the orchards of California. Michael Pollan describes how annual production of the standard McDonald's french fries from the russet burbank potato requires more than 17 different spraying cycles of at least eight different synthetic chemicals (2001: 218, 227) and results in biologically dead soil.

Synthetic chemicals were not only useful as weapons against "weeds" and "pests." They were also exploited for their remarkable stability and durability. For instance, PCBs' exceptional stability, inflammability, and high boiling point meant they were commonly used as insulators until they were banned in 1976 (Colborn et al. 1997: 89). Plastics proliferated into all areas of life in the latter half of the 20th century. Early plastics, such as ethocel, produced by Dow in 1935 to make use of excess caustic soda, were popularized in military products such as canteens and telephone headsets and as a water-resistant coating for tents and other gear (Doyle 2004: 146). After its initial use as synthetic rubber during World War II, styrene spawned Styrofoam for Navy flotation devices and quickly found other uses as insulating and packaging material. Vinyl chlorine has become ubiquitous in products like Saran Wrap (E. Brandt 1997: 218). Plasticizers such as p-Nonylphenol were added to these plastics to make them more flexible. This proved particularly useful for making tubes that could resist the intense gravitational forces produced by centrifuges in research laboratories (Colborn et al. 1997: 128).

The panoply of synthetic chemicals that Colborn researched was developed and used to shape and stabilize matter, such as crops; to produce homogenized environments, such as lawns; and to allow both people and things to operate in extreme environments ranging from those of combat to those of laboratory life (Saxton 2014). They were deployed to produce and maintain separations between categories of things (Pauly 1987). Crops were separated from weeds, human food from pests, wet from dry, hot from cold. Synthetic chemicals were the tools that allowed the production of high-modernist-controlled environments: factory farms, interstate transportation systems, pest-free homes, sterile laboratories, and manicured lawns. These environments and products served emerging national interests in controlling, organizing, and ensuring the success of their populations (J. Scott 1998; Russell 2001; Foucault 2007).

Colborn's fragments suggested a different plot for the story of these chemicals. Flowing from fields and factories, they entered the Great Lakes and diffused to concentrations too low for detection. They became part of the water, air, and mud and were eaten by passing plankton. As they entered the food chain, the disruptions began. Boundaries were transgressed and chemicals migrated from plankton to their predators. Some simply transfused into fishes' blood from water crossing their gills. Because many of these chemicals are lipophilic molecules, they dissolve easily and accumulate in animal fats. This begins the process of biomagnification in which each subsequent animal in the food chain consumes and concentrates the reservoirs of chemicals carried by their prey. PCBs biomagnify in the fat of herring gulls to 25 million times that of ambient levels (Colborn et al. 1997: 27). Wildlife biologists knew that lipophilic chemicals biomagnify in this way. However, the link between bioaccumulation and transgenerational health effects had not been made.

The species barrier was not the only boundary crossed by these chemicals. Colborn's collection suggested that they also traveled between generations—in the yolk of eggs, in blood across the placental barrier and in mother's milk, exposing offspring during development and infancy. This exposure produced intersex animals: feminized males and masculinized females. Created as border guards against weeds and pests, these chemicals traveled paths as yet uncharted by humans between species and across generations. Tracking the chemicals' scientific literature brought together research on reproductive, metabolic, immune, and neurological disorders. Substances in the environment that mimic estrogen, DDT and DES became linked with research connecting PCBs to immune suppression and cognitive deficiencies and studies showing anti-androgenic compounds such as plasticizers leaching into water.

A critical finding was the remarkable conservation of hormonal receptors across species. The estrogen-mimicking compounds DDT, DES, BPA, and p-Nonylphenol can bind to these receptors and produce abnormal responses in many species, including humans (Colborn et al. 1997: 74). Colborn realized that the diverse effects she was compiling—immune, metabolic, reproductive, and neurological—could all be produced by disturbances to the endocrine system, the interconnecting system of glands that links the body's major systems. Colborn's matrix suggested a new connectivity between the animals, humans, and their environment, between parents and their offspring and the disorders afflicting the Great Lakes region. A hidden, progressively concentrated flow of industrial chemicals was migrating between prey and predator, mother and child, producing an array of developmental disorders by disrupting hormonal signaling.

Colborn's new matrix revealed the inadequacy of the regulatory templates used to develop and deploy the chemicals. All classificatory systems are based on separations made between kinds of matter. As Geoffrey Bowker and Susan Leigh Star write, a "classification is a spatial, temporal, or spatio-temporal segmentation of the world" (2000: 10). The classificatory system in place at the time that these chemicals were developed played a role in how they were deployed. For instance, pesticides such as DDT were introduced when leading entomologists believed that pests were a separable category from other insects, and therefore that they ought to be or could be eradicated wholesale (Russell 2001: 156). Before the DES scandal, most toxicologists believed that the placental barrier separated the mother from her child, so they did not examine the effects of chemicals on the fetus in utero (Colborn et al. 1997: 49). Federal laws were derived from similarly rigid separations between environments and species. For example, the Federal Insecticide, Fungicide and Rodenticide Act regulated the occurrence of pesticides on foodstuffs rather than on other aspects of the environment because it was assumed that this was the only, or at least the primary, route of human exposure (Krimsky 2000: 61).

The petrochemical industries and synthetic chemists who developed these chemicals, the agricultural cooperatives that used them, and the federal agencies, toxicologists, and entomologists that regulated them all organized their actions according to boundaries they perceived from preexisting templates: the boundary between mother and child, humans and animals, genes and their environment. Regulatory responsibilities were allocated to disconnected groups because of the pervasiveness of these rigid categories. The Food and Drug Administration (FDA), Environmental Protection Agency (EPA), and the Fish and Wildlife Service and National Institutes of Health (NIH) are discrete organizations that seldom communicated with each other. Further, when toxicity is defined by templates such as "the dose makes the poison" and "toxic chemicals = cancer," only high doses of chemicals are tested, each individually, on adult bodies. In this scientific and regulatory structure, synthetic chemicals slipped through the cracks. Their powers of metamorphosis went unnoticed because they did not fit the toxicological and developmental models employed by regulatory agencies and researchers. These chemicals do not follow the boundaries perceived and mobilized by the state, boundaries defined to maintain clear (but misguided) differences between species, generations, and sexes.

Colborn's matrix showed chemical connections between supposedly separate categories precisely because she was not looking through a template. Part of beagling involved seeing things in relations of becoming: synthetic chemicals diluted into runoff becoming concentrated in fish fat; predator becoming prey becoming human fat and then breast milk. The chemicals she studied exposed these relationships by showing the leakages across regulations and research disciplines and how such categorization hid the leakages. Endocrine disruptors undo fixed categories such as sex, since they produce offspring of indeterminate sex. In Colborn's purview, sex is the product of concatenated relationships between hormones and genes, the bodies of mother and child and between a mother and her environment. These chemical disruptions suggest that the sex of a child is the product of all these relationships. Similarly, biochemical and evolutionary relationships between species and generations are exposed in the paths traveled by the synthetic chemicals in question. Chemical flow across otherwise seemingly fixed boundaries attests to their provisional nature exactly because it enacts the relationships between them: predator becomes prey, adult becomes child, and male becomes female. It is as though the chemicals themselves acted as agents in the discovery of endocrine disruption. As they disrupted Colborn's preconceptions as well as boundaries and bodies, these chemicals produced troubling new entities. Together with Colborn's nomadic embodied scientific approach, they also produced a new perspective on the world.[12]

TELLING A TALE

Colborn described how her research group director, Rich Liroff, picked up the matrix from the edge of Colborn's desk, asking, "What's this?" Colborn interrupted her writing to explain its significance: "Every single risk assessment we've ever done, every chemical out there, has never been tested for this health effect. Our whole approach to public health and risk assessment is wrong." Liroff was excited. He ran from the room, matrix in hand, and made a call to his Canadian office. Colborn was soon sharing her story at meetings with representatives from Health Canada, Environment Canada, and the U.S. Fish and Wildlife Service.

Colborn's data-as-story drew people into her collection who brought relevant research they had encountered. She lost sleep to late-night photocopying, reading more papers, and writing (phone interview with the author, August 2005). Becoming literally nomadic, she traveled between countries and cities, telling her story and building a network (Latour 1987). Colborn's message did not contain a new model or template. Rather, it was a localized tale, a story about the relationship between different health problems in the Great Lakes. Those who picked it up edited it to fit it into their own stories. Colborn's matrix of health effects in wildlife was merged with the political and economic data on the Great Lakes in a 1990 book produced by the International Joint Commission's team, *Great Lakes, Great Legacy?* Although noting the presence of intergenerational effects and stating that reproductive disorders rather than cancer were the primary threat to species in the Great Lakes, this book played down the new discoveries: "nowhere . . . [did] the authors explicitly link the reproductive problems of wildlife to a general phenomenon of endocrine system dysfunction" (Krimsky 2000: 20).

This was the first of many reports for Colborn, whose database and literature searches continued to expand into human health, chemistry, and historical records. Toxicologists in particular resisted her endocrine disruption hypothesis. In both volumes of *Toxic Chemicals in the Great Lakes and Associated Effects* produced by the group, "every bit of work [she] put in there that was associated with the endocrine effects [was] left out." Colborn explained that toxicologists are not trained to think in terms of developmental effects (Allan et al. 1991):

> I continually had to deal with toxicologists. . . . I wasn't a toxicologist. Toxicologists spend their energy on how much you can be exposed to before you show an obvious health impairment such as changes in organ weight and body weight, overt birth defects, or convulsions, etc. They were fooled for years because when animals were exposed to estrogen-like compounds . . . the babies were always big, [there were] no effects on litter size, and they were able to say that's a safe chemical. The toxicologists were not speaking in terms of development. (Phone interview with the author, August 2005)

In typical nomad fashion, Colborn met someone on her travels who would help her through her colleagues' resistance. Pete Myers, Director of Science for the National Audubon Society, was being wooed by the W. Alton Jones Foundation for the directorship position. He was so impressed with Theo's work that he agreed to take the position if he could recruit Colborn. Colborn accepted a chair with the foundation in 1990. The job provided a route to continue developing her hypothesis, but she was still without a recognized territory within the scientific community. She had an idea that would solve this problem:

> It was important to bring people together and discuss this. There were people there who had never heard the wildlife evidence. There were people there who had never heard the human evidence: the DES story, diethylstilbestrol, where the mothers shared that drug with their babies during their pregnancies, and how it affected the lives of those individuals whose mothers took that pharmaceutical. (Colborn, in Hamilton 1998)

Colborn knew that since she had only recently received her Ph.D. she was too junior to gain attention. Yet she "felt a tremendous responsibility to get this information out" (phone interview with the author, August 2005). A meeting with Jon Vondracek from the Johnson Foundation's Wingspread Conference Center solved this problem.[13] Colborn told Vondracek that she wanted to "bring some experts together, lock them up for three days and not let them go out and talk to anyone else . . . just have each one of them talk about their research" (phone interview with the author, August 2005). She hoped that this meeting would transform these researchers in the same manner that following and arranging their data had transformed her.

The meeting took place in 1991. Known as the Wingspread Work Session, it disrupted disciplinary boundaries and created a new environment for the interaction between people and fields of knowledge. Attendees and historians alike describe it as a turning point foundational to the field of endocrine disruption (Krimsky 2000: 25). Colborn organized a space, time, and process in which a diverse group of scientists, influenced by each other's research, could see themselves as related to one another. They emerged from the meeting as a new form, a pack.

Becoming a Pack

Colborn invited people from 17 different disciplines, including developmental biology, toxicology, psychiatry, wildlife biology, anthropology, cell biology, and endocrinology to the conference.[14] Representatives from three regulatory agencies, the EPA, U.S. Fish and Wildlife Service, and NIH, also attended. She arranged sessions to overcome participants' disciplinary and institutional separations. They would "all sit around and listen to each other and they can have discussions while they're talking with each other and then, near the end, [Colborn would] break them out into small groups" (phone interview with the author, August 2005). Importantly, she did not present her ideas. Rather, she orchestrated the order in which people would talk and to some degree the groups with which they would interact. She decided there ought to be "no more than four or five people in a group because if you get more than that you can't come to any conclusions in a hurry" (phone interview with the author, August 2005).

Colborn's organizational goal for the groups was to make them as diverse as possible in terms of disciplines with no more and no less than one wildlife biologist in each group and to separate those who happened to know each other. Attendees would interact with unfamiliar information and people. Posing three questions to each group stimulated conversation: "(1) Do you think there is a problem with chemicals that can interfere with the endocrine system? (2) What is your evidence for your answer to question one? and (3) What further research ought to be done?" (phone interview with the author, August 2005).[15]

Colborn arranged it so that the participants might hear the same data she had gathered in her travels. Her questions gave participants the scent to follow. But uncertainties remained. Colborn did not and could not control what people might say about their research. She did not know if they would take up the questions or converse in the groups. Finally, she could not know if they would answer these questions in the manner that she would have and whether they would subsequently come to share her problem.

The assemblage of people, process, and place served Colborn's cause well. The process of hearing about one another's work and the purposefully disruptive organization of the groups made possible the conferees' movement out of their traditional perspectives and along the path to a new one. Colborn said, "None of the groups [took] more than a few minutes to answer yes to question one." Each recognized this shared, disruptive problem (phone interview with the author, August 2005). Attendees' descriptions of the Wingspread session read like revelations. In the developmental biologist Frederick vom Saal's account,

> [It was] like a religious experience. . . . People realized the monumental information that was being put together there. The weekend put things into perspective for me. The magnitude and seriousness of the problem became very clear. People were stunned by what we learned there. It doesn't happen very often that you see things totally differently after a short period of time. It was a turning point. (Quoted in Krimsky 2000: 25)

The 1998 PBS series *Fooling with Nature*, about the endocrine disruption hypothesis, refers to the meeting as "groundbreaking" (Hamilton 1998). Colborn said an epiphany took place:

> Before the last speaker even had a chance to speak, they began seeing things that they couldn't believe: "Tell me about that again?" "Now what did you say?" They stayed up at night. They wouldn't go to bed. It was unreal. At two-thirty in the morning they were still arguing and it wasn't because they'd had a lot to drink—we didn't have that. This was genuine, genuine concern. (Phone interview with the author, August 2005)

Interestingly, this was not simply group scientific enthusiasm, a "eureka moment," but "genuine concern." It was an affective process in which scientists took up the hunt, sharing Colborn's passion and urgency to express, expose, and resolve this problem. A new unity developed among the attendees:

> By the second day there was a change in the way the individuals were behaving at the meeting. There was a bonding also, a tremendous amount of bonding. And by the third morning, these people were so moved

> by what they heard that they decided they wanted to produce what was called a consensus statement. They wanted the rest of the world to know what they had discovered that weekend. (Colborn, quoted in Hamilton 1998)

Wingspread's epiphany changed the disparate scientists' interpretations of their own work. It altered their worldviews and resulted in their "bonding" into a new group drawn together by their individual parts in making the disruptive flow of these chemicals collectively sensible. This bonding produced a second system, a heterogeneous pack of scientific disciplines and people who were able to see and speak the same problem when they combined. The scientists also rethought their relationships to their original disciplines. As Colborn told PBS,

> It was those individuals, then, who went back to their respective institutions and said, "We had better look at how we have been doing science in the past." We are seeing the results of those people rethinking how they did things in various universities and regulatory agencies. Even today, many of us have said that that meeting changed the whole course and direction of our lives. It really has. (Hamilton 1998)

The attendees became emissaries and diplomats for their problem, capable of transforming their own disciplines. The Florida ecologist Louis Guillette's interaction with a Wingspread attendee about the feminization of male alligators exemplifies the meeting's transformative potential:

> A colleague of mine came in and started talking to us about work he had done, and a meeting he had been to that summer, in which he had met Theo Colborn, and a number of scientists and said, "You know, environmental contaminants might be acting like hormones." And it was all of a sudden, "Bam!" It was one of these incredible experiences when you realize, I have hormonal abnormalities. I have possibly a contaminated lake. I know I have a top predator that accumulates contaminants, and then it all just kind of came together as a hypothesis. (Guillette, in Hamilton 1998)

A DES researcher, John McLachlan, who was hesitant to attend the conference, added, "Theo pulled us together. . . . We hadn't read each other's literature. Her main contribution is linking the wildlife and human effects" (Krimsky 2000: 26). By the end of the conference, the Wingspread group was not satisfied to simply return to research. They wanted to inform others by writing a consensus statement. Like the fearful quivering or barking recognition induced by the discordant barking of a pack, their consensus statement became an affective weapon that was "subsequently cited in many journalistic accounts and brandished by activists as a new call to arms against environmental contaminants" (Krimsky 2000: 29). It was instrumental in endocrine disruption becoming a field of study.

Colborn described the statement's revelation as "so shocking . . . that no scientist could have expressed the idea using only the data from his or her discipline alone without losing the respect of his or her peers" (Colborn et al. 1992: xv). It was published along with a book, *Chemically-Induced Alterations in Sexual and Functional Development: The Wildlife/Human Connection*, which contained a series of papers by 21 Wingspread researchers and provided not an ubiquitous model that applied across species but rather a "weight of evidence" that "was too great to be ignored" (Colborn and Clement 1992: xiv). While Colborn uses the term *weight of evidence*, the term *arch of evidence* may be more appropriate. Rather than replicating or repeating each others' finding, like the original matrix Colborn put together, the various chapters interconnected and fit together to support one another, like the stones of an archway. Together they produced a newly stable form, evidence of a systematic problem: endocrine disruption. This heterogeneous arch of evidence is suited to describing the problem it illustrates. Endocrine disruption may not be provable by a single experiment or test. Rather, it is multiple phenomena, the phenotypes of which are determined according to factors such as the species exposed, as well as its environment and lifestyle, the timing and mechanism of exposure, the type of chemical and the dose, timing, and length of exposure time.[16] Similarly, "no attempt was made to synchronize the literary style" of the statement's supporting papers. Rather "each chapter st[ood] alone, reflecting the discipline, research and opinions of the author(s)" (Colborn and Clement 1992: xii). In other words, the consensus statement and the book retained their heterogeneity and yet spoke together: "the group speaking as one, gave life to the idea in the form of the Consensus Statement" (Colborn and Clement 1992: xv; Mol 2002).

The consensus statement was ready to publish six weeks after the meeting. Colborn was about to head home to Washington, DC, from the biannual Great Lakes meeting when she received her first signed copy of it. Coincidentally she was traveling with the EPA administrator, who was also her former boss from the World Wildlife Fund, Bill Riley. She showed Riley the book: "He sat and read it, and he said, 'This is serious, isn't it?' And [she] said, 'Well, yes,' then he knocked [her] on the shoulder again. He was sitting behind

[her]. It was a little plane. And then he asked, 'Can I read that again?' And he took it and read it again. He said, 'Can I keep this?'" (phone interview with the author, August 2005). She promised him that she would have copies for his entire staff on his desk by 9:00 AM the next morning.

This story attests to the affective power of the consensus statement. It is not easy to ignore. This is another kind of matrix, a flow of force generated by a particular combination of heterogeneous people coming to see their work and themselves as interrelated. Formed by people drawn together from Colborn's initial matrix, this multitude of situated voices spoke together. It exposed a flow of synthetic chemicals that propagated previously unseen relationships of becoming between humans and other organisms, scientific disciplines, and regulatory agencies.

The possibility of endocrine disruption demands broad changes not only in the regulatory accounting of chemicals but also in scientific disciplines and agricultural and industrial practices. It calls for a concerted effort from entities like the EPA, FDA, and Fish and Wildlife Service, a rethinking, as Colborn said, of "our whole regulatory structure" (phone interview with the author, August 2005). The linkage between wildlife and human effects deterritorializes epidemiology, placing it in closer relation to ecology and wildlife biology. The intergenerational effects also affect the viewpoint and methods of epidemiology, requiring "multigenerational histories of individuals and their progeny and congener-specific chemical analyses of reproductive tissues and products" (phone interview with the author, August 2005). The potential influence of chemicals in the environment of the developing fetus requires developmental biology to produce relational biomarkers that signify the points of interaction between the developing embryo and its environment, thereby also placing the genome back in its developmental context.

The Aftermath of Wingspread

Wingspread's consensus statement, together with Colborn's matrix and its corollary database, became a map for further research and regulation of the newly created endocrine disruption frontier. Endocrine disruption research is now a recognized scientific field that has grown worldwide with conferences for endocrine disruption research in science,[17] policy,[18] social science,[19] and chemical industries,[20] and journals such as *Environmental Health Perspectives* covering the subject. In 2009 the Endocrine Society of America's national conference issued their first scientific statement: "The evidence for adverse reproductive outcomes (infertility, cancers, malformations) from exposure to endocrine disrupting chemicals is strong, and there is mounting evidence for effects on other endocrine systems, including thyroid, neuroendocrine, obesity and metabolism, and insulin and glucose homeostasis" (Diamanti-Kandarakis et al. 2009).[21]

The field's researchers are investigating questions such as whether the plasticizers BPA, phthalates, and nonylphenols could be related to preexisting problems in human populations such as autism, prostate cancer, and hypospadias. Different disciplines bring their own perspectives to these problems. Epidemiologists gather data on exposed populations and draw correlations between exposures and disease. Cell and developmental biologists investigate the effects of exposures in various animal models. Endocrine disruption research continues to bridge the relevant disciplines, with field sciences and laboratory sciences working on different aspects of the problem. The laboratory has become a field in itself, with researchers analyzing the contents of cage materials and animal foods that have been found to contain endocrine disruptors (Howdeshell et al. 2003; Koehler et al. 2003; vom Saal et al. 2005; Ruhlen et al. 2011). There has been an increasing recognition that, given the ubiquity of these chemicals, there is no unexposed place or population on earth (NHANES 2009).

Unlike many other fields of basic research, endocrine disruption has retained ties to regulatory institutions. In 1996 Congress's Food Quality Protection Act (FQPA) required that the EPA screen pesticides and environmental contaminants for endocrine disrupting effects. Wingspread group testimonies were crucial in the development of this bill along with policymakers' concerns about the link between endocrine disruptors and breast cancer (Krimsky 2000: 62–63). It was a quick success given that the Wingspread group had only published its statement five years before. Based on the FQPA, the EPA developed the Endocrine Disruption Screening Testing and Advisory Committee (EDSTAC) on which Colborn served. However, despite more than 10 years of development, no chemicals have yet been evaluated or regulated as endocrine disruptors based on this program. The Trump administration proposed cutting the entire Endocrine Disruptor Screening Program in 2017 (EPN 2017: 35). Colborn was displeased with the EDSTAC process, with both its foot-dragging delays and the final quality of its toxicological endocrine disruption assays, which she believed were grossly insensitive. She ascribed the insensitivity of the final screening program to the EPA not putting those who discovered endocrine disruption on the panel and subsequent panels.

However, Colborn's and Wingspread's efforts to generate a new frontier of research and regulation concerning endocrine disruption have been remarkably effective overall. In addition to the FQPA, Canada banned the commonly used plasticizer BPA for use in babies' bottles in 2008 and a similar bill went before the U.S. Congress a year later (Vogel 2008a, 2008b). The work of studying chemicals continues to be challenging, not merely because of the way that the sciences and regulatory structures that Colborn confronted are organized, but also because of the social, political, economic, geographic, and legal organization of petrochemical industries. These industries excel at producing new chemicals and creating uses and markets for them while also obscuring the hazards surrounding their production, use, and disposal. The next chapter explores how Colborn attempted to develop an institutional structure for endocrine disruption research by forming a nonprofit, science-based advocacy organization capable of countering the political, scientific, and cultural infrastructure of the chemical industry. Her databasing, nomadic scientific methods, and ability to cut data away from its traditional enclosure and recombine it to change its meaning all played a role in making visible another national public health problem: natural gas development.

HEIRship

TEDX and Collective Inheritance

TEDX fills in a large gap in public health protection. Drawing upon its computerized databases on endocrine disruption and coordination with researchers in the field of endocrine disruption, TEDX provides the very latest summaries of the state of knowledge and its meaning for human health and the environment. —NATIONAL INSTITUTE OF ENVIRONMENTAL HEALTH SCIENCES (2010)

Why did Colborn start an organization like The Endocrine Disruption Exchange (TEDX) after her research? Why leave an existing large nonprofit organization like the World Wildlife Fund and develop a small nonprofit around her growing research database? Most important, how does this small nonprofit, then run from a predominantly mining community in the Colorado mountains, fill "a large gap in public health protection," as the National Institute of Environmental Health Sciences (NIEHS) website argues? This chapter describes how Colborn developed a novel approach to environmental health research, which I term a health environmental impact response science (or HEIRship), suited to studying both the emerging health effects of endocrine disrupting chemicals (EDCs) as well as the public relations (PR), scientific, and regulatory strategies of the corporations that produce such EDCs.

TEDX and the Science of Sound Advocacy

The diminutive TEDX office on Paonia's small but scenic main street was a dramatic contrast to the organization's grand vision. Unlike in other American cities I had lived in, where town centers had been hollowed out by suburbanization, this main street was still Paonia's civic hub. TEDX was in the Harvester Building at the center of the street, just down from the single-screen movie theater, local radio station, and a small café. It was an oddity in its lack of relation and even relevance to other community activities. Tucked into the back of its building, visitors traversed a dimly lit corridor with offices on either side that house the local estate agent, an insurance agency, and a local wildlife photographer. A paper sign taped to the glass-fronted office door announced The Endocrine Disruption Exchange. Theo's office was a modest split-level space with an open floor plan. The top floor where I worked featured desks and a conference table.[1] Images from *Our Stolen Future* and Theo's talks decorated the stairwell. Three double rows of large gray filing cabinets dominated the ground floor. These held the paper version of "Monster," TEDX's ever-growing database of endocrine disruption research.

Monster is the heart of TEDX. At that time TEDX's database expert, Lynn, keyword-searched about 3,000 journals a week for articles related to endocrine disruption. She would then inspect the hits and add them to Monster. TEDX's monthly and yearly reports counted the number of searches performed and articles added. It was full-time work, particularly because of a backlog of articles from the years before TEDX had moved to Colorado. Lynn satirized her job stress in a large drawing taped to the gray filing cabinet closest to her horseshoe-shaped desk (see figure 3.1).

Knee-deep in a swamp that represents the Monster database and waving a shovel in distress, Lynn is surrounded by green crocodiles representing pressing database search requests. A stream behind her represents new articles continually dumping more information into the swamp. Lynn describes feeling overwhelmed by keeping pace with Monster's searches, data retrieval, and integration as well as ensuring the rest of the organization is working with the latest version. But this comprehensive database of published work, from many journals and countries, on endocrine disruptions brings together this far-flung and novel field of research. TEDX then combines these scientific publications with worldwide news articles and policy decisions related to the field, offering unique broad-ranging summaries of current work.

The staff provides an odd collection of expertise. Lynn, who came with Theo from DC, has a PhD in entomology. A part-time books manager, Kari,

FIGURE 3.1. Lynn's depiction of working on Monster. Courtesy of TEDX.

is an accountant by training. Mary, who worked on a range of issues such as aggregating searches from Monster into "products" for advocacy groups, scientists, and policymakers and providing information on their other projects such as Critical Windows of Development (CWD), has a background in psychology. I worked most closely with Mary, who was engaged in compiling a new database of oil and gas chemicals. Because TEDX aggregates published material rather than producing novel data, these disparate professional backgrounds did not unduly impede its research and Theo continually drew upon the scientific expertise of a growing worldwide network of endocrine disruption researchers.

Our meeting on my first day heard staff members' progress reports on the wide range of issues with which TEDX deals. Making a summary of TEDX's yearly progress for the board of directors and potential funders was the most pressing issue. TEDX is a nonprofit organization that survives on grants, donations from foundations, and private donors. The board ensures that the organization's finances are in order and that it is pursuing the guiding purposes in its founding mission statement:

> TEDX's core program is to gather, organize, and interpret scientific research relevant to chemicals called endocrine disruptors. Endocrine disruptors are chemicals that interfere with hormones, enzymes, growth factors, neurotransmitters, and any other signaling chemicals or processes that control development and function.
>
> TEDX's goal is to prevent exposure to and to reduce the production and use of endocrine disruptors.
>
> TEDX's strategy to achieve this goal is to provide customized scientific information to academicians, policy makers, government employees, community-based groups, health-affected organizations, physicians, the media and concerned citizens.
>
> TEDX's role is to assure the integrity of the science behind the endocrine disruption movement and to provide the foundation for *sound advocacy.* (Emphasis added)[2]

"Sound advocacy?" In contemporary parlance, "sound" is commonly attached to science in the phrase "sound science." Although multiple terms have been coined for science mixed with advocacy, none of them adequately describes TEDX's approach, which views the two as mutually constitutive rather than as opposed. PR strategists have developed the derogatory terms "activist" and "junk science" during the past 50 years to maintain or cast doubt on scientific certainty, particularly around public or environmental health issues that involve large-scale industries, such as the link between tobacco and cancer and the climate-change "debate."[3]

TEDX's connection of science with advocacy is part of a lineage of environmental health movements such as Woburn residents' "popular epidemiology" to diagnose their own cancer cluster through mapping community health complaints (Brown and Mikkelson 1997). These community responses to environmental toxics challenged expert and epidemiological modes of science. The toxics movement in particular is led by grassroots activists who employ health surveys and, increasingly, nonexpert environmental monitoring described as "street science" (Corburn 2005).[4] While TEDX works with community organizations and is a nonprofit performing informated environmentalism, environmentalism that relies on databases (Fortun 2004, 2011), TEDX does not primarily perform grassroots research. Rather, it provides weight of evidence studies and retains scientific credibility by routinely publishing papers in peer-reviewed scientific journals (Colborn 2004a, 2004b, 2004c, 2006; Colborn and Carroll 2007; Colborn et al. 2011, 2014; Vandenberg et al. 2012). Although Colborn was also a professor at the University of Florida, TEDX is not formally affiliated with an academic institution. It makes products that provide "the very latest summaries of the state of knowledge [in the field of endocrine disruption] and its meaning for human health and the environment."[5] What is this new form of science, expertise, and advocacy that is not for profit, not university based, not performed by a regulatory agency, not generated by communities or chemical companies and that is created by four individuals, a very large amount of paper, and a few computers and yet is so effective that the NIEHS trusts its ability to interpret the "state of knowledge" on EDCs for "human health and the environment"?

A NEW SCIENCE THAT ACCOUNTS FOR CORPORATE R&D

Now that endocrine disruption was socially and scientifically recognizable, Colborn faced the question of how to sustain this field and enable it to persist in the face of resistance from scientific, industry, and regulatory structures. Developing the field of endocrine disruption faced many of the same challenges as establishing the germ theory of disease, as described by the science and technology studies (STS) scholar Bruno Latour (1988). Latour describes how Louis Pasteur helped transform public health research by demonstrating the necessity of biological laboratories. Experimental sciences' relevance to public health crystallized during the mid-19th century. Robert Koch's postulates for proving disease causation by a microbe and Pasteur's public demonstrations in France of the efficacy of laboratory-produced vaccines were instrumental in reorganizing public health around the germ theory of disease. Pasteur convinced the hygienist movement that agents invisible to the naked eye, microbes, were behind contemporary public health crises such as cholera and anthrax outbreaks. In the laboratory, Pasteur was able to isolate, grow, and manipulate microbes, not only to mimic real-world illnesses but also to develop effective counter measures that could also be used to prevent those illnesses. Thus the laboratory became an "obligatory point of passage" for the public health world, the only place where cause and effect between microbes and disease could be established (Latour 1988).

Colborn and those researching endocrine disruption were also attempting to have endocrine disruptors recognized as a public health issue. But unlike microbes, endocrine disruptors have socially and economically powerful spokespeople in the pharmaceutical and chemical industries. TEDX's scientific research and involvement in advocacy were developed to respond to industry strategies of generating doubt about the "scientific certainty" of emerging hazards (A. Brandt 2007; Oreskes and Conway 2010). TEDX is an institutional form that responds to the largely unrecognized and unregulated impacts of corporations in environmental health discussions. Allan Brandt's penetrating history of the tobacco industry, *The Cigarette Century*, shows how PR strategies evolved in response to increasing evidence in the 1950s that smoking caused cancer. Moving beyond using doctors to promote their products (known as the third man technique) and staging public events to "make news" and build their markets, tobacco industry CEOs contracted the PR firm Hill and Knowlton in 1953, and it concluded that "scientific doubts must remain" about the link between smoking and cancer (A. Brandt 2007: 101). Consequently they founded the Tobacco Industry Research Committee and claimed that the committee would work in the public interest to investigate the cancer concern, while in actuality it supported research to perpetuate scientific controversy about the health hazards caused by cigarettes (A. Brandt 2007: 170).

The generation of scientific doubt became a theme in environmental health debates in the latter half of the twentieth century, sowing skepticism about the ozone hole, acid rain, climate change, and endocrine disruption (Oreskes and Conway 2010). The Competitive Enterprise Institute, which actively campaigned for tobacco and the oil and gas industries (about climate change), and other conservative think tanks such as the Cato Institute and the American Enterprise Institute launched a PR campaign to unsettle scientific consensus about DDT, the first example of an environmental endocrine disruptor (Oreskes and Conway 2010: 217). In 1996 a whistle-blower leaked internal documents from a Washington-based public affairs company, Mongoven, Biscoe and Duchin (MBD), that laid out "the battlefield for Chlorine" for the Chlorine Chemistry Council in order to combat increasing public concern that chlorine-based chemicals, including dioxin, DDT, and polychlorinated biphenyls (PCBs), were endocrine disruptors (Rampton and Stauber 2000: 134–35). The MBD documents expressed concern that EDCs' intergenerational effects and safety risks to children would increase the potential for precautionary regulations (Rampton and Stauber 2000: 135). MBD suggested that pro-industry scientists be identified and deployed against prominent endocrine disruption researchers when giving talks (Rampton and Stauber 2000). Theo reported the routine presence of an industry critic at her public presentations. Before *Our Stolen Future* was published in 1996, the industry-funded group the American Council on Science and Health obtained a galley copy of the book and prepared an attack (Rampton and Stauber 2000). Industry PR activities like these explain TEDX's tight security, the appeal of Colorado's remote location, and why the Monster database is only available on paper rather than online. Lynn always logged me in if I needed to perform Monster searches and I was never allowed to bring in my own computer or to work in the office alone or after hours.

A TALE OF TWO LABORATORIES

The United States has had two overarching modes of laboratory-based research in the twentieth century. One investigated the "natural" world and was based primarily in universities. The other used insights from that research to synthesize, stabilize, and sell novel configurations of materials in laboratory-based R&D in the pharmaceutical and chemical industries (J. Beer 1959; Travis 1993; Bensaude-Vincent and Stengers 1996; A. Chandler 2005). Unlike their university counterparts, industry laboratories were not geared primarily to uncover truths about nature, but to rapidly and reliably produce new and potentially valuable (i.e., commercially marketable) materials to transform the human world. These products were frequently used to control biological threats, from microbes and insects to human enemies (Russell 2001).

The first synthetic chemicals seemed innocuous. They were dyestuffs. But they transformed human environmental relationships. Dyestuffs, like indigo, were key colonial commodities, tying European powers to colonies in trade relationships. Ironically, coal tar–based dyes were innovated by an English chemist tinkering with this plentiful and cheap by-product of Britain's coal-burning industrial revolution. The first synthetic dye color, mauve, a color long associated with royalty because it was expensively made from mollusk shells, was popularized by Queen Victoria. Little did she know that the synthetic chemical industry would destabilize the economic foundations of her empire. Synthetic dye manufacturing took off in Germany because the country was not invested in colonial networks of dye production and lacked strong intellectual property protections. Chemical producers quickly called for strong intellectual property protections because once the process of making a chemical is discovered it is easily copied by competitors.

The German petrochemical industry produced industrial giants such as BASF, Bayer, and IG Farben (Travis 1993). The early German synthetic products also mimicked "natural" materials produced by competing nations' colonies: aniline-based colors replaced British red and blue sourced from madder and indigo that were grown in India, artificial fertilizer replaced guano exported by South American colonies, and early plastics and celluloid replaced ivory, bone, and horn exported from African colonies (Ester 2005). The chemicals produced by this industry were truly "something new under the sun," replacing fields with factories and spurring the growth of consumer-driven economies with brighter, cheaper products (J. Meikle 1995; Benjamin 1999; McNeill 2000). The first gas used in chemical warfare, chlorine was a by-product of the German dye industry. During World War II shortages spurred the development of synthetic rubber to replace natural rubber produced and exported by Brazil, Malaysia, and Ceylon (J. Meikle 1995).

Federal funding in the United States spurred the development of petroleum and oil as synthetic chemical feedstocks, creating a new industry in oil-derived petrochemicals (A. Chandler 2005: 23). The oil industry began to produce chemicals as well as energy. The major oil companies Shell, Standard of New Jersey (Exxon), and Standard of California (Chevron) were among the first companies to market petrochemicals prior to U.S. entry into World War II. By the end of the war they were leaders in producing feedstocks and polymers including polystyrene, polyvinyl chloride, polyethylene, and polypropylene (A. Chandler 2005: 23). Competition with chemical manufacturers prevented oil companies from fully capturing the downstream product markets for petrochemicals. However, a self-supporting oil-based industrial ecosystem had emerged by midcentury, stabilizing both the oil majors and petrochemical industries around western consumer and warfare-driven economies. By the 1970s, synthetic fibers accounted for 70% of the fiber market (A. Chandler 2005: 27), plastic production exceeded steel production in the United States (J. Meikle 1995), and an entire economic and physical ecosystem from clothes to transport and food production had stabilized around oil and natural gas.

Industry laboratories became factories in themselves, turning out new materials and processes to generate materials and then rapidly patenting them (J. Beer 1959). History of science research shows that scientific fields, like other institutions, take constant work to stabilize and maintain (Shapin and Schaffer 1985; Latour 1987). The tools used for sustaining scientific fields have been described as social, literary, and material technologies (Shapin and Schaffer 1985). Social technologies include annual meetings, college courses, and institutional departments. Journals and textbooks exemplify literary technologies. Material technologies are less easily recognized as social organizers. Examples include model organisms such as libraries of genetically mutant fruit flies and particle detectors (Traweek 1988; R. Kohler 1994; Galison 1997). Material technologies embed theoretical perspectives, questions, and means of generating evidence. They create communities around the shared skills, experience, and understanding of such tools (Traweek 1988; R. Kohler 1994; Galison 1997). Material technologies of science shape the kinds of questions a researcher is structurally able to ask and the kinds of evidence that can be generated (Clarke and Fujimura 1992; Rheinberger 1997).

Companies' research and development divisions developed their own material, social, and literary practices of science, including industrial laboratories and workbook conventions for the moments of discovery and patenting (Shapin and Schaffer 1985). In contrast to university labs built around individual researchers, corporate science worked to perpetuate the company's reputation rather than that of any one scientist. Company labs introduced open-lab designs where scientists could see each other, making it hard to hide a new discovery. Documentary techniques like the laboratory notebook remained, but with even stricter data recording protocols so that the moment of discovery could be identified for the sake of patenting the product or process. Intellectual property protection replaced research publication as the means of ensuring that companies reaped the benefits of their investments (J. Beer 1959: 70–90). All these practices were intended to generate a permanent tie between the company and its new material or process so that the company might internalize the profits of selling or manufacturing the materials that its R&D produced (J. Beer 1959).

Novelty, usefulness, and profitability have been the guiding principles in the design and development of chemical technologies. Chemical engineering and toxicology have separate histories, because chemists are not trained in toxicology and vice versa. Toxicological testing only takes place to meet regulatory requirements. It is a hurdle to be jumped, not a design principle, except when toxic chemicals, as in the case of pesticides, have markets.[6] Moreover, as chapter 2 argued, the dominant paradigms for toxicology were historically ill suited to distinguishing and assaying endocrine disruption's harms.

The legal, capital, and scientific ties between chemical manufacturers and their products dramatically change research on chemicals' health and safety, particularly because industry and industry-affiliated laboratories are able to construct science in their favor. The following discussion of a controversial Food and Drug Administration (FDA) decision in 2008 not to ban or further regulate a very commonly used plasticizer, bisphenol A (BPA), illustrates this point.

Rats and How the BPA Spreadsheet Deconstructs Industry Science

BPA was first synthesized in 1891 and recognized as a pseudo-estrogen in the 1930s by an English biomedical researcher, Sir Edward Charles Dodds (Dodds and Lawson 1936; Vogel 2009). Its hormonal activity was not considered (or possibly was unknown to industrial chemists) when it was developed in the 1950s as a component of epoxy resins by Shell Chemical (Vogel 2009) and of the now ubiquitous plastic polycarbonate by General Electric and Bayer (Doran and Cather 2013: 226). With uses in electronics, packaging, food linings, dental sealant, and water pipes, BPA is one of the most abundant synthetic chemicals worldwide (4.6 million tons produced in 2012 [vom Saal and Hughes 2005]).[7] A 1996 U.S. national health study of chemicals in human urine revealed that exposure to BPA is ubiquitous (Calafat et al. 2005, 2008).

Colborn and others were outraged when the FDA decided not to ban BPA in 2008, particularly given that it was recognized as estrogenic in the 1930s. They criticized the regulator's decision for contradicting the federally funded, independent, and cutting-edge science on BPA's health hazards and for relying on industry-produced data that created doubt about the effects of low-dose BPA. TEDX had been alerted to a brewing controversy over low-dose effects of BPA by three troubling occurrences. Theo's colleague from Wingspread, Fred vom Saal, published a data set in 1998 illustrating the hazards of low-dose BPA effects *in vivo*. However, two industry attempts to replicate vom Saal's results found no effects. His experiment was labeled as "junk science" and discredited. Secondly, the Environmental Protection Agency (EPA) argued in

2002 that the data on the risks of low doses of EDCs were too "premature to require routine testing of substances for low dose effects in the Endocrine Disruptor Screening Program" (EPA 2002). Finally, a Harvard Center for Risk Assessment study in 2004 funded by the American Chemistry Council found that there was insufficient evidence that low-dose exposures to BPA caused adverse health effects (vom Saal and Hughes 2005).[8]

Against this backdrop, TEDX gathered all the evidence available on low-dose BPA exposures and analyzed it. This process enabled TEDX to simultaneously show the problems of low-dose exposures to BPA via weight of evidence and to produce new ways of studying and confronting industry-funded research. The group's work began with a database search of Monster to build a spreadsheet on BPA research. TEDX reported research results to its board of directors in 2005:

> TEDX's first product was a Bisphenol A (BPA) spreadsheet. This is a user-friendly spreadsheet that lists every peer reviewed study published on the low dose effects of BPA using 1ppm [part per million] or less. Industry has managed to keep BPA one of the best kept secrets of the decade. We immediately shared an early draft with leading scientists in the field, which became the basis for four peer-reviewed papers to defend endocrine disruption research against attack by industry. (TEDX 2005: 2)

The BPA spreadsheet included each paper's citation, abstract, and a column for each effect that BPA induced. As this spreadsheet grew, the evidence became more and more certain that BPA is a very dangerous chemical. It revealed that practically every physiological and organ system is vulnerable to BPA during development, even at very low levels of exposure—within the range of everyday exposure. Importantly the spreadsheet also became a tool for deconstructing industry science: "The BPA Spread Sheet also revealed the flagrant use of unsound studies by corporate interests and their lies that BPA has never been proven to be harmful. We immediately distributed it to key scientists in academic and government laboratories" (Theo Colborn, personal communication, 2005).

Vom Saal, with whom TEDX shared an early version of the document, used it to analyze the results of industry-funded studies versus studies performed by non-industry-funded researchers. Although the 11 industry-funded studies showed no effects, 90% of the 104 non-industry-funded studies showed adverse effects (see table 3.1).

TABLE 3.1. Biased outcome due to source of funding in low-dose *in vivo* BPA research as of December 2004

	All Studies		CD-SD Rat Studies		All Studies Except CD-SD Rats	
Source of funding	Harm	No Harm	Harm	No Harm	Harm	No Harm
Government	94 (90.4)	10 (9.6)	0 (0%)	6 (100)	94 (96)	4 (4)
Chemical corporations	0 (0)	11 (100)	0 (0%)	3 (100)	0 (0)	8 (100)

Values shown are no. (%).

This figure is reproduced from table 1 in vom Saal and Hughes (2005), showing the difference in findings between industry-funded studies and nonindustry-funded studies and the effects of using estrogen-insensitive CD-SD rats versus other strains of rats.

Vom Saal also demonstrated that the researcher's choice of the strain of rat used was a significant factor in finding no effect. When the extremely estrogen-insensitive CD-SD strain of rat was used by either industry or non-industry-funded studies, no effect was found: "relative to human women, it [the CD-SD rat] requires 100- to 400-fold higher doses to produce effects" (vom Saal and Hughes 2005).

The discrepancy between industry and non-industry-funded studies became central to debates about whether BPA should be banned in the United States and how industry-funded science ought to be considered in the weight of evidence. Colborn and six Wingspread members went on to coauthor a paper criticizing the FDA for favoring industry studies that employed "good laboratory practices (GLP)" guidelines (Myers et al. 2009). The FDA developed these guidelines in 1983 after researchers at a Nalco subsidiary, Industrial Bio-Test, were found guilty by a federal investigation of doctoring scientific data (Myers et al. 2009). The GLP guidelines outline processes for the care and feeding of laboratory animals, facility maintenance, and equipment calibration as well as study protocols and data storage (Myers et al. 2009). Colborn and her coauthors argued that the GLP rules, when followed to the letter by industry, were paradoxically rendering endocrine effects imperceptible:

> Unfortunately, although GLP creates the semblance of reliable and valid science, it actually offers no such guarantee. GLP specifies nothing about the quality of the research design, the skills of the technicians, the sensitivity of the assays, or whether the methods employed

are current or out-of-date. (All of the above are central issues in the review of a grant proposal by an NIH panel.) (EPA, quoted in Myers et al. 2009)

Because the FDA promulgated GLP standards, it weighted GLP-implemented research over National Institutes of Health–funded and peer-reviewed research on the hazards of BPA as an EDC. Industry studies' GLP practices hid many of the subtle effects of BPA. For example, similar to the use of estrogen-insensitive rats, Rochelle Tyl and colleagues used a mouse strain that proved to be relatively estrogen insensitive as abnormally high doses of the positive control, estradiol, were required to show estrogenic effects in the mice (Tyl et al. 2008a and b; Myers et al. 2009: 312). Additionally the researchers did not use the most relevant biological endpoints: "a major limitation of the Tyl studies is the failure to measure more meaningful and sensitive end points in order to detect the effects of low-dose BPA exposure, which are often not macro-scopic in nature." The paper concludes, "Public health decisions should be based on studies using appropriate protocols and the most sensitive assays. They should not be based on criteria that include or exclude data depending on whether or not the studies use GLP. Simply meeting GLP requirements is insufficient to guarantee scientific reliability and validity" (Myers et al. 2009: 314).

While it is impossible to prove that industry studies intentionally selected research endpoints, animal feed, and a strain of mice that were unlikely to illustrate harms from BPA, null results using studies employing GLP guidelines had the strategic effect of forcing the FDA's hand in deciding not to regulate BPA (Myers et al. 2009; Borrell 2010). The FDA was compelled by its own definition of "sound science" into accepting industry-funded work over work funded by the National Institutes of Health. Here the Monster database became a material technology for evaluating the state of science and revealing important differences between industry-funded and non-industry-funded research. It enabled Theo and others to deconstruct industry research like sociologists of science, whose work reveals that scientific facts can be constructed (Shapin and Schaffer 1985) and that model organisms are manufactured and "non-natural" and therefore produce different results (R. Kohler 1994; Haraway 1997). Further research found inbred laboratory strains of mice and rats are less sensitive to EDCs than wild-type or non-inbred strains, raising major questions about how laboratory model organisms can be used to study EDCs, given that genetic diversity is the norm outside of laboratories (Vandenberg et al. 2012: 394).

Corporations attuned to how "objective" science can be subtly manipulated have become experts at producing favorable laboratory results (Conway and Oreskes 2010; Dumit 2012). Through PR and manipulation of the laboratory environment, corporations have developed social and material technologies of science that produce doubt about the hazards of their products. TEDX's Monster database is a novel material technology for responding to industry science. Remotely located but highly networked and nonprofit, TEDX is a novel social technology for organizing, interconnecting, and sharing research between scientists and advocates to more effectively respond to the PR activities of industrial interests. While there are structural parallels between TEDX acting as a scientific consultant for NGOs, activists and policymakers, and industry-funded think tanks, the vital difference lies in TEDX's focus on protecting public health and safety rather than sustaining corporate revenues.

CWD and New Science through Databasing

Chemicals generate surprising new relationships when they interact with the world after being released from the test tube. Inside a human being, a chemical's activity can vary, depending upon (1) its effect on different human (organ systems) and nonhuman (microbiome) tissues, (2) the other chemicals it encounters, (3) the dose and route of exposure, (4) the genetic makeup of the individual, and (5) the individual's history of previous exposure.

Diseases caused by environmental chemicals can be harder to recognize than illness caused by microorganisms. Bacteria, viruses, and fungi use their host organisms to replicate and reproduce themselves. If an organism is impacted by another disease-causing organism, there are likely to be large numbers of that organism in the host's body. It should also be possible to isolate the cause of disease in one infected individual and to use that population to induce further disease that demonstrates its provenance (Latour 1988). In contrast, chemical causes of disease have no unifying strategy (though some may build up in the host's body with repeated exposure over time). Unlike microbes, chemicals do not perpetuate themselves by replication inside their host. Chemicals are nonetheless bioactive agents, and the diagnosis and analysis of diseases caused by them require a science attuned to finding novel singularities that emerge from within human populations. Neither laboratory science, in its highly controlled environment, nor corporate research, primarily aimed at generating novel and commercial products, was designed for this.

Although tests may be developed for endocrine disruptor–induced diseases, chemicals can always exhibit environmentally dependent emergent behaviors. These are best identified by research that seeks singularities and accidents and that collects information across disciplines. Monster is a material technology suited for such activities. TEDX does not simply summarize research. Its database continually discovers new behaviors by aggregating data from different research fields (Daston 2012). Colborn's establishment of TEDX and maintenance of Monster continued the tradition she began with Becoming Beagle, in which a spectrum of health disorders was organized and understood as a family tree.

TEDX performs a new HEIRship approach to the study of hazardous chemicals. It views chemicals as agents of an epigenetic inheritance that fuses human and environmental health by focusing on human development and impacts across species (Landecker 2011). It reveals how chemicals form and transform relationships between humans, their environments, and other organisms based on shared evolutionary susceptibility. Practicing HEIRship is a kind of civic science developed to study endocrine disruption because it is simultaneously an ethical position, a rigorous research method, and a strategic approach to science that recognizes and responds to both chemicals and the companies that make them.[9]

TEDX's CWD project illustrates this potential by identifying stages along the timelines of fetal development in both humans and animals that are most vulnerable to chemical-induced disruption (known as critical windows). CWD (Critical Windows of Development) is a grand-scale project that took six years of work, and it has become a useful teaching and awareness tool for policymakers and scientists. CWD is a new way of showing and compiling the available data on a number of chemicals in order to give a holistic view of known and suspected ways they can impact a developing fetus.

While I worked at TEDX, CWD was a very large piece of paper stretched across an entire wall with a hand-written map of the fetal development stages broken out for each organ system. This project seemed so immense that I could not imagine it ever getting off the ground. Theo imagined users would visit the CWD website and be able to pick up any stage of human development in any organ system and see which chemicals were affecting development at what dose and how.

TEDX released CWD in February 2009. By then the wall chart had been replaced by a searchable online database and an interactive timeline combining available data on developmental stages with contemporary research on low-dose exposures to dioxin, BPA, phthalates, and chlorpyrifos (see figure 3.2).[10]

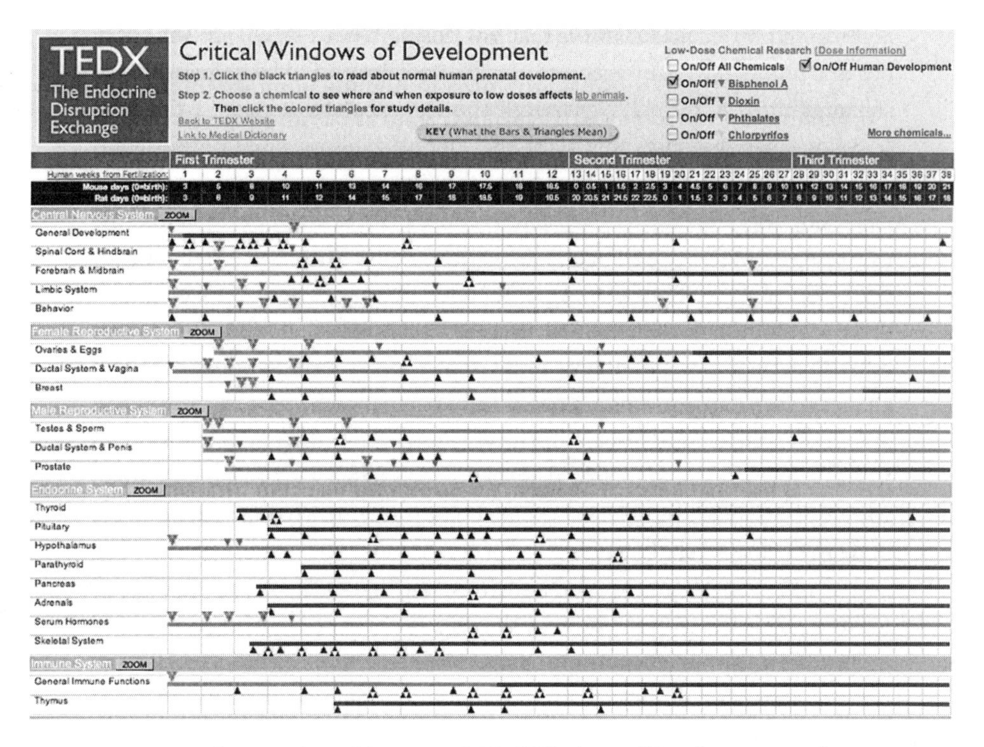

FIGURE 3.2. A screenshot of the TEDX Critical Windows of Development product from 2011, showing research on the impacts of low-dose BPA exposure. The triangles below the line show research on developmental stages; triangles above the line show research on BPA.

This screenshot of the CWD website illustrates the breadth and magnitude of the research behind the tool. The display window shows human, rat, and mouse development overlaid on the same time scale. Development is then broken out by system (nervous, reproductive [male and female], endocrine, immune), by individual organ, and by other developmental features (not shown in the image).

This timeline image exemplifies the data on low-dose BPA exposure. Green or multicolored triangles (indicating multiple clustered studies) above red and blue lines showed the BPA results. Clicking on those triangles would add a window with further information. For instance, Kawai and colleagues found that the offspring of pregnant mice fed with a low dose of BPA during the 11th and 17th days of fetal development have increased expression of the estrogen receptors alpha and beta in neural cells when they are examined at 5 and 13 weeks after birth (Kawai 2007). This finding is notable for scientists and regulators because it shows that low doses of BPA can change the number of a body's estrogen receptors, which could impact the whole hormonal signaling process. Can we even define this change as a disease?

Studies like this show that synthetic chemistry has released novel signals into cellular relationships forged through millennia of coevolution, sending ripples of unpredictable change across entire plant and animal systems. The estrogen receptor, for instance, is the earliest evolved hormonal signaling receptor (over 450 million years ago), therefore many species whose evolution diverged after its development share this signaling pathway and a susceptibility to similar EDCs (Guillette and Crain 2000; Thornton 2001). Random changes within organisms (genetic and epigenetic changes, for example), and in their environments, are the driving forces of evolution, often through novel changes in cell-signaling pathways. Hormonally based signaling pathways are particularly crucial in syncing up the numerous organ systems in the body in feedback relationships. How will developmental exposure of human and/or other organisms to agents such as BPA that change cell signaling affect the species' evolution? Beyond the difficulties in recognizing chemically induced diseases, we are entirely ignorant of EDCs' potential to alter our collective future. The only way to understand the consequences of our industrial biochemistry is to aggregate and analyze the results of experiments already in process, including those that are being lived across ecosystems by the generations of humans currently exposed to this complex set of inputs.

CWD opens a whole new way for a viewer to engage with this issue. The database does not collapse the results into a gross statistic. It displays individual data points that can be viewed next to each other with links to further explore scientific literature. This enables viewers to "double check" TEDX's summary for themselves. It also provides researchers new ways to analyze the data by viewing the results of different kinds of research together. For instance, behavioral data are accessible, along with biochemical data. For instance, one can access a 2008 study that found that feeding pregnant mice with low doses of BPA, during the human gestational equivalent of day 6 of week 6 to day 7 of week 10, results in behavioral changes in offspring: the time the pups spend nursing and in the nest decreases and the time spent alone increases (Palanza et al. 2008). Comparing biochemical and behavioral data does not confuse correlation with causation; rather, looking across such data allows researchers to begin to imagine novel connections between biochemical and behavioral outcomes. Rather than offering only a weight of evidence, CWD helps create new arches of evidence where findings from different fields fit together to explain and support one another (see chapter 2):

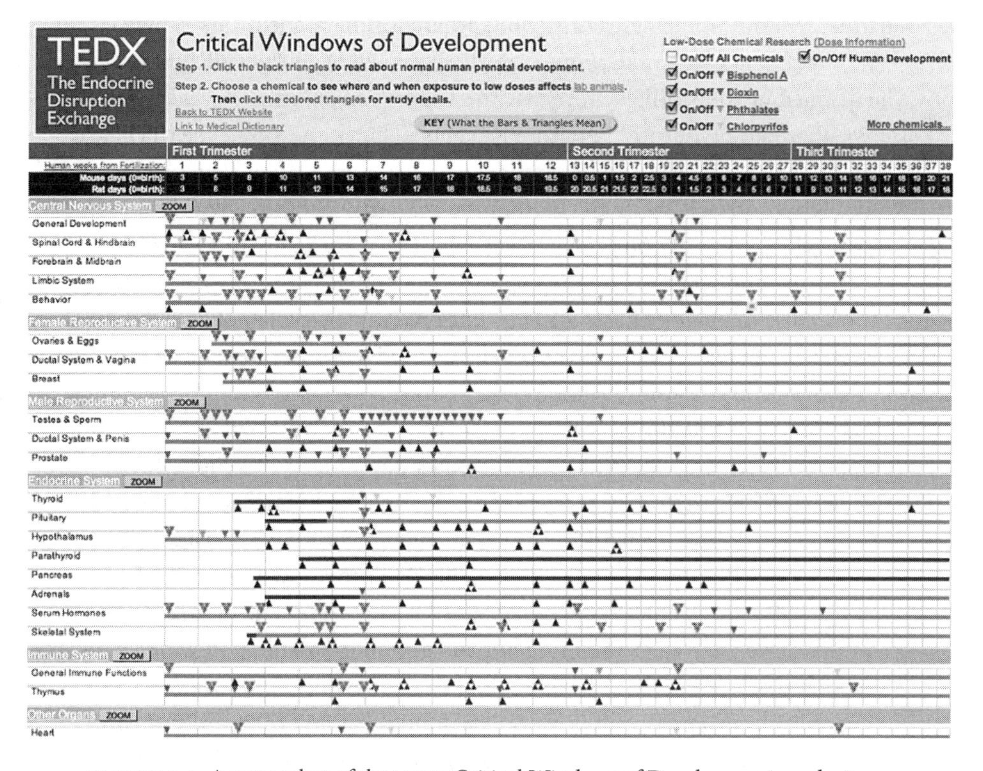

FIGURE 3.3. A screenshot of the TEDX Critical Windows of Development product from 2011, showing research on four endocrine disruptors: BPA, dioxin, phthalates, and chlorpyrifos.

Could physiological changes in the numbers of estrogen receptors be related to such behavioral changes? Without comparisons across disciplines, there is no way to identify a potential connection.

By allowing individuals to access and choose specific data and see it in the context of other available data, CWD changes how weight of evidence studies are presented and represented. A simple view of the whole database provides a visual sense of the scale of the endocrine disruption problem as well as illustrating that there are moments of particular vulnerability to it, based on the data from just four chemicals (see figure 3.3).

This page reveals that detrimental effects are clustered in the first trimester or very earliest developmental stages. Visual patterns suggesting systemwide effects also occur between weeks 6 and 7. The CWD makes interconnections between fields that are necessary to study a phenomenon like endocrine disruption. This approach potentially allows researchers to predict other chemicals' harms, identify developmental periods where testing a chemical's activity is vital, and understand more about basic development by illustrating how it can be derailed or transformed. A review of CWD for *Environmental Health News* noted that "one of the most important benefits is that the new timeline will help scientists instantly identify where more research is needed. 'For example, I was especially surprised to see the gaps in the phthalate data,' said Vandenberg, whose research focuses on how prenatal exposure to hormones disrupts rodent physiology and behavior" (Cone 2009). On the other hand, advocates using the database can easily illustrate that, while mechanistic explanations between diseases and fetal exposures have yet to be determined, there is ample evidence of biological impacts from fetal exposures (Cone 2009).

CWD transforms the process of engaging with environmental health issues. The sheer scale of information refutes any industry's simple lines of dismissal (for example, that there is no evidence for low-dose developmental effects of BPA). The format shows that chemicals should be considered together, rather than individually, if we really want to know their full effects. Imagine what this database could look like if it showed data on all 304 chemicals that have been detected in a body burden study by the Environmental Working Group?[11]

CWD did not attempt to achieve scientific certainty. It is a tool for hypothesis generation and precautionary decision making through HEIRship, identifying developmental moments in which the fetus is shaped by its environment and inherits a chemical legacy that shapes its future possibilities. CWD was developed by TEDX not only to highlight the unexpected impacts of chemical technologies at every developmental stage and at low concentrations but also to have an impact on regulation and science.

HEIRship—Practicing a Health Environmental Impact Response Science

Go and do good. —THEO COLBORN,
personal communication to Sara Wylie (February 2006)

The self-conscious use of PR to influence regulation, technology development, and market choices has ensured remarkable stability, resilience, and persistence for the central oil, gas, and petrochemical companies. While the petrochemical industry has derived a dizzying array of chemical products over the course of the last 150 years, the major manufacturers have not changed substantially since they were founded in the 19th century: there are Bayer (founded in 1863), BASF (1865), Dow (1897), DuPont (1802), Shell (1802), and Exxon Mobil, whose roots go back as far as Standard Oil. The usual economic explanation for this remarkable persistence is that entry-level investment requirements are too massive for most new companies. Thinking of such industries not as resistant to change, but rather as agents able to shape their regulatory and consumer environments helps to make sense of the need for organizations like TEDX that practice science-based, sound advocacy. Like that of the chemical industries, Colborn's science is a science of participation as well as observation. Industry fosters human-product relationships to benefit the producer, stimulating consumer desire. By contrast, Colborn's science uncovers the human-product relationships that are invisible, dangerous, and unhealthy.

Corporations recognize that we are capable of forming affective relationships with our technologies, hence the importance of branding. Branding encourages consumers to trust and form a relationship with a product by associating it with images, memories, and practices that are meaningful (Klein 2000; Lury 2004). Consumers do not simply buy a product based on its material content. From drugs to sneakers, commodities are relational in that humans create social and physical relationships with and through their consumption (Slater 1996; Lury 1996, 2004; D. Miller 1998; Klein 2000). Chemicals in particular, from antidepressants to house cleaners and makeup, are frequently advertised and used as ways of managing our human and environmental relations (J. Meikle 1995; Russell 2001; Dumit 2012). In these advertisements, the consumer is in control of the relationship. The commodities are advertised as tools for "objective self-fashioning" so consumers can optimize themselves and their environments (Dumit 2012). Products become the means for achieving control and presentation of both one's self and one's environment.

Converse to the advertising industry, which encourages consumers to make their own relationships with commodities, Colborn was interested in the invisible relationships that such chemicals forge that are not consciously controlled by the consumer. In a real sense, Colborn's work recognizes humans and nonhumans as inheritors of our chemical technologies such that our populations are literally heirs to our technologies. CWD embeds this perspective in a timeline by joining data on different chemicals and across species. This relational interconnectedness requires a frame shift at every social level, from individual consumer awareness and concern about workplace safety, to regulatory, scientific, and industrial practices that seek to account for this postindustrial condition. For Colborn we are all victims, even before birth, of "industrial terrorism in the womb" (Colborn, personal communication, 2006).

TEDX's products, *Our Stolen Future* and Monster, are part of an effort to generate cultural, regulatory, and scientific awareness necessary "to prevent exposure to and to reduce the production and use of endocrine disruptors." This involves working across scales, from an individual moment of consumption to changing scientific research strategies and regulations.

This connective and relational HEIRship approach accounts for the way TEDX works at so many levels at once. Web tools and products reach far-flung audiences, but TEDX is also very active in local debates. Weekly summertime "calendar spraying" of malathion (an organophosphate, cholinesterase-inhibiting pesticide) in order to control mosquitoes (T. Wills 2006) was a contentious local issue during my time in Paonia. TEDX actively supported ending it by hosting teach-ins about the biological hazards of malathion and testifying before the town council (T. Wills 2006: 5). This local issue produced versions of Monster's database that were tailored to information about malathion.

TEDX is also connected with a large number of NGOs such as the Pesticide Action Network and the Collaborative on Health and the Environment, allowing it to contribute to numerous issue areas.[12] At another level, Colborn's Wingspread collaborator, Rich Liroff, started an advocacy organization that works with shareholders to change company behavior.[13]

TEDX also works at national and state levels to transform regulation. It advocated in 2009 for an alternative approach for formulating regulatory action on endocrine disruptors. In 2009 then Massachusetts senator John Kerry and Congressman Jim Moran of northern Virginia introduced a version of a bill initially formulated by TEDX entitled the Endocrine Disruption Prevention Act (Kerry 2009, 2011).[14] The act would have authorized, according to a press release,

> an ambitious new research program at the National Institute of Environmental Health Sciences to identify EDCs and establish an independent panel of scientists to oversee research and develop a prioritized list of chemicals for investigation. If the panel determined that a chemical presented even a minimal level of concern, it would compel the federal agencies with established regulatory authority to report to Congress and propose next steps within six months. (Office of Congressman Moran 2009)

This press release criticized proposed EPA testing as "limited to only a handful of pesticides and are based on science that many consider outdated." Under the new bill, a panel of scientific experts within NIEHS would perform weight of evidence studies rather than base regulatory evaluations on arguably insensitive assays. The bill proposed an HEIRship approach to regulatory evaluation that, rather than setting standard laboratory conditions to recognize toxicity, allows a flexible approach where new methods of science can emerge to capture emergent toxic phenomena. Kerry reintroduced the bill in 2011 as the Endocrine-Disrupting Chemicals Exposure Elimination Act of 2011. However, despite the support of the Endocrine Society and American Academy of Pediatrics among others, both the House and Senate bills were referred to committee and "died" (ES 2011).[15] Nonetheless, support from significant senators and representatives and the director of NIEHS was a testament to TEDX's ability to scale from a rural Colorado town to Congress, generating effective, differentiated responses to industry activities. The next chapter shows how an HEIRship approach proved to be extremely effective in addressing a new and equally powerful technology, fracking, and social actor, the oil and gas industry.

Stimulating Debate

Fracking, HEIRship, and TEDX's Generative Database

This chapter describes how The Endocrine Disruption Exchange (TEDX) HEIRship approach, a health environmental impact response science, provided a new and powerful means of resisting the frontier of gas extraction. Making a database to identify and respond to fracking chemicals' potential environmental health impacts turned the information vacuum around natural gas expansion into a productive demand for answers. It generated a novel map of this terrain that revealed unrecognized relationships between human and wildlife illnesses potentially produced by these chemicals' travels. The database is "generative" not only because it countered the secrecy surrounding oil and gas chemicals, but also because, when combined with landowners' stories, it generated new scientific, social, and regulatory tools to investigate the potential health impacts of natural gas extraction. Combining a spreadsheet of chemicals used in gas extraction with a listing of the available scientific studies for each chemical's health effects, the database became crucial in developing in-depth journalistic coverage of the issue. A Pulitzer Prize–winning investigative online newsroom, ProPublica has published more than 100 articles since 2008 in its ongoing investigation of natural gas extraction called "Buried Secrets." Theo Colborn, TEDX's database, and the stories of landowners such as Laura Amos in Garfield County figure prominently in this series (Lustgarten 2008a).[1] The investigation's title, "Buried Secrets," implicitly accepts the framing that TEDX's database helped to generate—that natural gas extraction has been undertaken in a way that has consciously obfuscated public health risks. The *New York Times* followed ProPublica's lead with investigative reports that highlighted other unrecognized risks, including concerns that drilling waste brings radioactive material to the earth's surface (Urbina 2010, 2011a–g).

Colborn testified before the House Committee on Oversight and Government Reform in 2007 (Colborn 2007), and she also prompted congressional and regulatory investigations (Waxman 2007; Environmental Protection Agency Region 9 [EPA] 2009; House Committee on Energy and Commerce 2013). The fact that the database could not be contested without companies releasing information has been instrumental in forcing companies to reveal data about fracking chemicals. The second largest natural gas producer in the northeastern United States, Chesapeake Energy released a database in 2011 of chemicals used in fracking.[2] Schlumberger, the largest oilfield services provider and the second largest provider of fracking, also declared a need for openness about these chemicals (Wethe 2009). Even Halliburton, which had threatened to pull out of Colorado if forced to release information on its products' use in fracking (Hanel 2008a), began providing limited information on fracking chemicals' composition (WGAL.com 2010).[3]

Whereas Theo's previous work of studying endocrine disruption made use of databasing to open a new frontier of research, she and TEDX used the database of fracking chemicals to thwart the natural gas industry's reliance on information sequestration in order to expand its interests.

Informating Environmentalism

The oil and gas database was the online counterpart of Colborn's beagling process, described in chapter 2. The database's development by a small nonprofit company is a testament to how "informating" environmentalism is transforming possibilities for advocacy. After the Bhopal disaster in 1984 (in which a Union Carbide pesticide factory in India released a toxic gas in the sleeping city of Bhopal, killing and injuring hundreds of thousands of people), federal passage in the United States of the Community Right to Know Act in 1986 transformed what the general public could know about the chemical industry (Fortun 2001). The Toxic Substances Release Act ushered in a new phase of "informating environmentalism," as it required for the first time that companies report toxic releases to the government, making them publicly accessible via the Toxics Release Inventory (TRI) (Hadden 1994; Fortun, 2004, 2011).

New kinds of environmental investigation became possible with access to this information. For example, Scorecard, an online database that acts as a portal for TRI data, allows individuals to track their local pollution sources. It was started in 1998 by the Environmental Defense Fund (EDF), a nonprofit organization first organized through a successful movement to ban DDT.[4] Designed in collaboration with toxicologists and an MIT computer scientist, Phillip Greenspun, Scorecard compiled data from more than 150 publicly available databases on the toxicity, location, and quantity of chemicals released by specific U.S. companies.[5] Greenspun convinced the EDF to release Scorecard as a website rather than as stand-alone software, a paradigm shift for ensuring free public accessibility. Kim Fortun argues that such databases have become key to knowing and relating to our environment (Fortun 2004, 2009, 2011).

Companies were so surprised to see how much pollution they released that some voluntarily agreed to reduce emissions (Fung and O'Rourke 2000; Karkkainen 2001; EPA 2003). Scorecard was enormously successful at its inception, aggregating data from sources such as the U.S. Agency for Toxic Substances and Disease Registry, California's EPA Air Contaminant database, and International Program on Chemical Safety's International Chemical Safety database in order to allow users to become researchers about the hazards of pollutants produced near them. Scorecard crashed on its first day after receiving 40 hits per second rather than the 20 per second for which it was designed (Foster, Fairley, and Mullin 1998). In its first two months, the site received more than 10 million hits (EDF 1998). An industry executive described it as "the worst thing that could possibly happen to the industry" (Foster et al. 1998).[6] Continuing that statement's legacy, Scorecard was vital to TEDX's construction of a database of the health effects of chemicals used in gas development. That it is now possible for a small organization like TEDX to develop a database with such a compelling impact on national discourse about the safety of oil and gas development has demonstrated the change in environmental advocacy enabled by web-based environmental information.

Historically, databases and maps have been preeminent means used by national governments for building and managing territories and populations. Because maps and databases are necessary for connecting citizens with property to assign voting rights, assess residency, and store and calculate tax information, they are vital to modern state organization and function, as in the collection of census data (Foucault 1979; Deleuze and Guattari 1987; J. Scott 1998). They are vital to the process of knowing and constructing individuals as citizens (B. Anderson 1991) and to the bureaucratic processes of a state (Weber 1968). Databases can be defined as information-storage mechanisms that separate objects and allow users to relate them to each other on the basis of an established system of classification. Geoffrey Bowker (2005) notes that making a database is engaging in a memory practice, a way of deciding what is worth remembering, and that memory practices change over time and place. As described below, Scorecard and TEDX's database exemplify how databasing software and networked, publicly accessible databases change who, what, and how information can be collectively remembered. They enable changes in the social process of knowledge generation (Bowker 2005).

Starting Small: The Beginning of the Database

TEDX began to realize the power of linking chemicals used in fracking to potential human health effects in 2005 from the process of systematically analyzing chemicals in products that might be used on nearby Forest Service (FS) land at Spaulding Peaks (FS 2005). The FS had drafted an environmental assessment report on the impact of drilling in this area by the Gunnison Energy Corporation (GEC), and it called for public comment (FS 2005). GEC, historically the company that runs the two remaining coal mines on the western fork of the Gunnison River, had filed an environmental impact statement (EIS) regarding its plans to drill for natural gas in the National Forest area on October 24, 2005 (FS 2005). Colborn wanted to comment on the EIS for the Spaulding Peaks plan after connecting Wes Wilson's letter to Representative Diana DeGette (Dem.-CO), critiquing the 2004 EPA report, with Laura Amos's health problem. Wilson had told Colborn that the EPA had been directed not to look into fracturing. Colborn contacted the Western Slope Environmental Resources Council (WSERC), a group that she helped found. The WSERC and Theo decided to write extended comments on the Gunnison Energy Corporation's EIS.

TEDX's role was to find information about the chemicals that GEC planned to use in fracturing. Theo was already aware of some of these chemicals, including 2-Butoxyethanol (2-BE), but the products listed in the EIS raised more potential health impacts (see table 4.1). For example, LoSurf-300 Non-Ionic Surfactant produced the results shown in table 4.2.

TABLE 4.1. Potential chemicals that may be used for fracking fluid

Chemical Product Name	*Application*
LoSurf-300 Non-Ionic Surfactant	Surfactant
CAT-3 Activator	Activator
Sand-Premium Brown	Proppant
LGC-36	Liquid gel concentrate
Nitrogen-N_2	Analytical/synthetic chemical use
Sodium Persulfate Breaker	Breaker
AQF-2 Foaming Agent	Foaming agent
BA-40L Buffering Agent	Buffer
GBW-3 Breaker	Breaker

Source: This table is reproduced from a Gunnison Energy Corporation environmental impact statement (FS 2005: table 2-16).

TEDX's analysis showed that LoSurf contained three chemicals: one carcinogen and two neural and developmental toxicants. TEDX submitted comments on the Spaulding Peaks EIS to the FS, the Bureau of Land Management, and the EPA (WSERC 2005). Their comments spurred the regional EPA office to lodge its own formal objection to the proposed drilling plan. The document's success in influencing the EPA's decision spurred TEDX to expand its research into chemicals used in natural gas extraction.

The 2005 Energy Policy Act's exemption of fracturing from the Safe Drinking Water Act made gathering information difficult because there was no requirement for companies to disclose the chemicals they used.[7] Although one could find products used in fracking under their trade names, there was no public information about their chemical compositions. TEDX's research had two aims: to identify products used in fracturing and to name their chemical constituents. This effort involved the Oil and Gas Accountability Project (OGAP; now Earthworks/OGAP), a small national organization dedicated to making the oil and gas industry accountable on many issues, including its usage of potentially toxic chemicals. The Freedom of Information Act requests made by OGAP provided lists of products as well as some of their hazardous components. OGAP asked TEDX to further investigate the chemicals' health impacts.

When I arrived, TEDX was working out a format in which information from different sources could be integrated to build a master database. Becoming sleuthing beagles on the Internet, we expanded the database of products and their constituents used in oil and gas production from two binders of 20 or so products, to a seven-binder set covering 280 products. Building

TABLE 4.2. The components and known effects of LoSurf-300

LoSurf-300, non-ionic surfactant (liquid). May cause eye, skin, and respiratory irritation. May cause headache, dizziness, and other central nervous system effects. May be harmful if swallowed.

Components and known effects:

	Ethoxylated Nonylphenol	*Aromatic Naphtha Type 1*	*Isopropanol*
CAS#	9016-45-9	64742-95-6	67-63-0
% of product	5–10	10–30	30–60
Volatile			
Water soluble/miscible			
Carcinogen		X	
Mutagen			
Cardiovascular/blood toxicant			o
Developmental toxicant	X		X
Endocrine toxicant	X		
Gastrointestinal/liver toxicant		X	o
Immunotoxicant	X		
Kidney toxicant		X	o
Neurotoxicant		X	X
Reproductive toxicant	X	X	
Respiratory toxicant		X	o
Skin/sense organ toxicant		X	o

Note: X = confirmed in scientific literature, o = confirmed on Scorecard.

Source: Reset from WSERC (2005); reproduced with permission.

this database was very similar to Theo's process of gathering the literature that helped found the field of endocrine disruption. Awash in a sea of data, I had no sense of where exactly we were heading. In the beginning, I spent my days with *ols* and *eths*—half-recognized numbers and words.

One of my first projects was to synthesize a coherent approach to this evolving project. A lack of standardization, both in terms of the mode of data storage and in the nomenclature used to describe each chemical, was a major early roadblock to establishing the database. For example, the Spaulding Peaks work existed in several forms: a list of comments on the site, an Excel spreadsheet of the products and their constituent chemicals, and a black binder that was organized alphabetically by the products' brand names.

There was also a white binder containing printouts of product searches, arranged alphabetically by chemical component rather than product name. The different organization of the two binders made it hard to relate each chemical back to the product of which it was a constituent. Beyond these two folders, there were two other product lists arranged in a grid on white printout paper, containing the product name, its composite chemicals, a brief note on their health effects, and a unique chemical identification, or CAS, number.

Combining everything by product on a master Excel spreadsheet yielded 54 products whose constituents made up 339 rows. The language used to describe the categories of toxicants also required standardization, since, in the evolution of the database, different TEDX staffers had developed different nomenclatures. After discussion, the definition of a toxicant was standardized to that used by Scorecard, as chemicals that have been "identified as recognized toxicants based on the hazard identification efforts of authoritative national and international scientific and regulatory agencies." Suspected chemicals, on the other hand, had yet to be identified as a recognized toxicant by the above bodies:

> Chemicals are identified as suspected toxicants based on reports in the scientific or regulatory literature, or on information abstracted from major toxicological databases. . . . Suspected toxicants possess evidence that they can cause specific adverse health effects, but no authoritative hazard identification is currently conducted by regulatory agencies or scientific organizations for that health effect.[8]

We used X, denoting Scorecard's "Recognized," for recognized weight of evidence studies; the red S category was for chemicals noted as "Suspected," according to Scorecard; also included were those chemicals with health hazards additionally supported in PubMed searches of the literature (denoted by a black S). We continued using a material safety data sheet (MSDS) for something already coded unless further data showed that a chemical should be counted as a recognized toxic, then signified with an X.

Database Parasites

After the new nomenclature was in place, the real work started; the mind-boggling, eye-blurring, quotidian work of searching for data on the composition of each drilling product and then looking for information about the

FIGURE 4.1. The structural formula for 2-Butoxyethanol.

toxicity of each chemical component. During this process it became clear to me that TEDX's database was merely a subset of many other databases, each of which was itself also a subset of still others. The process gave experiential meaning to Fortun's statement that "to be informated is to be beset by possibilities for constant re-ordering and re-visualization. . . . They are a place where change happens, sites of transaction between the past, present and future" (2009: 21). TEDX and I displaced information from databases while searching for information on chemicals. We reordered and revisualized them to give them new meaning.

There is a long history of efforts to keep information on chemicals coherent. The numerous names and signs that are available for representing chemicals include molecular formulas, structure diagrams, systematic names, generic names, proprietary or trade names, and trivial names. For example, 2-BE, the chemical that potentially caused Laura Amos's adrenal tumor, has seven different common names: 2-Butoxyethanol; butyl cellosolve; butyl glycol; ethylene glycol monobutyl ether; Dowanol; Bane-Clene; Eastman EB solvent; and BH-33 industrial cleaner. It can also be represented by its structural formula (see figure 4.1).

2-BE can also be represented by its chemical formula: $BuOC_2H_4oH$ ($Bu=CH_3CH_2CH_2CH_2$). Given the array of names and ways of representing each chemical, it can be hard to find comprehensive information for any given chemical. The Chemical Abstract Service (CAS) has been tracking chemicals since 1957. The "unique" number that it assigns sequentially to each newly "described" chemical has no scientific significance, but rather links to a wealth of information about each chemical. As the CAS website describes, "When a chemical substance, newly encountered in the literature, is processed by CAS, its molecular structure diagram, systematic chemical name, molecular formula, and other identifying information are added to the Registry and it is assigned a unique CAS Registry Number."[9] The Registry as of 2017 contained "more than 130 million unique organic and inorganic substances and more than 67 million sequences."[10] Thus the number system created by the CAS became the linchpin, the first, most useful, and clear way for tracking and organizing the plethora of identities of each individual chemical.

TEDX's database was enabled by the availability and standardization of information provided by the CAS database. However, that database was not produced with an entity like TEDX in mind. This is one of the excellent properties of databases accessible on the Internet—they can be used in unforeseen ways by people for whom they were not designed. Parasitizing the CAS Registry was the first step in the larger work of building the database by extracting, recombining, and integrating data from many other online sources.

I used the CAS numbers to search for any information on each chemical in other databases such as Scorecard, Google, and PubMed; in the available MSDS; and in databases specializing in identifying components of pesticides, including those of the National Institutes of Health, the EPA, and the Centers for Disease Control.[11] A document tracking every website, search term, and result created the paper trail behind each search. This was duly printed and added to the growing paper version of the database. This practice gradually evolved into the Excel spreadsheet.

By the end of my first winter month at TEDX, we had researched 138 chemicals. Our data organization became more refined and a series of Excel spreadsheets emerged: one that showed only the chemicals, another listing the products and their constituents, and a series of sheets for each kind of product used in oil and gas extraction. These products included biocides used to clean the wells of bacteria, corrosion inhibitors used to protect the pipes, cross-linkers used to thicken fracking fluids during injection, and emulsifiers to break down those fluids—a total of 18 different kinds of products used in all stages of gas-well development.

THE PROMISE OF PREDICTION

Each potential problem chemical led to deeper literature searches or searches in new toxicant categories. Mary, one of the TEDX staffers, and I called excitedly to one another across the office to describe what we found. Because our job was to seek out potential harms, discovering them encouraged us to search further. We shared particularly interesting findings with the entire group. I remember encountering a study carried out by Greenpeace on Soltex, a drilling mud or shale stabilizer, after a large quantity had been dumped in the North Sea. Soltex had looked relatively benign before I uncovered this study (see table 4.3).

The paper, entitled "Muddied Waters: A Survey of Offshore Oilfield Drilling Wastes and Disposal Techniques to Reduce the Ecological Impact of Sea Dumping" (J. Wills 2000), referenced an article written in 1995 by Greenpeace

England that described the practice of weighting Soltex drilling mud with heavy metals (see table 4.4). This document changed the profile for Soltex dramatically, as heavy metals can be extremely toxic. Table 4.4 shows the Soltex TEDX entry, including the Greenpeace material, where the S signifies "Suspected," according to Scorecard (see table 4.5). When I circulated the data, the staff were excited by the find. Such moments were vital to keeping our momentum.

Some might argue that the process of trying to find evidence of potential threats to human and animal health is not truly scientific. In an ideal world, the scientist is agnostic about research outcomes. Certainly, researchers seeking to show no potential harms would have had a very different mind-set from those of us who worked on TEDX's database, and they would likely have obtained different results. Does that make TEDX's approach unscientific? Not necessarily. The precautionary approach of identifying possible harms that might emerge or be latently occurring is proper to environmental justice and a key aspect of HEIRship.[12] It attempts to elucidate, predict, and track the emergent relationships produced by technologies, because the human and environmental health stakes are high. HEIRship's immediate goal is to determine if there is the potential for harm. This requires actively seeking evidence for concern rather than using an epidemiological approach.

Traditional epidemiology often fails to account for the needs of chemically exposed populations, who must become ill or continue to be exposed in order for causal inferences to be made between their exposures and illnesses.[13] Theo's experience with a local epidemiologist hired to study the health of people in the Garfield County region illustrates this. The epidemiologist had been hired with money from a settlement with Encana, the Canadian oil and gas company discussed in chapter 1.[14] Encana gas production had contaminated Divide Creek in Rifle, near Rick Roles's and Laura Amos's homes (Heiman 2004).[15] Concerns about chemicals were supplanted when natural gas began bubbling up in the creek, leaving the water ignitable and contaminated with benzene (Webb 2004). The epidemiologist heard about TEDX's efforts and became concerned that information published from the database might affect how the community reported their symptoms and thereby skew the data. Such concerns sideline a community's right to know about potentially hazardous exposures, in pursuit of maintaining "pure research subjects" who can be objectively studied without the study interfering with their lived reality. Research that objectifies subjects in such ways has been strongly criticized by research participants and sociologists of science for creating power asymmetries and for putting research goals before human rights (Epstein 1996; Allen 2003, 2004).

TABLE 4.3. The components and known effects of Soltex

Components	*CAS #*	*% of product*	*Volatile*	*Water Soluble/miscible*
Sodium asphalt sulfonate	68201-32-1			
[Polyglycol [mono] oleate]	9004-96-0			
[Polyglycol P400,..P2000, ..P-4000,.. Polyglycol type P1200,..P2000,..P250,.. P3000, . . . P400,..P750 (all==Actocol 51-530)]	25322-69-4			
[1,2,4-Trimethylbenzene (==1,3,4-trimethybenzene; 1,2,5-trimethylbenzene, UNS-trimethylbenzene. Pseudocumene)]	95-63-6			
[1,3,5-Trimethylbenzene (SYM-trimethylbenzene; Mesitylene)]	108-67-8			
Dipropylene glycol [*sic*; see original list]	25265-71-8			
Nitrogen	7727-37-9			
["Light aromatic solvent" (sweetened hydrotreated light aromatic solvent naptha)]	64741-87-3			

Note: S = suspected toxicant scorecard, X = confirmed toxicant scorecard.

Source: Reset from the TEDX natural gas development database, 2006; reproduced with permission from TEDX.

Theo also thought the epidemiological study itself was flawed. She reiterated criticism made in environmental health literature that the health study itself was misguided, as it investigated the whole county rather than only those living around oil and gas wells (Allen 2003; Corburn 2005). Against the background of whole-county data, exposed populations would be statistically lost. Another major flaw in this study was that the population primarily exposed to potentially harmful toxicants, the oil and gas employees, were transient workers and thus not included in results. Furthermore, the study only

Known Effects											
Carcinogen	*Mutagen*	*Cardiovascular/blood toxicant*	*Developmental toxicant*	*Endocrine toxicant*	*Gastrointestinal/liver toxicant*	*Immunotoxicant*	*Kidney toxicant*	*Neurotoxicant*	*Reproductive toxicant*	*Respiratory toxicant*	*Skin/sense organ toxicant*
		S						S		S	
								S		S	
S									X		
										S	

looked at health data recorded by medical professionals and was unlikely to catch any of the health issues occurring in the environment. It examined only broad-scale endpoints like cancer and mortality rates that were unlikely to change immediately. Such gross statistics mask the subtle, potentially intergenerational effects that TEDX had highlighted for endocrine disruptors. Theo argued that, given all of these problems, the epidemiological study was unlikely to show any health hazards and thus might provide evidence that drilling was safe, thereby forestalling further study.[16]

TABLE 4.4. Soltex components

Components	*Concentration (mg/kg)*
Antimony	6.0
Arsenic	0.4
Barium	16.0
Cadmium	0.6
Chromium (total)	1.2
Cobalt	2.0
Copper	1.3
Fluoride	200.0
Lead	3.0
Mercury	0.2
Nickel	11.0
Vanadium	16.0
Zinc	2.1

Source: Table 6 from J. Wills (2000), referenced in an article written in 1995 by Greenpeace England that described the practice of weighting Soltex drilling mud with heavy metals (Reddy et al. 1995, cited in J. Wills 2000: 55). Courtesy of *Sakhalin Environment Watch.*

Chapter 6 investigates how many of Theo's predictions came true. Overall, she thought that such a limited study was not equal to the task of illuminating the scale, diversity, and particularity of potential health impacts from the array of chemicals that TEDX uncovered. She argued that it would be far better to begin such a study knowing what to look for. This would give exposed communities a way to recognize the relatedness of the diverse health symptoms experienced by them, their pets, and surrounding wildlife. It was only after we began summarizing the database for release in the summer of 2006 to individuals and collaborating NGOs such as OGAP that I realized what a powerful tool it was for reframing perceptions of oil and gas extraction as a serious long-term threat to public health.

A Generative Database

OGAP published the first public summary of the database on its website in June 2006 (OGAP and Baizel 2006; Sumi 2006; TEDX 2006). It was disseminated as part of a larger press release in which OGAP, TEDX, and affiliates called for the full disclosure of chemicals used in oil and gas extraction in

Colorado. The summary began as follows: "This project was designed to *explore* the health effects of the products and chemicals used in drilling, fracturing ('frac'ing'), and recovery of natural gas. It provides a *glimpse* at the *pattern(s)* of possible health hazards for those living in proximity to gas development" (TEDX 2006: 1 [emphasis added]).

This exploratory approach is telling because it establishes the idea that the relevant information is incomplete. There was an unknown terrain out there that TEDX had just begun to investigate. The use of *pattern* suggests the database was a template—like a map, it suggested a group or set of health effects that could now be searched for in oil and gas communities. The term *glimpse* further suggests that the map was incomplete and that the database was future-oriented. This summary rhetorically argues that the TEDX database may be used to link disorders back to gas development. It suggests that, because there is no publicly available record of what chemicals are used, or their amounts, location, and monitoring by the gas companies, it is impossible for TEDX or anyone else to show where the chemicals used in gas extraction travel or where contaminated water and bodies might be. Therefore, the TEDX database sought to aggregate and represent the *available* information and to link it to health problems that could be found in communities.

The summary states that "the four most common adverse health effects for the chemicals in the spreadsheet are skin/sense organ toxicity, respiratory problems, neurotoxicity, and gastrointestinal/liver damage" (TEDX 2006: 1). Unlike an epidemiological study that would be unlikely to have raised any concerns (for the reasons noted above), the TEDX database could be used as a map through which the process of natural gas development could be investigated, for example, by examining whether these health effects were occurring when fracking took place.

State-by-state summaries reinforced this intent of the TEDX database. The first public summary discussed above "addresse[d] only those chemicals and products for which there is evidence that they are or have been used in western Colorado" (TEDX 2006: 1). In this way, the database bridged the informational and physical landscapes of gas development, reorganizing them in reference to each other. It suggested new avenues of action that might change the informational landscape and prompt further research in the physical one. Although the TEDX database is clearly not a literal map, that is, a representation of a place that relates objects to be found there spatially, it is used as a map; it serves as a new frame through which to see and explain health problems in the gas patch that provided new routes for action.

TABLE 4.5. Soltex entry from the TEDX gas development database (2006), which now includes the Greenpeace data from 1995

Components	*CAS #*	*% of product*	*Volatile*	*Water soluble/ miscible*
Sodium asphalt sulfonate	68201-32-1			
[Polyglycol (mono) oleate]	9004-96-0			
[Polyglycol P400,..P2000, ..P-4000,..Polyglycol type P1200,..P2000,..P250,..P3000, . . . P400,..P750 (all == Actocol 51-530)]	25322-69-4			
[1,2,4-Trimethylbenzene (=1,3,4-trimethybenzene; 1,2,5-trimethylbenzene, UNS-trimethylbenzene. Pseudocumene)]	95-63-6			
[1,3,5-Trimethylbenzene (SYM-trimethylbenzene; mesitylene)]	108-67-8			
Dipropylene glycol [*sic*; see original list]	25265-71-8			
Nitrogen	7727-37-9			
["Light aromatic solvent" (sweetened hydrotreated light aromatic solvent naphtha)]	64741-87-3			
2-(2-Methoxyethoxy) ethanol (*see* Diethylene glycol monomethyl ester)	111-77-3		X	X
Sulfolane (http://wlapwww.gov.bc.ca/wat/wq /Beguidelincs/sulfolane.htm)	126-33-0			
Antimony	7440-36-0			
Arsenic	7440-38-2			
Barium	7440-39-3			
Cadmium	7440-43-9			
Chromium (total)	7440-47-3			
Cobalt	7440-48-4			
Copper	7440-50-8			
Fluoride	16984-48-8			
Lead	7439-92-1			
Mercury	7439-97-6			
Nickel	7440-02-0			
Vanadium	7440-62-2			
Zinc	7440-66-6			

Note: X = "Recognized" toxicant from Environmental Defense Fund Scorecard.
S = Chemicals noted as a suspected toxicant according to Scorecard.

Source: Courtesy of TEDX.

Carcinogen	Mutagen	Cardiovascular/ blood toxicant	Developmental toxicant	Endocrine toxicant	Gastrointestinal/ liver toxicant	Immunotoxicant	Kidney toxicant	Neurotoxicant	Reproductive toxicant	Respiratory toxicant	Skin/sense organ toxicant	Wildlife toxicant
		S						S		S		
								S		S		
S									X			
										S		
			S	S	S		S	S	X			
		S										
		S						S	S	S	S	
X		S	X	S	S	S	S	S	S	S	S	
			S					S	S	S		
X		S	X	S		S	S	S	X	S		
S					S	S	S		S		S	
X		S	S		S	S	S	S	S	S	S	
		S	S		S		S		S	S		
					S		S	S		S		
X		S	X	S	S	S	S	S	X		S	
		S	X	S	S	S	S	S	S	S	S	
X		S	S			S	S	S	S	S	S	
						S	S			S		
		S	S			S			S	S	S	

The Natural Gas Development Spreadsheet broke down the percentages for potential harms from the 214 chemicals listed into 13 categories of toxicant:

- 58% can cause skin/sensory organ toxicity
- 55% can cause respiratory problems
- 45% are gastrointestinal/liver toxicants
- 36% are neurotoxicants
- 26% are kidney toxicants
- 25% are cardio/vascular/blood toxicants
- 23% are carcinogens
- 23% are reproductive toxicants
- 18% are developmental toxicants
- 15% are immune system toxicants
- 10% are wildlife toxicants
- 9% are endocrine disruptors
- 7% cause mutations (TEDX 2006: 2)

The summary then separated out chemicals by their potential mode of travel, that is, whether they were soluble or volatile, and determined the following harms for 50 (23%) of the chemicals on the list that are soluble, or miscible:

- 80% are skin and sensory organ toxicants
- 70% are respiratory toxicants
- 70% are gastrointestinal/liver toxicants
- 40% are neurotoxicants
- 36% are kidney toxicants
- 34% are cardiovascular toxicants
- 24% are reproductive toxicants
- 22% are developmental toxicants
- 22% are immune system toxicants
- 16% are wildlife toxicants
- 16% cause cancer
- 12% are endocrine disruptors
- 4% cause mutations (TEDX 2006: 3)

The fact that soluble chemicals contain higher percentages of toxicants than insoluble ones suggests that water contamination ought to be a concern, because toxicants are likely to dissolve in water. It is a compelling way to frame the data for western audiences who regard water as an extremely important resource. A similarly worrying pattern also holds true for chemicals that can vaporize, as industry commonly stores wastes in open-air pits. The report offers the following data on volatile chemicals, particularly 62 (29%) of the chemicals on the list that can vaporize:

- 65% are skin and sensory organ toxicants
- 65% are neurotoxicants
- 63% are respiratory toxicants
- 58% are gastrointestinal/liver toxicants
- 45% are cardiovascular/blood toxicants
- 42% are kidney toxicants
- 39% are reproductive toxicants
- 31% are developmental toxicants
- 29% are carcinogens
- 19% are immune system toxicants
- 16% are wildlife toxicants
- 11% are endocrine disruptors
- 6% cause mutations (TEDX 2006: 3)

This holistic method of data presentation through percentages could be somewhat misleading because the same chemical might appear in several categories (e.g., as a neurotoxicant and as a skin and sensory organ toxicant). TEDX's general goal was to raise questions about the body of chemicals used in fracking as a whole. By illustrating a range of toxic effects, the report reaches across audiences: those concerned about wildlife, water, air quality, cancer, and endocrine disruptors.

The database grew dramatically over time and TEDX profiles also began to break out those chemicals specifically used in fracking (see figure 4.2). TEDX's 2009 analysis of the 246 products believed to be used in Colorado summarized the data in pie charts (see figures 4.3, 4.4, and 4.5). Here the viewer can see that 93%, or 228, of the products have health effects and, of these, 86% have health effects in 4 to 14 of the toxicant categories. TEDX also related the data on harmful chemicals back to research on endocrine disruption, showing that 43% of them are endocrine disruptors (2009a: 3). The pie-chart presentation form visually conveyed TEDX's message that it is the chemicals/products as a whole used by this industry that are health concerns.

The database was used as a map in part because the possibility of actually "mapping" these chemicals in the field had been made impossible by the process of information sequestration described in chapter 1. Fortun argues that informated environmentalism is transforming environmental debate through the ability of databases, combined with users' experience, to draw

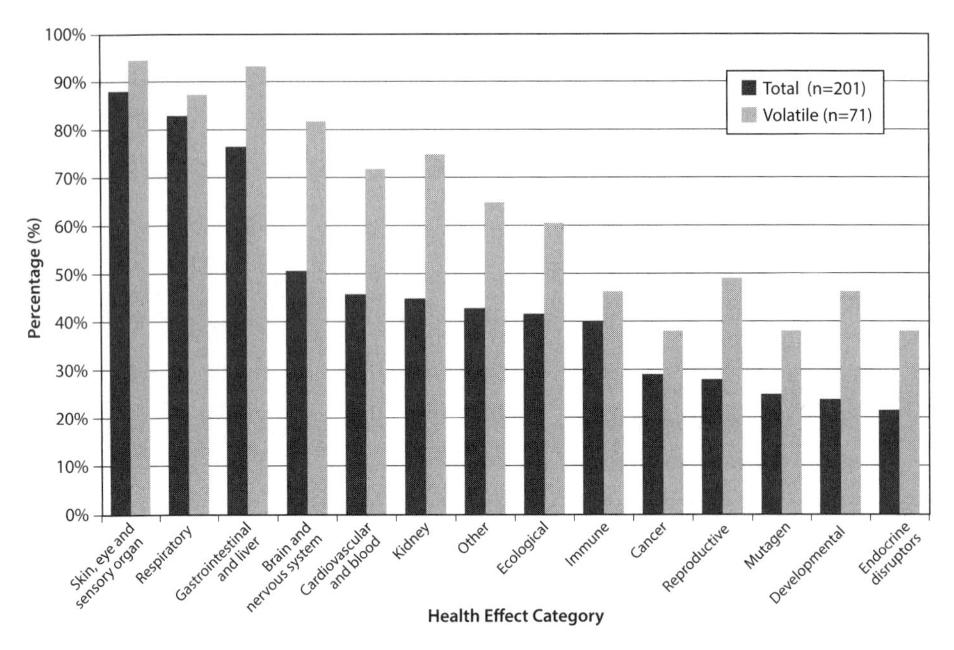

FIGURE 4.2. TEDX profile of fracturing chemical toxicity (2009b: 4).

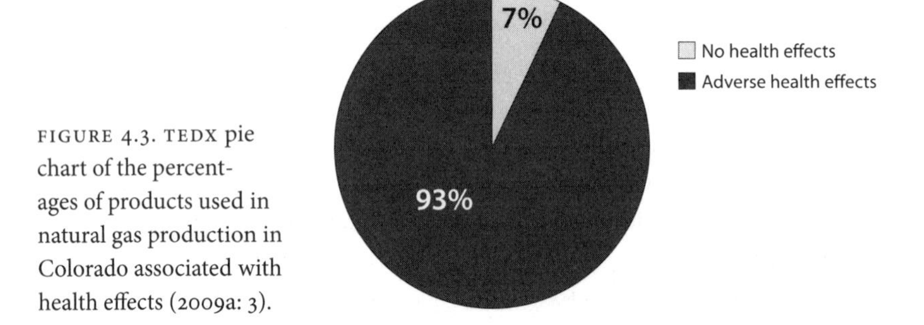

FIGURE 4.3. TEDX pie chart of the percentages of products used in natural gas production in Colorado associated with health effects (2009a: 3).

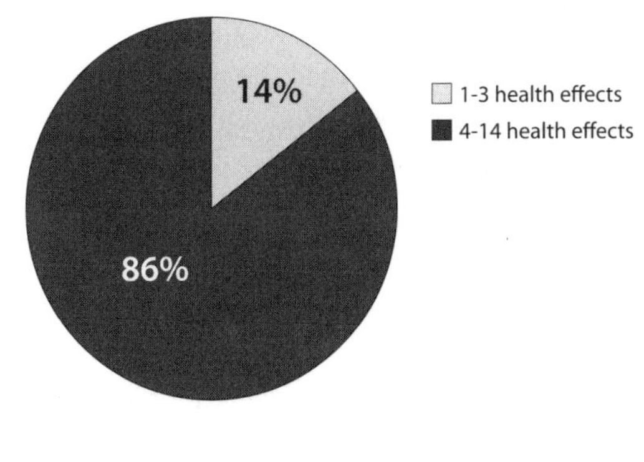

FIGURE 4.4. TEDX pie chart of number of health effects associated with products used in natural gas production in Colorado (2009a: 3).

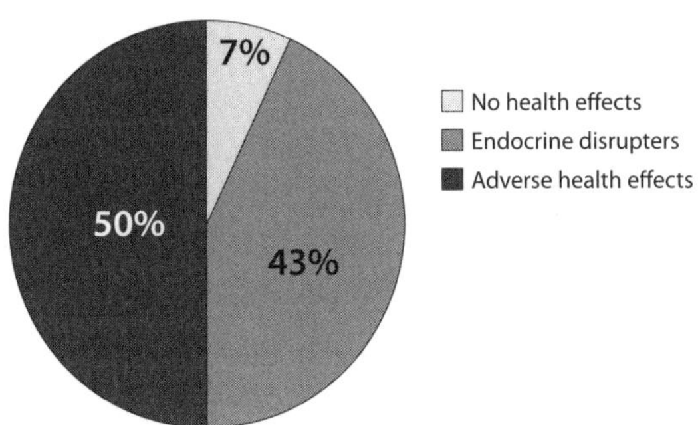

FIGURE 4.5. TEDX pie chart of the percentages of products used in natural gas production in Colorado that contain endocrine disrupting chemicals (2009a: 3).

attention to "discursive gaps" "in what discourses can say or even recognize" (2009: 9). TEDX's database tactically transformed the paucity of public information to its advantage. It insisted that "we make no claim that this list is complete," and that "only a fraction of the chemicals used by the natural gas industry is covered in this spreadsheet" (TEDX 2006: 1). Fortun notes that this admission of limited certainty is a frequent trope in informated environmentalism: "This double gesture—providing information as the basis for action, while qualifying the validity of the information—is characteristic of environmental information systems. Often, they produce working knowledge, which is not claimed to be perfect knowledge" (2009: 17). As there was no master list available to compare with TEDX's, it could only be shown to be incorrect if the gas companies decided to reveal the chemicals they use.

TEDX admitted that it did not know the exact chemicals used in Colorado and therefore could not determine how well the database reflected health hazards on the ground:

> Several reasons led to the lack of data about the health effects of some of the products and chemicals on the spreadsheet:
>
> (a) We found no health effect data for a particular chemical or product.
>
> (b) Some product labels have no ingredients listed.
>
> Some product labels provide only a general heading such as "plasticizer," "cross-linker," etc. Some product labels state only "proprietary" or provide only the name of one or two ingredients plus "proprietary." (TEDX 2006: 1)

TEDX also critiqued the utility of one of its major information sources, the MSDS, saying that it provided information only on acute toxicity for a product in general without necessarily breaking it down by the toxicity of each constituent chemical in that product.

Furthermore, the summary adds, the spreadsheet could not elucidate the actual quantities of each chemical used: "This spreadsheet provides no clues to the volumes of material injected underground during natural gas development. However, typical drilling and stimulation activities use up to 350,000 gallons or more of fluid at each frac'ing. What chemicals constitute these liquids, and at what concentrations, is unknown" (TEDX 2006: 2).

And the spreadsheet could not speak to the impacts of chemical mixtures to which people living around gas development were likely exposed: "This spreadsheet provides only a hint of the combinations and permutations of mixtures possible and the possible aggregate exposure" (TEDX 2006: 2). Furthermore, much of TEDX's chemical data dated back to the 1960s and 1970s, because more recent studies could not be found (TEDX 2006: 1):

> Many of the chemicals on this list have been tested for lethality and acute toxicity. The majority have never been tested for low dose, long term effects that may not be expressed until long after exposure. Nor have adequate ecological studies been done. For example, most of the chemicals have not been tested for their effects on terrestrial wildlife or fish, birds, and invertebrates. It is reasonable to assume that the health endpoints listed above could very well be seen in wildlife, domestic animals, and pets. (TEDX 2006: 3)

Rather than undermining the authority of the database, these unknowns became a primary achievement. Each knowledge gap displayed the paucity of scientific data on low-dose health hazards and the inadequate reporting by companies. These issues supported TEDX's argument that the potential gap between the database and exposure to chemicals used on the ground is the result of the lack of regulatory oversight, monitoring, and public disclosure of information. The database thus became a map of known unknowns that allows the reader to imagine the compendium of information hidden from the public, infusing the landscape of natural gas development with spaces of uncertainty that generate grounds for demanding "full disclosure . . . to protect our watersheds and public health," as "proper monitoring of air and water cannot be designed without knowing what to look for" (TEDX 2006: 4). TEDX has become more sophisticated in representing and presenting these known unknowns as its major finding (see figures 4.6 and 4.7).

The pie-chart representations of data separate the problem of knowing the chemical components of products from knowing the amount of each chemical in a given product. Both are necessary to understand a product's potential toxicity. According to TEDX, full chemical constituents are known for only 6% of the 246 products used in Colorado. Moreover, for 33% of the products the chemical composition is basically unknown, as less than 1% of the information is available (TEDX 2009a: 2).

Theo eagerly accepted a request from the New Mexico Oil and Gas Regulatory Commission to review its test results of gas drilling waste's solid content, thinking it might improve public information about fracking chemicals. However, when TEDX reviewed the laboratory results, it found that most of

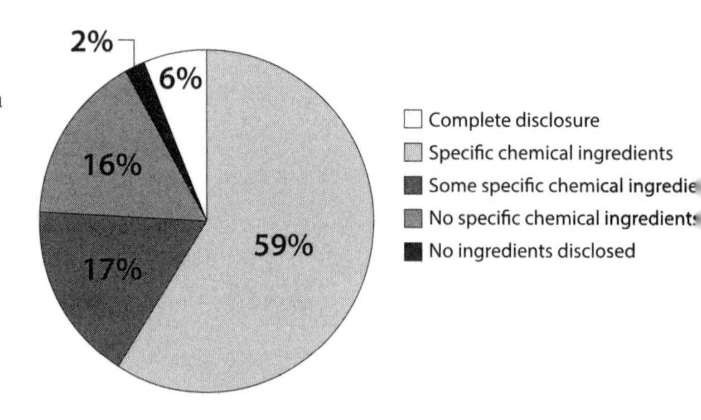

FIGURE 4.6. TEDX evaluation of the information available about 246 products used in natural gas extraction in Colorado (2009a: 2). The pie chart shows the degree of chemical disclosure for the products used in natural gas production in Colorado.

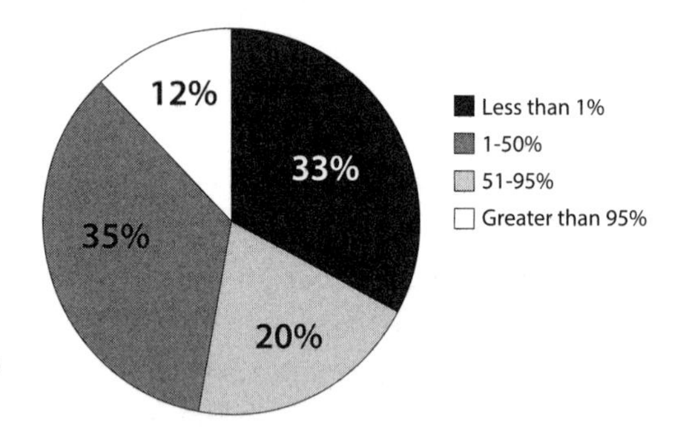

FIGURE 4.7. TEDX evaluation of the information available about 246 products used in natural gas extraction in Colorado (2009a: 2). The pie chart shows the amount of information disclosed about the composition of the products used in natural gas production in Colorado.

TABLE 4.6. Toxic chemicals lists and the 42 chemicals detected

List	*# of Chemicals on List*	*Percentage*
CERCLA 2005	39	93
EPCRA 2006	26	62
EPCRA List of lists	29	69

Note: "CERCLA 2005" is the Comprehensive Environmental Response, Compensation and Liability Act Summary Data for 2005 Priority List of Hazardous Substances. "EPCRA 2006" is the Emergency Planning and Community Right to Know Act Section 313 Chemical List for Reporting Year 2006 (including toxic chemical categories). "EPCRA List of Lists" is the Consolidated List of Chemicals Subject to the Emergency Planning and Community Right-To-Know Act (EPCRA) and Section 112(r) of the Clean Air Act.

Source: The information on these lists comes from a TEDX analysis of chemicals found in New Mexico natural gas drilling waste pits, based on a national list of hazardous chemicals (2007: 1). Reproduced with permission.

the chemicals in drilling waste solids were not included in the TEDX database at all. TEDX found instead that the vast majority of the chemicals qualified as already regulated hazardous substances (see table 4.6; TEDX 2007, 2008).[17] This finding led to speculation that the lack of disclosure rules around extraction enabled industry to unsafely dump waste already known to be toxic. Given that every drilling operation produces such pits, Theo warned that the gas industry was potentially leaving thousands of hazardous single-acre Superfund sites across the country.

TEDX also found a range of chemical constituents that do not even possess CAS numbers, precluding the possibility of collecting information on their toxicity: "It was impossible to link 49 of the chemicals without CAS numbers to any health category aside from the health data reported on an MSDS. The limitations of MSDS data for possible health effects are noted above" (TEDX 2009a: 4).

These information gaps were foregrounded as reasons why full disclosure is required, and they also became avenues for further investigation. For instance, chemicals without CAS numbers, which were often described as petroleum distillates, have become the subject of extended investigation by TEDX and another science-based environmental advocacy organization, Environmental Working Group (EWG). EWG produced a report, "Drilling around the Law," which showed that petroleum distillates contain benzene, toluene, ethyl benzene, and xylene (BTEX). In 2004, industry had agreed to ban the use of diesel as the primary fluid in fracturing, precisely because it contained BTEX chemicals (Horwitt 2010). EWG argued that the industrial use of these distillates constituted avoiding the voluntary ban on using diesel.

PERSONAL STORIES AND THE DATABASE

The first release of the summary of TEDX's database, combined with an OGAP report documenting incidents of frack fluids contaminating streams and water wells and of air-pollution complaints, galvanized individuals and communities resisting gas development. It started a ball rolling that has changed perspectives on gas development for many impacted individuals and at state and national levels. The first publication by TEDX/OGAP reframed local incidents as human health hazards. It documented an explosion of fracking fluid that covered a field with fluid half an inch thick and that contaminated irrigation ditches. The regulatory commission, Colorado Oil and Gas Conservation Commission (COGCC), hardly responded and classified the spill "as non-significant" (Sumi 2006). Yet TEDX's data argued that "the fluids contained a host of potentially toxic substances, including ethoxylated nonylphenol (15–40%); trimethylbenzene (3–7 %), light aromatic naphtha (3–13%); oxyalkylated phenolic resin (15–40%); ethylbenzene (0–2%); xylene (3–13%); and isobutyl alcohol (10–30%)" (OGAP and Baizel 2006: 4). These data provided new grounds for viewing this spill as significant to human health. Read against the background of TEDX's research, companies' and regulatory authorities' responses to these claims of contamination seemed grossly inadequate, if not negligent.

TEDX and OGAP also documented a complaint to the COGCC that a landowner's well water was contaminated. First the commission told the resident, John, that it was probably due to an earthquake or drought, not the unlined pit located about 350 feet above his house. After five months of taking samples, however, the COGCC concluded that the contamination was coming from the pit. The COGCC requested information on the potential materials in the pit, and some product information arrived from the company, Maralex, a few months later. The agency did not require Maralex to test for hazardous chemicals even after receiving MSDSs that showed them to be present: "For example, products contained ethoxylated nonylphenols, 2-bromo-2-nitropropane-1,3-diol, acrylamide, heavy aromatic petroleum naphtha and dipropylene glycol monomethyl ether" (Sumi 2006: 3). The agency did not require remediation of the water even when at the time the OGAP/TEDX incident report was published, John reported that the water still tasted metallic. Moreover, even requests for MSDSs made by the COGCC failed to get full disclosure about what was used in the drilling (Sumi 2006). Against the background of the TEDX report, such conduct appeared to be negligent.

A third incident related to air-quality issues. A group of 10 residents in Rifle complained repeatedly about a nearby drilling operation that produced so much air pollution that residents found themselves gagging and unable to breathe outside. They made repeated complaints, yet only one air sample was taken, despite residents' reports of headaches and other ailments. Residents were told the contents of these fluids were mostly harmless—guar gum and water—but could be odiferous. Eventually one air sample was taken nearby that showed "that benzene and xylenes exceeded the U.S. Environmental Protection Agency's 'non-cancer risk levels' for these compounds—at 67 $\mu g/m^3$, benzene was present at more than double the risk level. Other detectable compounds included acetone, toluene and ethylbenzene" (Sumi 2006: 7). Barrett, the company responsible, was found to be illegally moving condensate to pits that were overflowing (after workers had been informed of the COGCC rule that it was illegal to keep filling pits after they were already full). Barrett then attempted to dispose of the excess by igniting the fluid, which resulted in clouds of smoke. When questioned about this violation by local reporters from the *Post Independent*, the head of the COGCC said that "burning condensate does get rid of the odor problem quickly, which we thought was beneficial." The newspaper also reported that the Garfield County environmental health manager said he "felt pretty good that there was [*sic*] no direct impacts to anybody in the area" (Sumi 2006: 7).[18] The company's disregard for residents' complaints occurred against a background story that the chemicals used in drilling were safe. The TEDX database attacked this narrative by making a weight of evidence argument for the potential toxicity of the vast majority of drilling products and by focusing attention on what cannot be known and why.

Most of the individuals described in these examples could tell with their senses that something in their environment was unsafe—strange smells, burning sensation in the eyes, strange tastes—and they identified changes in their bodies such as headaches, nausea, difficulty breathing, and rashes that indicated a problem. But until TEDX's work, they did not have a scientific basis from which to argue that more research, regulation, and monitoring were urgently needed. They also could not express their concerns that the potential contamination could have long-term impacts on their health and their land. TEDX's work has given such individuals a place to stand and fight against the narrative that drilling is safe.

A Changed Landscape of Perception

As the primary impact of the gas development frontier is borne by those who live and work within it, it is important to examine how the TEDX database reframed and changed individuals' experiences of their health problems and of gas development. Perceiving the landscape of gas extraction through TEDX's map leads one to look for its patterns and to experience its unknowns. I follow this effect in one landowner's story, that of Rick Roles. After completing the TEDX/OGAP summary statement and remembering the list of Rick's symptoms (paralysis, immune dysfunction, recurrent pains throughout his body, and unusual levels of hydrocarbons in his blood), Theo decided to give him the summary of TEDX's findings.

When we visited Rick a few weeks later, he had fused this new information with his own narrative. He had tried to give the summary to the mother of a child whose preschool was having a rig installed next door, but she refused to take it. He also told us he intended to give the TEDX report to his neighbor who had developed a tumor, along with the name of a lab where the tumor could be analyzed. Rick now interpreted the eclectic illnesses occurring in the area as related to each other via gas development. For instance, he connected his own poisoning, his neighbor's tumor, and the neurological and behavioral impairments of his friend's children to shared exposure to gas development. His narratives were structured around known unknowns. By threatening to violently force an industry representative to drink condensate, Rick was expressing a desire to expose the worker to the same uncertain, untreatable, and unrecognized space that Rick inhabits.[19] Rick described the industry's strategy for dealing with sickened residents with reference to his friend Susan, who could only leave the house with a cylinder of supplied air: "No one knows when someone in the valley gets sick, the oil companies keep it quiet, and they're so sick they don't get out to talk to people." Combining the database with his personal experience, Rick perceived a large, partially hidden, yet interconnected group of ailing landowners to whom he felt responsible as a spokesman and advocate to bring them and others out of the dark.[20] He acted as a sentinel, an expression of and spokesperson for the toxic environment produced by natural gas extraction.[21] The TEDX database gave new legitimacy to the uncertainty and disorientation he experienced and gave him a new way to relate it to people around him. Combined with landowners' stories, it transfigured the residents' experience of place. This transformation did not arise from making concrete connections, but rather by mapping gaps and patterns in which landowners like Rick could situate themselves. The uncertainties that the database revealed did not shut down criticism. They became the basis of its cultivation.

As news of TEDX's work spread and Theo traveled widely to discuss the data, Rick's story has been repeated across the country. For example, a woman in Pavilion, Wyoming, heard Theo's talk and then linked the strange and diverse health problems occurring in her family. She already believed that her seizure disorder was related to the contamination of her well water during a frack, but after hearing Theo's talk, she realized it could also be related to her son's liver damage, her daughter's recent miscarriage, her granddaughter's kidney failure, and the respiratory issues occurring in her family and community. These people had all drunk, showered with, and cooked with the water after the company had made perfunctory efforts to clean up the well.

The TEDX database does more than raise questions of whether health problems are linked to gas development. It provides a platform for linking diverse health issues to chemicals used in oil and gas extraction as a whole, rather than on a chemical-by-chemical basis. If regulatory efforts, or even scientific investigation of contamination by oil and gas chemicals, were handled on a chemical-by-chemical basis, there might be no way to perceive connections between these illnesses. The database turns the logic of the frontier on its head: by the very mechanism of information sequestration that otherwise makes tackling this issue impenetrable through other forms of advocacy and science, the database becomes an obstacle to companies that want to argue with its findings. It took the frontier's vacuum and mapped into it an extremely compelling story about the chemicals' potential health effects. The TEDX database was useful to individuals in the immediate term, unlike the epidemiological study described earlier, which could only be useful years down the line to determine whether or not health had declined (Coons and Walker 2005; see also chapter 6). It gave people grounds to re-describe health problems as potentially related to oil and gas practices. The database has also helped to shift state and federal policies. TEDX has become a hub for data about chemicals used in oil and gas development. Other sources and states send in information to be analyzed from Emergency Planning and Community Right-to-Know Act Tier II reports. The New York City Water Board reviewed all of TEDX's data, prompting a deep inquiry into the health and safety of the practice, which resulted in the 2014 statewide ban on fracking (Lustgarten 2014).

Building on the hypothesis that TEDX and other databases are powerful in their capacity to generate new knowledge and identify discursive gaps, the practice of HEIRship transforms field sciences by taking boundary confusion and the inherent messiness and lack of researcher control as unique strengths that counter the limitations of laboratory research and traditional modes of epidemiology (Epstein 1996; Kuklick and Kohler 1996; Kroll-Smith 2000; R. Kohler 2002; Allen 2003, 2004; Murphy 2004, 2006). The TEDX database uses HEIRship by attempting to identify potential illnesses related to chemicals used in gas extraction, thereby providing a foundation for interrelating illnesses that might be produced by the travels of these chemicals in the environment. It was designed to respond to the lack of available data on fracking due to lack of regulatory oversight, but it not only outlined a set of risk-laden chemicals but also aggregated a large range of health problems as potential outcomes of chemicals used in natural gas extraction. The database mapped and illustrated a "family tree" of potentially interrelated illnesses through the diversity of toxic and potentially toxic chemicals employed in fracking. In so doing, it acted generatively as a map for investigating this discursive gap.

Industrial Relations and an Introduction to STS in Practice

The critic is not the one who debunks, but the one who assembles. The critic is not the one who lifts the rugs from under the feet of the naive believers, but the one who offers the participants arenas in which to gather. —BRUNO LATOUR (2004: 246)

The previous chapters illustrated how Theo Colborn assembled new arenas for the scientific and social critique of both endocrine disruptors and chemicals used in fracking. The HEIRship approach used by The Endocrine Disruption Exchange (TEDX), building and using databases to draw together diverse health effects, enabled affected people to perceive each other as related. The evolution of "informated environmentalism" (Fortun 2004, 2011) produces not only new modes of environmentalism but also new data for scientific understanding, aiming to advance research and to respond to chemical-producing companies. The next four chapters examine how the social sciences, particularly science and technology studies (STS), are adopting novel approaches to research that actively construct new ways of doing sciences rather than simply deconstructing the present problems surrounding science and engineering.

STS has a rich history of examining the "social construction" of scientific facts and technologies (E. Martin 1987, 1994; Traweek 1988; Marcus 1995b;

Rabinow 1996; Galison 1997; Haraway 1997; M. Fischer 2009, 2010, 2012; Dumit 2012). From the 1970s onward, STS, gender studies, and critical histories of science and technology all coalesced around questioning the objectivity and "naturalness" of scientific facts and technical objects. They deconstructed how scientific knowledge and technology were made, in order to show their cultural and historical contingency.[1] Latour's *Laboratory Life* (1979, with Steven Woolgar) and *Science in Action* (1987), Sharon Traweek's *Beamtimes and Lifetimes* (1988), and Karin Knorr-Cetina and colleagues' *Determinants and Controls of Scientific Development* and other works (1971, 1975, 1980) pioneered anthropological/sociological approaches to question the naturalness of scientific facts by taking ethnographic methods into the laboratory. Laboratories became places where facts were made rather than simply found.

The most compelling STS research brings power relationships implicit in the structure of research and technologies to the foreground. For instance, Steven Shapin and Simon Schaffer's *Leviathan and the Air Pump* (1985) showed how, in the 17th century, access to and participation in science were initially limited to landed gentlemen, because their social status enabled them to credibly bear witness to scientific findings. Scientific societies were warranted by royal charters and in turn supported the social structure of power.

When sciences became organized around laboratories, this introduced a new element into social power relationships: to argue with the "matters of fact" produced in laboratories, the scientist must first establish his own lab and participate in the experimental program. The sciences thereby organized a powerful new terrain for generating evidence, beyond the established biblical canon, legal history, or logical theorizing (Shapin and Schaffer 1985).

In his 2004 essay "Why Has Critique Run Out of Steam?," Latour examines how the STS-style deconstruction of evolutionary theory, climate research, and endocrine disruption is increasingly employed by climate-change deniers, public relations professionals, right-wing fundamentalists, and left-wing conspiracy theorists. Deconstructing factual matters can lead to a postmodern relativism in which every *truth* is devalued as constructed. This can impair critical theorists' ability to address matters of concern, ethical and values-based issues. Feminist science studies have long advocated for engaged forms of research that build alternative political possibilities based on ethical principles. Donna Haraway's "Cyborg Manifesto," in *Simians, Cyborgs, and Women* (1991), inspires women to embrace the blurring between organism and technology brought about by assisted reproduction, biotech, computation, and nuclear warfare. Arguing against simplistic interpretations that technology is antifeminine because women are associated with the natural rather than the "man-made" world, she encourages women to take an active, constructive role, to build human-machine relationships that are ethical, mutually respectful, and nonviolent. Colborn's database can be viewed similarly, as a constructive critique of science based on an ethical interest in recognizing, reducing, and preventing harm.

While chapters 2 and 3 have illustrated a *classical* STS approach, that of deconstructing how Colborn built a scientific field by making endocrine disruptors and chemicals used in fracking visible as subjects of scientific study, the following chapters illustrate constructive approaches to STS that build upon a deconstructive analysis in order to offer new possible power relationships.

In 2007, Chris Csikszentmihályi, an artist and technologist at MIT's Media Lab, and I founded a research group named ExtrAct. The ExtrAct team built participatory online databasing and mapping tools that aimed to interconnect previously disparate gas-patch communities. These tools were inspired by Colborn's databasing methodologies. They attempted to respond to the structural asymmetries between regions of extraction and the oil and gas industry. This chapter describes how these asymmetries arose, through the technical and informatics systems that enable fossil-fuel extraction. ExtrAct experimentally responded to these informatic asymmetries by using digital media to network gas-patch communities and academics. As a "constructive" mode of STS research, ExtrAct not only critiqued the industry but also actively participated in constructing material technologies of science that facilitated alternative social, technical, environmental, and scientific relationships. I call this research methodology "STS in practice." It aims to move from deconstruction to building more just, ethically accountable and reflexive science and technology.[2]

Mapping Subsurface Academic and Technical Networks

The erasures that make gas development a difficult terrain for surface residents to map and study can be contrasted with the subterranean capital and informatics systems that make the subsurface visible and accessible to natural gas companies. The magnitude of the engineering that reveals gas and oil deposits became apparent to me after I heard a lecture delivered by a representative of Schlumberger, the world's largest oil and gasfield services company, at the MIT Media Lab in the spring of 2008 (IHS 2014). The Media Lab is unique. It started in 1985 as a space for artists, technologists, and scientists to create collaborative research "inventing a better future."[3] Corporate funding supported Media Lab's research, though researchers had latitude to work on their own problems. Schlumberger was a major lab sponsor, one of the 21 donors that contributed to its endowment.[4] Sponsor representative visits were frequent, and twice yearly every research group gave sponsor demonstrations. Although Schlumberger did not sponsor ExtrAct, we had to demonstrate its tools for community oil and gas monitoring to company representatives as part of our obligations as Media Lab researchers. Students were also expected to attend Schlumberger's annual sponsor's lecture.

A small conference room, one of the few well-lit, windowed rooms in the older of the two Media Lab buildings, was crowded with students arranged around a long corporate boardroom table. The Schlumberger representative, Claude Baudoin, stood at its head next to a projection screen. Baudoin was an Information Technology and Knowledge Management Advisor with over 30 years of experience at Schlumberger, having directed technology practice, in the Network and Infrastructure Solutions division, and software engineering and IT, in the Test and Transactions division. Baudoin had also been influential in establishing Schlumberger's relationship with the Media Lab.[5]

Schlumberger has a much closer relationship with MIT's halls of invention than with any of the gas-patch communities I have visited. As Bill Readings argued in 1996, universities have moved away from their modernist moorings as "cultural projects of maintaining their respective nation-states" to become instead "transnational bureaucratic corporations" (quoted in Fischer 1999a: 456). Readings noted that, rather than being pure publics of rational communities, universities are places where the complexity and diversity of postmodern inequalities of culture and capital are becoming impossible to ignore. In many ways, MIT pioneered the integration of corporate sponsorship with university research in its 1919 "Tech Plan":

> Tech Plan rebuilt MIT's entire operation around corporate patronage. It created a centralized Division of Industrial Cooperation and Research—the forerunner of today's ubiquitous technology-transfer offices—to facilitate corporate-funded research projects on campus, open the institute's libraries to industrial sponsors and share alumni records with corporate recruiters. (Kaiser 2011)[6]

At Baudoin's lecture, I realized that although I was at my home institution, I was still technically "in the field," as the informational and technical networks influencing the lives of those I studied in Colorado had been extended into the Media Lab. As an MIT student rather than an ethnographer of science driving around the gas patch, I was closer to the "inside" of the industry than I had ever been before. The cultural anthropologists Michael M. J. Fischer and George Marcus write about the "complicities of all sorts [that] are integral to the positioning of any ethnographic project" (1996: xviii). The Schlumberger representative's lecture about the company's technology research and development reminded me that anthropologists, like scientists, cannot stand outside of their subject of study. Marcus and Fischer argue that recognizing such complicities can generate awareness of already-existing alternative or resistant actions:

> And it is here that the power of ethnography as cultural criticism resides: since there are always multiple sides and multiple expressions of possibilities active in any situation, some accommodating, others resistant to dominant cultural trends or interpretations, ethnography as cultural criticism locates alternatives by unearthing these multiple possibilities as they exist in reality. (116)

Awareness of my multicomplicities raised the question of how I would address them analytically and methodologically. A critique of Schlumberger and MIT's relationship would necessarily involve the complex problem of interrogating the culture of my home institution.[7] Graduate students are told cautionary tales of self-reflexive anthropologists who trespass into the dangerous territory of studying their home institutions. It gets complicated. However, rather than tracing the complicities inherent in MIT and Schlumberger's corporate relationship and my role as a student both listening to Baudoin and inviting Colborn to speak at an upcoming MIT conference, I sought to leverage this complicit, complex site and social position into a more nuanced understanding of how we all came to be together. I asked myself, "How might our relationships be reconstructed" (Readings 1996)? How might I become a critic that constructs? Could I practice STS in a way that materially offered new, less extractive, and more responsible relationships between my research work, MIT, Schlumberger, and gas-patch communities?

It is important to understand exactly why MIT and Schlumberger were prospering from their close financial, intellectual, and educational relationships. Why was Schlumberger interested and invested in MIT and the Media Lab in the first place? Baudoin described Schlumberger's technical challenges and innovations. He stressed that although oil and gas are mundane objects, visualizing and extracting them is an exciting scientific and technological challenge. Engineering schools have long been wedded to corporate capitalism (Noble 1977). Historically, they have trained the workforce for research and development jobs within corporations (Noble 1977; Downey 2007). Schlumberger aggressively recruits graduates from the world's top universities: 7,000 new recruits every year or about 8% of the company's total workforce (*Innovation Quarterly* 2008). This effort relies on deep and lasting links with research institutions. The company's "Ambassador Program" states that it

> defines around 50 elite universities around the world to which Schlumberger wants to be particularly close. The program includes the usual choices, such as MIT, as well as less obvious ones, such as universities in Khartoum and Saudi Arabia. The company appoints senior managers, preferably graduates from those universities, to manage the relationship, by attending careers fairs, meeting faculty, organising internships and arranging teaching visits. (*Innovation Quarterly* 2008: 2)

Thus Baudoin's graduate student audience was a stock of potential future employees who might hear something that could spark their interest in a corporate career.

Furthermore, Schlumberger was attracted to participating in and developing research programs and technology that both assisted its technical development and improved its efficiency. Its primary research and development headquarters lies across the street from MIT (Tuz 2004). MIT's work on artificial intelligence helped it build "expert system" software: programs that simulate expert decision making that proved applicable for determining optimal well spacing in Schlumberger's target gas or oil reservoirs (Winston and Prendergast 1986).

Baudoin's lecture provided a different view of the gas and oil industry and its institutional relationships from the one I had acquired living and working in the gas patch. The surface hardly appeared in the technologically minded view offered at MIT, and the people living in gas-patch communities were not mentioned in the talk or in Schlumberger's promotional materials. I understood then that the industry is a vast technical system that generates not just new software and physical technologies in pursuit of fossil fuels. It also develops informational systems for "seeing" the subsurface that, in their design, disadvantage surface communities. Chapter 1 described how information is the key product marketed by the company, and Baudoin reiterated this insight. It was the Media Lab's excellence in designing digital data analysis, interaction, and visualization that attracted Schlumberger.

Simulating the Subsurface

Mapping the subsurface has required the technical development of alternatives to the human senses and expanding them with technologies such as electromagnetic and acoustic seismic imaging. Schlumberger made its market niche by black-boxing one mode of sensing, the differential electrical conductivity of geological layers, and linking this with potential fossil-fuel reservoirs (Bowker 1994).[8] It maintained its market edge by dramatically expanding technical sensing. The company's seismic imagining subsidiary, WesternGeco, has the largest seismic vessel fleet in the world (Offshore 2005). Schlumberger's sophisticated technical and social system for "knowing the subsurface" is in stark contrast to the paucity of informational infrastructures available to gas-patch landowners. The company's technical, scientific, and social networks that bring the subsurface into relief mask the surface and its inhabitants in three ways: (1) by myopically focusing on workers and fossil fuels; (2) by extracting and exporting information about the subsurface to distant sites, a practice that dislocates surface owners from the planning process behind resource extraction; and (3) by simulating data in a manner that excludes the surface from the calculation.

VAST MYOPIC MACHINES AND VITRUVIAN WORKERS

Schlumberger's seismic imaging has a worldwide reach. Its fleet of ships uses "Q-Marine technology," which is recognized by experts as "the most advanced technology available in the seismic industry where individual hydrophones spaced at 3.125 m intervals [on kilometer-long streamers (chains of connected hydrophones)] sample and transmit data continuously back to the vessel" (Carton et al. 2006: 1). "With their trailing gear [seismic vessels] are easily the largest man-made moving objects on Earth," according to a vice president of marketing for WesternGeco (Viner 2013). WesternGeco's newest vessel design for arctic seismic operations, the *Amazon Warrior*, has the capacity for 124 miles of streamers (Kliewer 2014).

In parallel to Q-Marine systems, Schlumberger's Q-Land system for land seismic imaging lays sensors in networks across tens of kilometers of land. In Algeria, the Q-Land system was used to map "44 km^2 [17 square miles], with a dense grid of sensors, equating to a density of 20,000 sensors per km^2" (Messaoud et al. 2005: 51). The sensors listen for vibrations produced by machines, thumpers, which generate sound waves by hitting the surface (Bozman 1993). The Q-Land system sensors gather real-time digital data from up to 30,000 different sources, generating extraordinary quantities of raw data (Papworth 2009). A pilot test of the system in Kuwait in 2004 produced "16 terabytes of 384 million uncorrelated single-sensor seismic traces, which equates to a single-sensor trace density of over 16 million traces per km^2—a densely sampled wavefield by any definition" (Shabrawi et al. 2005: 68). Moreover, the sensors operate in various environments, from arctic to desert or down hole, and can link up in chains as tall as the Eiffel Tower (George 2000). The on-land sensors have been "ruggedized" to resist frost and repel sand, and each individually configurable sensor is equipped with a GPS unit and time-stamping so that it can transmit data in real time (Papworth 2009).

Schlumberger's land and sea sensing systems give the company's seismic analysis its global reach. Paul N. Edwards (2010) developed the concept of a "vast machine" in his history of meteorology's global but piecemeal efforts to make global climate science by gathering, standardizing, and aggregating measurements from discrete points across the earth, in order to approximate global weather. Schlumberger's vast machine of seismic data collection differs from meteorology's public scientific system by interiorizing vital subsurface knowledge within proprietary systems. Contrasting the two systems raises the pressing civic question of who should own data about the subsurface.

Schlumberger's vast subsurface data-production machine relies on cybernetic systems that are ever more finely designed to ensure the seamless interoperation of humans and machines (Mindell 2002, 2008; Edwards 2010).[9] Before the *Geco Eagle* seismic imaging vessel crew set foot onboard, they had completed 40 training courses, using virtual reality (Swinstead 1999: 60). Designers modeled human operator behavior in three-dimensional computer programs that were then evaluated by experts (Swinstead 1999). Such simulations are common in the industry. For example, researchers from the University of Texas and Cornell University developed "a simulation tool to evaluate the impact of crew sizes (people and equipment), survey area, geographical region, and weather conditions on survey costs and durations. Schlumberger uses it to obtain and profit from a larger portion of the global seismic survey market" (Mullarkey et al. 2007: 120).

The researchers estimate that "based on the number of surveys that Schlumberger conducts each year, it should save about $1.5 to $3 million each year" (Mullarkey et al. 2007: 120). The *Geco Eagle* and its crew training are the results of this approach; the ship and its humans were designed, simulated, and entrained so that they interoperated as seamlessly as possible (Swinstead 1999).

Schlumberger's representation of Leonardo da Vinci's Vitruvian Man perfectly expresses the company's totalizing design ideal of human-machine interoperability. The *Encyclopedia Britannica* states that da Vinci "envisaged the great picture chart of the human body he had produced through his anatomical drawings and *Vitruvian Man* as a *cosmografia del minor mondo* (cosmography of the microcosm). He believed the workings of the human body to be an analogy for the workings of the universe."[10] Schlumberger transformed this iconic male into a gender-ambiguous, fully clad laborer, complete with a protective construction hat, gloves, and boots (see figure 5.1).

Here, Schlumberger's naturalized laborer redesigns humanity's relationship with the universe through machines. This idealization of seamless control and engineering to protect the worker makes an interesting counterpoint to gas-patch landowners, who feel that they have lost control of their bodies through their inability to control their environments.[11] While vast sums of capital, management, and design have gone into constructing the relationship between workers' bodies and machines, landowners' bodies, affected by the outcomes of this work, are marked by an absence of any conscious design and modeling of how their bodies might interact with the industry. On the surface, the relationship between landowners' bodies and the gas industry appears to be completely undesigned. Landowners are frequently not even made aware that seismic imaging is occurring in their area, yet data from beneath their feet are drawn into information architecture that is the key to planning and organizing resource extraction (Finley 2013b; Junkins 2013).[12]

EXTRACTIVE INFORMATIC SYSTEMS, DATA DISLOCATION, AND IMMERSIVE SIMULATION

Data go to Schlumberger's supercomputing hub in Houston, Texas, via the company's satellite and digital networked data system called DeXa.Net. This private satellite system, orbiting over 23,000 miles above the earth, allows Schlumberger to send and receive information among its worldwide operations (BW 2002a: 1, 2002b). In 2002, the Hollywood producer/director John David Cameron and his brother James used the system for a real-time TV broadcast of an expedition to the site of the sunken *Titanic*: "'For the first time ever, we brought live footage from the bottom of the Atlantic Ocean to viewers around the world,' says John David Cameron. 'The whole point of this show was to offer the viewer a real-time you-are-there experience'" (Perin 2002). The STS scholar and cultural anthropologist Stefan Helmreich argues that this "you are there experience" is produced by efficiently masking a titanic technical system (2009): the research vessel, the submersible filming, the satellite circling the earth, Schlumberger's inland supercomputing

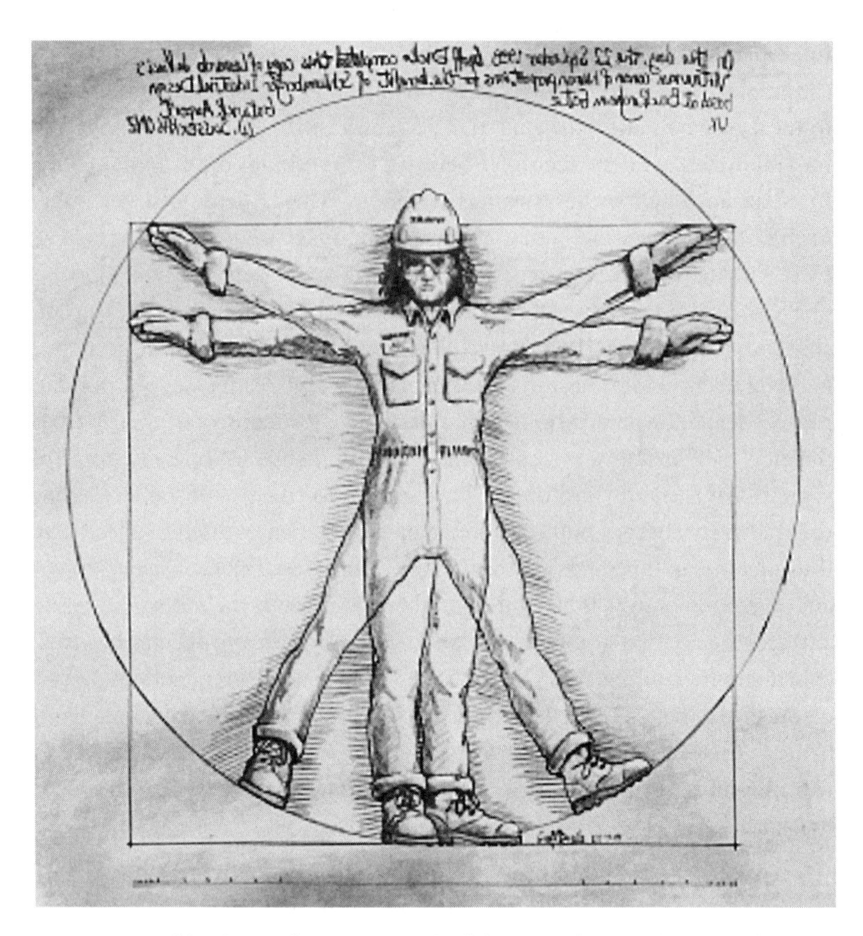

FIGURE 5.1. Schlumberger “Vitruvian Worker” (Swinstead 1999: 46). Copyright Schlumberger, used with permission of *Oilfield Review*.

hub, the LA film stage, the cable TV network, and the viewer’s TV, to name a few of the parts, are all required to simulate the experience of “being there.”

Technical systems’ ability to recede from attention when creating a simulation is an intentional result of Schlumberger’s system architecture, the final goal of which is to create 3D simulations of the subsurface. Schlumberger has been perfecting processes to combine data sets of oil and gas reserves over time so that clients can virtually walk through the subsurface (BW 2001). The company has used such technology for over a decade. A company memo from September 1998 read as follows:

> Schlumberger advanced the use of high-performance 3D data visualization in the oil and gas industry through the introduction of GeoViz

> software and the Alternate Realities Corporation's VisionDome system, for which Schlumberger is the exclusive licensed reseller to the industry. This combination provides geoscientists and engineers with the first fully immersive, portable, virtual-reality environment for constructing 3D models of subsurface reservoirs, selecting drilling targets and designing well trajectories to maximize oil and gas recovery. (Schlumberger 1998: 4)

Baudoin's MIT lecture described how reservoir models could be accessed remotely from virtual-reality suites in Houston, Tokyo, New York, and Paris. From these sites, Schlumberger employees and their clients could walk through renderings of oil and gas deposits miles below distant seas or half the world away:

> Inside Reality [Schlumberger's proprietary virtual-reality (VR) software, released in 2003] is the industry's only virtual reality commercial software for interactive well planning and real-time geosteering. Inside Reality integrates all data types required for full reservoir characterization. Users navigate and interact with the data by natural body movements like walking, pointing and grabbing or by using a 3D data wand instead of the traditional keyboard and mouse. Its unique human interface and the level of immersion it offers encourages team collaboration and brainstorming, allowing teams to be more creative. (BW 2003: 1)

For those unfamiliar with the process, simulations can obscure the mechanisms by which they are constructed and can encourage users to mistake them for reality (Masco 2006; Turkle 2009). They make accessing oil and gas reserves seem relatively simple, because the viewer can manipulate and walk around in spaces that are in fact physically inaccessible. Joseph Masco, an anthropologist who studies the U.S. development of nuclear weapons and simulation technologies, examined how Native Americans, living in what he called the "nuclear borderlands" exposed to nuclear testing, were obscured for research scientists and the military. The ban on subsurface nuclear testing led to the development of 3D VR suites called CAVES. This technology is the basis for that employed and modified by the oil and gas industry today (Turkle 2009). In these 3D virtual spaces researchers could step inside a simulation of the explosion without risking their bodies. Masco worried that the outcome of the researcher's simulated union with such a destructive force concealed the bomb's potential for devastation, as well as the very real bodies that still bear radiation damage from nuclear explosions and weapons production.

Schlumberger's VR suites arguably produce the same effect, reducing concern about impacts to surface life by immersing viewers in a map that highlights the "treasure" and makes it seem palpably accessible. The architecture of remote sensors and satellites makes it unnecessary for engineers and clients to actually visit a site. Those who process and interpret the data are in Schlumberger's supercomputing hubs often thousands of miles away from the site of data collection. Maps have been analyzed as "immutable mobiles" that travel from peripheries to "centers of calculation," allowing for those at a distance to plan futures for peripheral locations (Latour 1990).

Historically, the development of geography was central to defining, developing, and administering extractive relationships with colonies to support metropolitan centers. Similar asymmetrical power dynamics between regions of extraction and corporations are now achieved through the export of seismic data. Moreover, subsurface simulations sideline representations of surface life by encouraging users to focus primarily on the reserves of fossil fuel. Artifacts that interfere with mapping fossil-fuel reserves, such as subsea mountains or surface buildings, like railroads or highways, are identified as sources of noise and written out of the picture. The Schlumberger systems' greatest strength is their ability to reduce the signal to noise ratio: "All seismic images include noise. . . . What the Q system does is minimize the noise that is recorded. And the noise that the system picks up is recorded such that it can be removed from the final product so that the resulting image is sharper. It is a high fidelity image" (Maksoud 2005: 67). Because noise is a serious problem for on-land seismic studies, the Q-Land system is initially run merely to pick up background noise so that it can be edited when the seismic data are gathered later (Bisdorf and Fitterman 2004; Maksoud 2005).

Baudoin's lecture illustrated that, although humans now have the ability to gather and model vast amounts of data about the world, this huge, distributed technical sensorium and information-processing system is narrowly focused, calibrated to identify only fossil fuels. Schlumberger's system is an information architecture built to find the proverbial needle in a haystack by eliminating the haystack. The needle becomes an easy find if one uses a magnet, but a magnet is a very poor informational system for describing the haystack, because it makes the haystack irrelevant. For clients, Schlumberger's vast informational system coalesced into an event, the experience of seamless subsurface data analysis produced by removing any signals that might confuse the image or complicate the issue. Schlumberger's visualizations of oil and gas reserves delete from consideration the teeming political, social, and ecological life on the surface. Moreover, the system actively dislocates surface owners from representations of the subsurface and excludes them from the process of planning the future of resource extraction. Schlumberger and MIT's collaboration helps produce a vast myopic machine that extracts and exports seismic data while informatically, economically, and politically undermining the relationship between surface owners and the land beneath their feet. Thus companies are able to formulate and begin to execute plans to dramatically reorder surface life around subsurface riches, without the knowledge or participation of surface residents.

Industry's close relationship with universities tends to produce technical infrastructures that align with corporate development goals. Students are rarely taught to question the social and cultural aspects of these industries or to study their history and development. They are trained instead with technical skills (Downey 1998). A historian of science, Langdon Winner, famously argued that technological development should be taken as seriously and analyzed as thoroughly as new laws, because of the tendency to build political and cultural assumptions into technologies (1980).[13] The development of automatic tomato pickers by the University of California state extension services is a classic example of this myopia (Noble 1977; Hightower 1978). Engineers asked to solve the purely technical problem—that of how tomatoes could be gathered most efficiently—designed a fully automated process that omitted social questions about laying off tomato pickers. Although automation was initially more expensive for growers, its benefit for industrializing farms was not to cut costs but to weaken workers' organizations (Hightower 1978). This technical approach to research and development that removes social or political issues from consideration has led to many of America's land-grant universities—universities (such as MIT) established with a mandate to help agricultural communities—playing a vital role in both eliminating small farms and promoting the industrialization of agriculture (Noble 1977; Hightower 1978; Kloppenburg 1988; D. Fitzgerald 1993, 2003; Stoll 1998; Stine and Tarr 1998). Schlumberger and MIT as well as other engineering universities continue this legacy by approaching natural gas extraction as a technical practice rather than as a political practice with social ramifications.

FROM EXTRACTIVE INFORMATIC PRACTICES TO EXTRACTIVE ENCLAVES

Seismic imaging is the first step in developing oil and gas resources in a fashion that systematically sidelines the surface owners' control over the extraction process. Everywhere the oil and gas industry goes, dreams of unimaginable riches emerge: flows of liquid gold, fantasies of fabulous future wealth (Peluso and Watts 2001; Watts 2003). For inhabitants of resource regions, however, these dreams are inevitably shibboleths, as a scramble for resources creates social rifts. Extracting resources requires a rewriting of relationships regarding the ownership of surface property, beginning with the leasing or purchase of mineral rights. Historically, corporations have exacerbated existing social stresses in order to serve their interests (Sawyer 2004; Santiago 2006). The tradition of national sovereignty over subsurface minerals (inherited from English common law) means royalty wealth goes to national bodies rather than to those living where resource extraction will occur. This does not hold in the United States except on federal lands. However, the federal government is the largest landholder in the western United States, and it has actively leased the subsurface to facilitate the shale gas boom.[14] To extract resources, mineral rightsholders (nations, states, or individuals) must enter into a "Faustian pact" with oil and gas companies that have the technology and infrastructure (Peluso and Watts 2001: 205).

The "technical and social" processes of extractive industries develop and perpetuate structural inequalities between resource producers and consumers, and they benefit from and assist in producing social volatility in regions of resource extraction.[15] The absence of either a "rule of law" in regions of extraction, or of strong social ties, supports the development of extractive industries that "need only directly control a fairly limited piece of ground and secure access to the relevant external market" (Ferguson 2005: 206). For example, rather than establishing lucrative regions of extraction in relatively peaceful and stable African nations such as Zambia, oil extraction thrives in the most violent and undemocratic regions of Congo, Nigeria, and Angola: "Such [capital] investment is not occurring (as World Bank doctrine would suggest) where what they call 'governance' is good and the rule of law strong. Rather, the countries that (in the terms of World Bank and IMF reformers) are the biggest 'failures' have been the most successful at attracting foreign capital investment (Ferguson 2005: 380).

Extractive industries secure "enclaves of extraction," establishing barriers to insulate the companies and resources (Ferguson 2005, 2006). Private roads are developed to travel to and from company-controlled zones of production. Migratory labor camps are organized and only accessible to workers, young men who work in 24-hour shifts. Such infrastructures create an inaccessible interiority. Andrew Barry (2006) calls them "technological zones," which have reached an extreme form in the offshore oil rigs (Appel 2011). These zones appear to be technically justified but hide their political and economic functions: securing networks that direct capital and resources to flow favorably to oil and gas companies with minimal "leakage" of resources and capital to surrounding areas. Such disengaged or, as Hannah Appel (2011) describes them, "modular," technical arrangements provide for the perpetuation of an oil and gas industry in which populations in producing regions are made structurally irrelevant to the success of resource extraction.

As I began to have a sense of Schlumberger's world-spanning, cybernetic seismic system as the first step in the extractive process, I wondered what kind of cultural critique might effectively bring the surface and its inhabitants back into view? How might the "technical" and the "political" be conjoined to include discussions with individuals like Rick, Laura, and Theo? How could I conduct research to do more than critique the absences in Baudoin's model? An alternative sociotechnical discourse was required to improve the social, political, and technical ability of people with whom I had worked in Colorado to shape the industry at MIT and Schlumberger.

From Critical Technical Practice to STS in Practice

CRITICAL TECHNICAL PRACTICE

Schlumberger's use of MIT's artificial intelligence (AI) research is a good point from which to start developing a constructive approach to STS. The term "critical technical practice" was coined by the MIT researcher Phil Agre. While he was working on AI, Agre began to question the relationship between the models of human behavior embedded in AI systems and actual human perception (Agre 1997).[16] He realized that assumptions about perception, such as, that there is no difference between a thing's name and the object itself, were built into his own AI programs. People built technical systems that presented the world to them as they assumed it is, rather than how it is actually experienced.[17] These problems are hard to notice within the discipline because AI programs successfully work within their technical parameters despite being based on flawed assumptions. This success recursively confirms the model's accuracy. Agre developed a self-reflective process he called critical technical practice to resolve this problem, and he encouraged engineers to unearth the assumptions that they had tacitly built into their machines (Downey 1998; Helmreich 1998; Forsythe 2001).

Chris Csikszentmihályi introduced me to Agre's work. Csikszentmihályi's research merged STS, engineering, and art, not only to engage in critical technical practice but also to develop technologies that expose the politics within and social power dynamics established through machines.[18] Chris's "Computing Culture" research group in the Media Lab was a niche within MIT where such self-reflexive design flourished (Yoquinto 2014).

TACTICAL MEDIA AND ANTHROPOLOGY AS CULTURAL CRITIQUE

Csikszentmihályi is part of a larger movement within art and engineering that is generating artifacts or practices that reveal and question the politics built into technical and scientific systems. In order to critique the military development of unmanned drone weapons, he designed a set of "unmanned" protest robots that could be operated remotely by protesters so they would not have to physically face the police. Ironic and witty, the project called attention to the structural violence of weaponized drones that turn killing into a virtual practice where the body of the assailant is structurally hidden (Yoquinto 2014). Csikszentmihályi's project went beyond academic analysis of the practice. It disrupted the logic of drone warfare and generated a space of cultural critique for the audience.

The movement has been loosely organized under the rubric of tactical media, a term that refers to the critical usage and theorization of media practices that "draw on all forms of old and new, both lucid and sophisticated media, for achieving a variety of specific non-commercial goals while pushing all kinds of potentially subversive political issues."[19] Tactical media practitioners embraced Michel de Certeau's concept of "tactics" as "weapons of the weak," used to unsettle normally invisible power relationships by creatively misusing power structures to momentarily provide opportunities for critique (J. Scott 1985).

The artist Natalie Jeremijenko's tactical media work was salient for my interest in chemicals and environmental justice issues. Jeremijenko's Feral Robotic Dogs project taught inner-city youths how to wire cheap robotic dogs with sensors for volatile organic chemicals so that they could be released in brownfield sites to track pollution.[20] Pack releases of robot dogs were mediagenic events. The dogs enabled public participation in both generating scientific data and creating the means of gathering that data by turning toys into tools. The Californian artist Beatriz da Costa took this idea further. Inspired by a picture of pigeons carrying spy gear in World War II, da Costa placed tiny nitrogen oxide sensor units on pigeons to map real-time urban air pollution (da Costa and Philip 2008). Da Costa's and Jeremijenko's works expand tactical media into questions of how science operates and how scientific data circulate to change how we study environmental and human health. They turn junk technologies and pest species into new partners in relating to our urban environments (Haraway 2003). Tactical media and tactical biopolitical performances borrow from the subjects they are critiquing; they mimic, displace, or expand some item, to defamiliarize our experience of the object. This work embodies the spirit of anthropology as cultural critique in media other than the usual written word. Anthropology's power has previously been argued to rest on its ability to make the familiar strange, and the strange familiar. Critical technical practices also invert these cultural experiences to foreground their contingency.

ANTHROPOLOGY IN DESIGN AND THE LIMITS OF TACTICAL MEDIA

While methods for cultural critique similar to those practiced in anthropology flourished in art, anthropology was being discovered as a design tool. Anthropology has developed into a kind of brand in technological development: a discipline able to bring the exotic user into the picture. Lucy Suchman, who pioneered ethnography in centers of technology design and development, has described the perpetual rediscovery of anthropology as a design tool: "[The journalist's] email message explained that he was doing 'a big story on the rise of ethnography and anthropology in business.' I replied that I felt that I could not bring myself to do another interview on this topic, which I pointed out had been 'news' for over a decade, including in *Business Week*'s own pages" (Suchman 2007: 2).

Anthropological and ethnographic methods have become standard parts of corporations' user-centered design, as companies seek to understand how consumers interact with products (Cefkin 2008, 2010). However, anthropology's strength as "cultural critique" is often overlooked in corporate contexts. Design theorists argue that ethnography's focus on consumer choices ironically promotes sustainable design centered on consumers' awareness of their environmental impact, rather than political economic systems of production (DiSalvo, Boehner et al. 2009; DiSalvo, Brynjarsdóttir, and Sengers 2010; Dourish 2010).

Focusing on consumers ignores the political and economic questions associated with production processes. Environmental historians have criticized consumer-centered sustainable design as "green liberalism," the attempt to address harms from inherently wasteful industrial processes that, by design, externalize production costs, through encouraging environmentally conscious individualized consumer choices or actions driven by culturally manufactured feelings of guilt (Steinberg 2010: 13). Recycling, for instance, is the ingenious product of mass-market beverage companies for whom cheap canning enabled low cost, mass distribution, which replaced local bottle collection. The "disposability" of cans allowed long-distance shipping and undercut local beverage firms, which had collected and reused bottles (Steinberg 2010: 14). Using rhetoric like the "litter bug" made consumers the guilty party for the increased waste of "disposable" containers. The notion of recycling, emerged from industry efforts to prevent the passage of ordinances banning disposable bottles and cans. Recycling effectively passed onto the public the costs of a waste stream that the industry had previously borne by picking up and reusing containers (Steinberg 2010: 15).

Annina Rüst critiqued this transfer of guilt by corporations onto individual consumers with her creation of an art object called the Thighmaster, "a form of carbon penance."[21] Consumers use the Thighmaster to punish themselves for overconsuming electricity. The device is based on the *cilice*, a self-mortification tool worn around the thigh by members of the conservative Catholic religious sect Opus Dei. Linked wirelessly to technologies in the owner's household, the Thighmaster grips its wearer's leg with metal spikes if electricity is consumed above a certain level, disciplining them to turn off electronic devices. Like many of the products emerging from Chris's group, the piece's humor and unstable position as art object/product circulated it far more broadly than Theodore Steinberg's illuminating analyses of recycling. It was featured in the *New York Times*' 2008 "Year in Ideas" series.[22] Unfortunately, the *Times* reportage missed the deeper irony of the piece, that attending to climate change with yet more consumable electronics is as unlikely to solve the issue as a single monk is to rid the world of sin by punishing himself. The Thighmaster's tactical success is undercut to some extent by the *Times* coverage, suggesting that it might actually be a useful consumer device, particularly if wearers are alerted to overconsumption with a noise or other warning, rather than being harmed.

The reception of Rüst's work suggested the limitations of tactical media's serious jokes and raised questions about how to design critical technologies that can lead to structural, political, and economic change. What kind of infrastructure, academic-regulatory-corporate, is required to take the Feral Robotic Dogs project or the Pigeon Blog out of art circles into daily life? Ironically, my first idea for a gas-patch tactical media project was to develop a cheap and easy-to-assemble sensor unit called iSense that residents could wear or put around their property.[23] I thought such sensors could be a new way to gather data on individual exposures. Our discussion about this project's unwitting acceptance of green liberalism's individualistic frame led to ExtrAct. ExtrAct drew from the rich lineages I've discussed here—critical technical practice, tactical media, and anthropology as cultural critique—to build material rather than written critiques within STS that hoped to offer possibilities for changing the power dynamics between surface communities and extractive industries.

FROM TACTICAL MEDIA TO CIVIC MEDIA

Counterposed to tactics, strategies can be enacted by those who have the power to occupy and delimit a physical space that is recognized by others, whether they are enemies, collaborators, or competition (de Certeau 1984: 36). In their heyday, newspapers, as an information source, and journalism, as a profession, held such strategic power, protected as necessary for a democratic public sphere. The "free press" enjoyed its own rights, as well as legal and social protection (36), occupying a powerful strategic position that defined what counted as knowledge and who had the authority to produce it (36). The newspaper's loss of its strategic hold over civic discourse, due to the emergence of the Internet, allowed MIT to enter the discourse of defining what "civic media" meant.

In 2007 and 2008 Csikszentmihályi collaborated with two other Media Lab faculty members, Henry Jenkins and Mitch Resnick, to establish a Center for Future Civic Media (C4FCM) to develop the next generation of journalistic tools, which they called "civic media": media that enable collective contemplation and action in the name of shared interests. C4FCM was inspired by a quote from John Knight (the founder of the Knight-Ridder newspaper empire): "Thus we seek to bestir the people into an awareness of their own condition, provide inspiration for their thoughts and rouse them to pursue their true interests."[24]

Rather than devising ways to save the newspaper, C4FCM sought to address civic sociopolitical issues through using digital media to restructure information production and circulation (Csikszentmihályi et al. 2007). Political and economic questions about how to connect civic media to collective actions were at the heart of C4FCM and it won the largest award to date from the Knight Foundation, $5 million, to establish a center at MIT.[25]

In this context, Csikszentmihályi and I began discussing health issues in the Colorado gas patch, a problem that had received little mainstream media coverage at the time (2007). It was a systematic yet distributed problem that could not produce enough news value to gain coverage. We wanted to address the underlying structural informational and technical asymmetries between communities and companies by building databases and maps similar to those developed by TEDX. Communities were disadvantaged by technical systems that exported data beyond their control and that excluded them from directing or overseeing the process of gas and oil development. How could we create systems that brought landowners back into view, that enabled their oversight of extraction, and that helped them to act collaboratively on shared issues such as water or air contamination?

Science and engineering shape power relationships in two ways. First they set what research questions are investigated, who investigates them, and how they are investigated. The questions we ask about the world, and how we ask them, shape what we know. For instance, regimes of imperceptibility made the study of fracking and endocrine disruptors challenging. Second, those who have access to the knowledge and technologies that the sciences and engineering generate will shape power relationships. Practicing STS could work to reconfigure one or both of these aspects of how science and engineering shape power dynamics.

STS has some history of applied work to reshape inquiry and culturally biased assumptions. A mathematician and critical theorist, Ron Eglash, created mathematical modeling software based on non-western mathematical practices, documented in his ethnographic work on the common use of fractals in African mathematics (Eglash 1999; Eglash et al. 2006). Eglash drew on ethnography as cultural critique to embed fractal-based math in software that could expand this alternative approach to teaching math. However, similar to the distinction between studying biological forms of disease and chemical forms of disease, Eglash's African fractal math, while vital for developing postcolonial mathematics, did not challenge vested interests with the exceptional economic, political, and social power of the oil and gas industry. The oil and gas industry is structurally designed to be legible within itself (through creating and interiorizing flows of information and resources) and to be illegible for those outside the industry. In the case of chemicals used in fracking, companies were actively trying not to disclose any information that was necessary to study the problem.

Although it is impossible to grasp the world in its totality through databasing or mapping, it is also impossible to be in the world without leaving traces. Science and the social sciences share the work of determining how we detect and connect those traces (Traweek 1988; Galison 1997). Haraway argues that "there is one fundamental thing about the world—relationality. Oddly, embedded relationality is the prophylaxis for both relativism and transcendence. Nothing comes without its world, so trying to know those worlds is crucial" (1997: 37). Relationality was at the core of both Schlumberger's and Colborn's work. Colborn attempted to chart the relationships unexpectedly generated by chemicals, while Schlumberger depended on the ability to build informatics networks across great distances in order to bring the subsurface into view, for corporate centers of calculation, by underdeveloping (1) their relationships to surface inhabitants and (2) the inhabitants' relationships to their subsurface environments. How could increasing the ability of gas-patch residents to interrelate with similar informational databases and mapping tools construct new perspectives on the relationships this industry generates and technologically obfuscates?

Chris introduced me to the "experimental geographer" Trevor Paglen, who devotes his research to peering into black holes on America's national maps, black holes composed of military systems such as secret bases and classified units and communications installations (Paglen and Thompson 2006; Thompson 2009; Paglen 2010). Paglen develops methodologies on the fly. He used a camera developed for satellite imaging to see secret bases from the ground, and he collaborated with amateur plane spotters to track CIA rendition flights (Paglen and Thompson 2006). His acquisitional, context-based, methodological innovations are suited to tracking objects that are either trying not to be seen or are consciously hidden.[26] They share nomad-like qualities with Theo's methodological approach. Experimental geography begins from the premise that to map something is to take a position. It is an intervention. Scholarship, written or nonwritten, is also an intervention that has the possibility of changing or stabilizing the status quo in tiny and major ways. I avoid the term "applied" in defining "STS in practice" because I see both written and material practice as an intervention. The question for me is whether the intervention seeks to deconstruct and reveal existing power relationships or to offer and develop new possible power relationships.

ExtrAct builds on Eglash's practice of creating tools based on existing resistive practices, anthropology's strength as cultural critique (Marcus and Fischer 1996), civic media's attention to connecting information to action, and tactical media's situational borrowing to transform the objects of critique. As the ExtrAct team,[27] we engaged in "STS in practice," with the aim of transforming the political, social, and technical relationships the industry forms with regions of gas extraction, through these actions:

1. Analyzing and consciously developing tools with different power dynamics than the industry's present informatics systems by designing them for and with communities.

2. Encouraging academic participation in the development of tools for nonprofit ends to break down the model of corporate-centered research identified by David Noble as leading to academic work that turned political issues into technical problems, or what the anthropologist James Ferguson has analyzed as "the anti-politics machine" (1994).

3. Developing a platform for alternative modes of grassroots data gathering and research, modes that are inherently multivocal and allow for emergent cultural critiques from various participants. As the cultural anthropologist Michael Taussig argues, "From the represented shall come that which overturns the representation" (1986: 135). We wanted to collaborate with gas-patch communities to generate alternatives to present representational practices, practices that consistently wrote them out of the system.

4. Developing those alternatives in such a way that they could be maintained and expanded within gas-patch communities and nonprofit organizations, in the same way that research from MIT can be used by Schlumberger.[28]

ExtrAct

A Case Study in Methods for STS in Practice

In the Beginning There Were Para-sites

Planning for ExtrAct began in earnest in January 2008. We started a research group at MIT comprising people with programming and web-design backgrounds who were interested in the social and political aspects of technology design and would be capable of codeveloping a web-based project with communities impacted by oil and gas. The Oil and Gas Accountability Project's (OGAP) research director, Lisa Sumi, was the first person we invited to join the project. Sumi sought novel approaches to monitoring and reforming extractive industries, and she had experience attempting to coordinate mining activists across the Americas, a networking effort that failed due to technical difficulties. She was also well connected with both landowners and advocacy organizations.

Chris Csikszentmihályi was also attracted to the work of a Harvard history of science undergraduate Christina Xu, who had organized an extremely successful conference on Internet culture, ROFLCon (ROFL is Internet slang for "rolling on the floor laughing"). The child of Chinese immigrants, Christina grew up on the Internet. As a teenager unable to drive in suburban Ohio, her social life was online. She understood Internet culture and could reasonably be called an expert user. Christina also had experience with science and technology studies (STS) because her undergraduate thesis on computer interface design was supervised by the STS theorist Chris Kelty. His study of open-source and free-software communities later informed ExtrAct's development (Kelty 2008). Xu became an essential project member and went on to direct ExtrAct when I left to write my dissertation. Dan Ring was also a logical choice for the ExtrAct team because he began programming at age 6, thoroughly understood computers, and had worked at one of the few critically oriented Media Lab spinoffs, a group called Thought and Memory, that aimed to build a database for tracking company environmental, health, and labor practices. An academic's son and native of South Dakota, Dan had yet to graduate from MIT because he tended to sideline his assigned work in favor of his own projects. Matt Hockenberry came aboard to develop the "front-end" look and feel of the websites. Matt was a Media Lab graduate with excellent web-design skills and a history of working on environmental issues. One of his projects, Sourcemap, developed alternative modes of product manufacture by mapping supply chains.[1]

Dan, Christina, Matt, Lisa, Chris, and I became ExtrAct's core group. Our first goal, in the spring of 2008, was a research visit to explore the gas patches in New Mexico and Colorado to introduce the MIT team to the gas patch and vice versa, to gather and discuss ideas for what to create, and to form potential partnerships to develop and pilot these tools. In one week we met with three citizens' alliances, three environmental health scientists, one prominent lawyer, one national advocacy group, one regional environmental newspaper editor, and numerous individual landowners—as many kinds of actors involved in struggles over natural gas in as many different places as possible in the shortest amount of time. We also toured gas patches in Farmington, New Mexico, and Garfield County, Colorado, and took a flight over the gas patch in Colorado's wilderness. We wanted the team to collectively experience the complexity of these places rather than strip it away. Our fieldwork moved much faster than traditional ethnography in the hyperfast world of product design because the Knight News Challenge conference, where we intended to present ExtrAct's first digital tool, was scheduled for June, just six weeks away.

ExtrAct aimed to design tools for Lisa's network of NGOs and aggrieved landowners, who were ready for tours and had frequently taken journalists around the gas patch. While it was important for our group to hear landowners' messages, we also needed to open a space beyond these scripts to discuss collaborations and dig into the cultural critiques and tactics informing their advocacy. These meetings needed to be "para-sites," nontraditional field sites where ethnographers and their collaborators create a space, meeting, talk, conference, or presentation that is "outside conventional notions of the field in fieldwork to enact and further certain relations of research essential to the intellectual or conceptual work that goes on inside such projects. It might focus on developing those relationships . . . whereby the ethnographer finds subjects with whom he or she can test and develop ideas" (Holmes and Marcus 2008: 100).

Traditionally, para-sites involve meeting collaborators elsewhere, such as at an academic campus.[2] Here our para-site would be in the "field." We wanted to develop from ethnographic concepts a collaborative project based on critiques of the present information-sharing and technical infrastructure of this industry. These cultural critiques would rise to the surface through para-ethnography, the process through which participants in a culture become reflexive about their own experiences.[3] Advocacy itself is often a kind of para-ethnography, as advocacy organizations are based on cultural critiques that address community concerns (Fortun 2001). Each meeting to develop our para-ethnography aimed to move past the habits of presentation, the script and practices that activists employ to change the world, toward a self-reflexive conversation: How did these groups organize initially? Which of their tactics worked and why? How did they select tactics and hold together as organizations? Through these discussions we hoped to take imaginative leaps to decide what we might develop collaboratively.

Creating new tools required all participants to imagine new possibilities, what the design theorist Carl DiSalvo calls "projections" for future organizing (2009). DiSalvo built on John Dewey's conclusion that publics are actively constructed. He identified two tactics that help build publics through design: projection and tracing. Projections help constitute or reflect on consequences of future possibilities by offering potential visions of the future, not fully realized but plausible enough to inspire conversation about how that future might be realized or transformed (2009: 52). Our meetings hinged on moments of projection that would move us from critical reflection into a conversation about constructing possible futures. These projections helped us imagine how discourse around natural gas would be transformed if we could create the means to trace the industry's diverse and distributed actions and impacts.

Tracing powerfully constitutes publics because tracking an artifact brings places and people whose existence is shaped by and shape that artifact into relationship with each other (DiSalvo 2009: 55). In Michel de Certeau's terms, tracing is a tactic that produces a new way of organizing, perceiving, and inhabiting physical space that generates strategic power.[4] Theo Colborn's databases, deployed as maps, had traced new publics—the endocrine disruption research "pack" as well as the community of ill people related by and to gas development.

The following sections illustrate ExtrAct's development through para-ethnography paired with participatory design. It is helpful to enumerate these steps, although in practice we intermingled them:

Step 1. Use ethnographic techniques to generate para-sites for reflection and the articulation of cultural critiques with collaborators from the field.

Step 2. Employ projections to stimulate collective discussions of imagined possible futures.[5]

Step 3. Move iteratively between steps one and two to understand how tracing technologies might be developed and deployed to generate strategic spaces for publics shaped by critical technical practice.

The next section illustrates the process of articulating cultural critique through para-sites and the use of projections. It is organized around the cultural critiques that emerged during our research and interspersed with descriptions of how we began collectively "projecting" ideas for further collaborations.

Fieldwork

CULTURAL CRITIQUE 1: LESSONS FROM THE LANDSCAPE

While the last section focused on discourse between people, it is important to remember that landscape is also a vital actor in this discourse. Natural gas-rich basins brought the gas industry, and eventually the ExtrAct team, to New Mexico and Colorado. Our first stop was the San Juan Citizens Alliance (SJCA) offices in Durango, Colorado.

Known as the Million Dollar Highway, the Red Mountain Pass to Durango takes you more than 11,000 feet up through the San Juan Mountain Range. The road was cut in the 1880s to reach rich mining areas that are now sites of major environmental remediation efforts because of arsenic and heavy metal contamination in the water. Red Mountain's bald, stark, rust-colored peak speaks for itself; it proclaims its iron riches in vivid shades of burnt red against blue sky. Ochre-colored slag heaps and crumbling mining machinery litter the landscape around Silverton and Ironton, testaments of the region's history. The towns are all but empty in winter. In summer, tourists ride up the mountain from Durango on a quaint, coal-fired steam engine. Passing through these lands of industry past, I wondered what legacy the shale gas boom would leave. Would tourists someday photograph the gas patch?

The road winds down the southern side of the range, through groves of quaking aspen. The turning leaves, top sides bright green and under sides a dusty olive, create a shimmering canopy that trembles and changes with the wind. Aspen is rhizomatic. Each group of trees shares an underlying substructure. In *A Thousand Plateaus* (1987), Gilles Deleuze and Félix Guattari use rhizomes as a figure to challenge the modernist tendency to make divisions between objects. Rhizomatic structures, linking apparently independent gas-patch communities through their relationships to the gas industry, became a useful theme and operating metaphor during ExtrAct's development.[6] Chris drew on this metaphor in his presentation at the inaugural Knight News conference later that summer. He connected ExtrAct's mission to Manuel Castells's critique of network societies ([1996] 2000), describing how networked corporate entities are able to generate flows of information, people, objects traveling between and within physical spaces. Communities grounded in place are unable to control and order these flows. The rhizome metaphor helped us focus on interconnective moments between disparate groups and interests, to consider where they stood at present and could be interwoven in the future.

CULTURAL CRITIQUE 2: GEOGRAPHIES OF RESISTANCE

Durango's historic Main Street has a quaint feel, carefully curated for the lucrative tourism industry. The street is lined with gift shops selling curios such as minerals, fossils, and expensive Native American art works from the nearby Southern Ute and Navajo tribes. A range of upscale restaurants serves anything from French cuisine to sushi. Bars and microbreweries cater to every thirst. During the summer the street is bustling with families, outdoors folks, and locals. The door to the SJCA office is squeezed beside a brewpub and is easily missed.

The SJCA has three adjoining offices up one flight of stairs. The five of us trooped up them at 10 AM on a warm, bright May day to meet with the alliance leaders, Josh Joswick and Steve Kaiser; an alliance member, Amy; as well as OGAP's then executive director, Gwen Lachelt, and its new director of research, Renee.[7] While OGAP focused on national oil, gas, and mineral issues and has offices close by, the SJCA has a broader but more regional purview. The organization addresses issues from immigrant rights to forest and wildlands protection, as well as water and energy concerns that include gas development and local proposals for a coal-fired power plant. OGAP's and SJCA's interests overlap and they keep in close contact over oil and gas issues, but they are both autonomous organizations. SJCA is a grassroots group with a regional member base and offices in Cortez, Colorado, and Farmington, New Mexico, as well as the central office in Durango.

OGAP, SJCA, and their landowner members all encounter the gas industry through different social and physical geographies. Each organizes and brings their advocacy to bear in a different way. The alliance serves communities in the San Juan oil and gas basin. OGAP is a national organization with paid staff as well as thousands of members nationwide. It is a hub for both coordinating legislative action at national and state levels and supporting the growth of local citizens' alliances. Landowners are concerned with their own property and counties. These sites intersect and diverge to form three different spaces from which to change the process of gas development (Gerlach and Hine 1970). Understanding how these different geographic and organizational foci are tactically deployed and interrelated was vital to contextualizing how ExtrAct could help strengthen their intersections, without eliminating the differences in their approaches to issues.

OGAP is organized around policy issues and by political geographies, in that it seeks to influence state laws and federal regulations. It has helped change landowner laws and other state regulations in Colorado and numerous other states, by promoting and organizing citizens' alliances (OGAP 2007a, 2007b). OGAP depends on small citizens' alliances like the SJCA to organize local grassroots coalitions. It supports such groups by training organizers, sharing successes between states, and drafting legislation. In regions unfamiliar with gas development, OGAP is often the first point of contact for concerned citizens. The most dedicated of these citizens often become organizers in new areas. OGAP facilitates this process by providing informational resources and holding meetings where landowners from impacted areas can educate those in new drilling areas.

The SJCA's organization unifies the Four Corners area in the San Juan Basin rather than dividing it along state lines. The basin is composed of sedimentary rock rich in coal and uranium that had been deposited in the Mesozoic era. The Fruitland Formation coal bed traps methane (the primary energy source of natural gas) and runs unevenly through a 46,000-square-mile area, encompassing northwest New Mexico and southwest Colorado. In 2007, this coal formation produced 1.32 trillion cubic feet of natural gas, making it the country's largest coal-bed methane natural gas producer (EIA 2009). The San Juan Basin is also a drainage system that acts as a water filter that eventually drains into the San Juan River and then the Colorado River. The SJCA representatives explained that organizing around the basin environment, rather than the four states, allowed it to manage wildlands, river, and energy issues in a more comprehensive way that considered the region's geography and organisms and their shared environments.

We met two of the landowners who were part of the SJCA's coalition of landowners and residents living in the San Juan Basin. The Fitzgeralds are rancher activists who live outside Durango and are close friends with Gwen Lachelt, who founded OGAP in 1999. They became active on oil and gas issues in the 1980s when drilling began in the basin, in what Lachelt described as the "good old direct action days." They blocked bulldozers attempting to enter San Juan National Forest's HD Mountains because Amoco was beginning to drill before completing an environmental impact statement. We drove to their place along dusty, rutted, single-lane roads, passing through scrub brush and over arroyos, tiny waterways that carve deep winding grooves in the yellow and gray parched earth. Dogs bounded up to our car as we parked by the horse-drawn buggy the Fitzgeralds occasionally use to take their produce to the market. Jim and Terri have lived on their 300-acre property for the past 40 years, growing veggies for the local market and raising sheep, a few milk cows, horses, and donkeys. We squinted, even in the mild May sun, as we emerged from our car laden with enough backpacks, tripods, cameras, computers, and recording equipment for a major expedition. Jim and Terri met us, and their rough hands shook our pale tailored-for-typing counterparts. "Manure-movers of the world" read Terri's dusty sweater, an appropriate, ironic slogan for this lifetime activist, who wears her life in the sun in the lines on her face. The excited dogs shepherded us out of the dusty parking area and over a small irrigation ditch toward the Fitzgeralds' nineteenth-century homestead of dark worn wood, where we were ushered inside with a somewhat cautious welcome.

A wood stove made it surprisingly warm inside. Light streamed from a single stained-glass window above the door, diffusing the sun to yellow and purple. One uncolored pane at the lower right let in a column of light to frame Terri, as she scratched her cat's back and reflected on life and activism in the gas field. We settled awkwardly on the edge of an old sofa and an eclectic mix of chairs beside the stove. I arranged my iPod at the center of an indistinctly colored and well-worn rug and Chris asked if he could set up the video camera. Professionalism had been easier for the ExtrAct group at SJCA's office, where wall sockets and wireless Internet were readily available. Here it was more comfortable, but we were out of place. Yet we knew this scene was the frontier of gas development around which we had to design, one of many homes across the country where cohabiting with gas extraction was becoming inevitable. Various sites were already connected in the gas-patch communities. If we wanted to join that conversation and create discussions about how MIT could help, ExtrAct had to collaborate to build web tools that would function smoothly across environments and be adaptable to each place.

CULTURAL CRITIQUE 3: TECHNOLOGY WON'T SOLVE THIS PROBLEM, PEOPLE WILL

After pleasantries and coffee, we made our first attempts to explain exactly why we were crowded into the Fitzgeralds' living room. I told them about how the project came together while I was working with Theo as an anthropologist of science. As I retraced ExtrAct's emergence, I became conscious of how much my role was shifting. In normal ethnographic mode, I would have been scribbling notes and listening intently, but here I was making eye contact back and forth between the MIT group and the Fitzgeralds, craning forward with my hands in motion in what felt uncomfortably like a sales pitch. I wanted to find the hook that would open this conversation and bring these activists onboard.

Jim and Terri told me that they also knew Theo. They wanted to ask her opinion about a current rule-making discussion on toxics; they had also found that the hazards of fracking chemicals, revealed by Theo and OGAP's work, had given them new leverage in a water-use dispute in which they were involved. My backstory over, I introduced Chris to talk up the possibilities for online organizing. For a moment, it seemed that the little momentum gained from our conversation about toxics was lost. Jim frowned and Terri looked away. "I've never sent an email," Terri said. "I wouldn't know where to start. Jim sends our emails and checks them once a week." Jim operationalized their discomfort: "Technology won't fix this issue. People will," he said, and promptly stood up to hold forth on the importance of face-to-face interaction.

I had a sinking feeling, followed swiftly by a moment of self-criticism, and then confusion. Shifting between the positions of ethnographer and participant is continual and subtly negotiated in micro-moments. When I saw plainly that Jim had erected a wall between technology and people, the salesperson in me panicked. I thought, *I'm going to have to try another tack to get him to see this differently*. It was not as though Jim was entirely technology-averse; they got their electricity from solar panels and he uses email. But as an ethnographer I thought, *That's interesting. What does face-to-face interaction have that the technology doesn't? What's lost? What is technology to Jim?* As an ethnographer, my care was to understand the particular makeup of Jim's mind and how that might explain our interaction. But as a codirector of a project we were trying to get off the ground with people like Jim, both as participants and important sources for additional community support, our success rode on his favorable opinion of whether it could work. The two roles are not as incompatible as they may seem, because finding a way to make what is said recognizable and sensible to someone else's point of view depends on connecting with their point of view in the first place. Instead of challenging Jim on whether technology could replace face-to-face interactions, I agreed with him and asked why they were important to him and where he had been successful with such interactions. Turning the conversation over to their stories in this way broke open the discussion. "That can't work," an apparently impenetrable barrier, was replaced with "That would be interesting," and then "Well, what if it could do this?" and an excited "What if we did it this way?"

The Fitzgeralds thought that landowners' personal stories were the most effective tool for transforming gas development, because shared stories led to collective concern and anger. Jim illustrated this point with a story about the frustrations they experienced when making complaints about gas extraction. County meetings, organized to hear landowners' complaints, used to provide opportunities for people to come together over similar issues. Landowners could speak about their problems and others in the audience were able to add their own. Landowners were able to recognize that their issues and anger were shared. Then the county imposed a rule that barred repeat complaints, so people were prevented from speaking if their issues were similar to earlier concerns. Federal agencies also prevented such collective recognition by using "open houses" at which landowners had to wait in line to present their problems individually to county representatives, with the result that stories were shared even less. Landowners' ability to recognize themselves as a collective to demand action from the state (COGCC) was further diminished as complaints were automated and landowners were reduced to recording their complaints by video instead. This was the "technology" Jim and Terri doubted. There was a vast difference between sharing your outrage with a video camera and expressing it face-to-face in a community setting.

The common thread raised para-ethnographically by meetings with the Fitzgeralds, SJCA, and OGAP was that their work depended on sharing stories. Most of their activities involved coordinating and collecting these stories. For the Fitzgeralds, storytelling had been tremendously successful when they had traveled on OGAP-organized trips to new gas fields in Alaska and Trinidad, Colorado, to educate novice communities by relating their experiences of living in the gas patch. Jim explained that these face-to-face interactions were important because companies develop gas wells by making individual contracts with landowners, a process that turns neighbors against each other. Just as the county had actively separated landowners so that they could be managed as individuals rather than as a collective, the industry employs these divide-and-conquer tactics[8] at the very beginning of gas development. For instance, when a drilling company wanted to sink wells in their area, company landmen from Louisiana went from house to house providing and gathering information about drilling. During their initial "outreach," the landmen asked about who lived next door and what they were like. After a few visits, they started telling neighborhood tales: "Do you know what Judy down the road said about your husband?" They spread gossip. They warned people not to chat with their neighbors and risk missing out on the best deal that, of course, was being offered by their company and only right at that moment. One of the Fitzgeralds' neighbors was offended by the gossip and ignored this ploy, contacting Jim and Terri to see what they made of the situation. Together they realized that they had all been rushed and isolated. Outraged, they called a neighborhood meeting and the whole community got together to discuss the rumors they had been told. They realized that the landmen had been subtly dividing them to find out who was the "softest touch" and might accept a well on their property.

When the next gas company began calling, the residents were wise to this tactic and tried to call a neighborhood meeting to collectively meet with the corporation's representatives. The landmen initially refused, but eventually agreed to attend, and the Fitzgeralds notified their neighbors and scheduled a meeting date. At the last moment the landmen changed the date. They tried a second time but the landmen again changed the date. Finally, a third meeting was arranged but residents were faced with a new ultimatum. Those gathered were told that if they refused to negotiate a drilling contract, one of them would simply be bonded on; that is, the company would give a bond to the state as insurance against property damage and go ahead with development, without the owner's consent. Trapped by this threat, one of the neighbors broke down and made a deal. The Fitzgeralds were forced to suffer a massive engine running outside their house around the clock for five months.

After a four-hour discussion at the Fitzgeralds, they took us to see the well on their property. The gas company had made many errors drilling the well, failing to properly replant trees or control run-off. The construction crew had even attempted to stop the Fitzgeralds' filming on their own property. Cohabiting with gas development clearly put landowners in a good position to be aware of gas industry tactics and report on the development process. The challenge for ExtrAct lay in designing a means to collect and circulate landowners' knowledge, regardless of the industry and regulatory agency actively undermining the emergence of collective structures.

PROJECTION 1: GATHERING STORIES

Lisa asked the Fitzgeralds whether ExtrAct could help collect stories more formally, describing how she had long wished she had a log of them to which she could turn when working for OGAP. She explained that so many tales were forgotten or lost because people with the knowledge moved on or did not write them down. "It would be great to have a way to collect those stories," she said. "But who would actually do the work?" Jim asked.

Chris wondered if we could set up a system for gathering and entering gas-patch stories into a shared database similar to a project he had worked on called Speak Easy. Chris's group had developed Speak Easy to facilitate communication between nonnative and native English speakers in Boston's Chinatown. Children who had translated for their non-English-speaking parents answered a phone and offered callers translation services for a few hours a week. Jim said the SJCA had about 8–12 really active members and 50 to 60 who were very reliable if their work was directed. The tide of the conversation had shifted. He was now actively imagining how we could work together and the ways in which the collaboration would benefit the SJCA.

Jim lamented the decline of grassroots activism since the gas industry had become entrenched. OGAP was increasingly focused on legislative issues and had hired more full-time specialists to deal with these regulatory matters. It had moved away from organizing community meetings and visiting regularly with landowners. He wondered if our digital tools might help revive grassroots activism. We said that story collecting could certainly be tailored to particular campaigns. Terri had asked if a story database could be organized "so you could search something like 'water' and find all of the cases related to that." In the water dispute in the courts she had mentioned earlier, landowners using tributary water downstream from gas drilling were not being properly informed about how their water supplies would be affected.

Our conversation brought ideas to life. Chris suggested that using the web to find out where state agencies had issued new drilling permits would allow citizens' alliances to alert residents before the landmen arrived. People could be empowered with information preemptively and become connected to OGAP. Like OGAP's community-to-community education trips, the stories database would connect communities new to gas development with those who could teach them about what to expect from gas development and how to deal with it.

PROJECTION 2: TRACING COMPANIES THROUGH CITIZEN MONITORING

A similar process of projection happened in our conversation with OGAP and SJCA representatives (DiSalvo 2009). Chris's Macbook presentation was almost entirely lost on the SJCA representatives; Josh Joswick described it as a "one-man techno rave." However, interest picked up when Chris described his group's critical uses of technology: how the politics of technologies and systems could be exposed by using them in unexpected ways. His examples around this theme produced speculation about how tactical media could affect the technical and social asymmetries surrounding gas extraction.

Chris's presentation about the Government Information Awareness (GIA) database led Gwen to consider how digital tools could be used to trace company activities. Chris's student, Ryan McKinley, built GIA in response to the Defense Advanced Research Projects Agency's Total Information Awareness (TIA) program, which fed everyday information on citizens (including bank records and credit-card use) into algorithms to determine a person's "terror quotient" (Bray 2003; McKinley 2003). GIA was an ironic tactical media response to TIA. It was an online database that gathered information on members of the House and Senate, combining public record information with users' tips to calculate politicians' "government quotients." In its first 12 hours online, GIA posted the Department of Justice's internal phonebook as well as the home address of the CIA director. Chris showed a video of Ryan being interviewed by Fox News, which concludes with the interviewer saying, "This seems like a perfect opportunity for someone with an axe to grind to sandbag a politician who is just simply trying to do their job" (McKinley 2003: 87–91).

Ryan responded, "I'm looking at this more as a process, as a way to give us a hope of maintaining a democracy. The technologies we are developing are building a very dangerous asymmetric balance of power." ExtrAct's project also sought to redress these "dangerous asymmetric balances of power."

Chris's presentation introduced TXTMob,[9] Tad Hirsch's workaround to overcome the chain-linked fence "free speech" zones created to contain demonstrators at the 2004 Democratic and Republican National Conventions in Boston and New York (Hirsch and Henry 2005). In the 1990s, organizations responded to police containment at protests by creating listservs, for better coordination through media centers for sharing information (Hirsch and Henry 2005; Juris 2008).[10] However, by the 2000 Los Angeles Democratic National Convention, law enforcement had become "savvy" to listservs, shutting down the independent media centers' satellite uplinks and arresting protestors simply for having walkie-talkies (Hirsch and Henry 2005). Tad realized protesters needed a technology that everyone possessed so that individuals could not be arrested for "possession." He hit upon the idea of using cell phones. Inspired by the flashmob phenomenon, which gathered groups to artistic events through text-message instructions, Tad applied the flashmob idea to political protests and built a text message-sharing system via phone lists. This system facilitated more flexible, smaller-scale protests that could emerge and dissipate very quickly (Rheingold 2002; Hirsch and Henry 2005).

TXTMob was so successful that the *Wall Street Journal* published a story on it (Bialik 2004). During research to put the story together, the *Journal* reporter told phone companies that TXTMob was using their networks and the companies began to filter the service. Rather than abandon the program, Tad asked hacker communities to help find a way to reroute the texts. People from all over the world responded by opening browser pages that acted as relays, covering up the initial server from which the text was generated and successfully preventing the majority of phone-company filtering attempts.[11] Both of the projections Chris presented suggested the potential for disruptive digital technology that could be adapted for gas-patch use to transform possibilities for grassroots political actions and responses.

ExtrAct's digital-media tools could allow OGAP and SJCA to focus attention on particular companies and build grassroots industry-monitoring networks, potentially solving both groups' concerns that there was currently no way to view the oil and gas extraction industry in aggregate. Steve, the SJCA organizer, described problems with infilling in the San Juan Basin where companies were interested in maximizing the number of productive wells in an already explored and developed formation. In his region of Farmington, New Mexico, seven active gas formations overlapped at different subsurface depths. State regulation set well spacing for each formation in terms of square-mile, or 640-acre, "sections." At one formation, eight wells per section might be allowed, but in another, four might be the maximum allowable. Although each formation was leased and contracted by different companies, all wells could be accessed from the same surface area. Well densities of 25 wells per square mile had emerged in Steve's area as a result. Since each well had to have an access road, water lines, electricity, and pipelines, his region had become "a vast industrial complex" that dominated the landscape. Because there was no planning for aggregate surface impacts and no one took responsibility for analyzing gas development in aggregate, Steve said his area was now a "national sacrifice zone for gas development."[12] His concerns echoed those of de Certeau: landowners, SJCA, and OGAP lacked a strategic space for tactical action. They did not "have the options of planning [a] general strategy and viewing the adversary as a whole within a distinct, visible, and objectifiable space" (de Certeau 1984: 37).

Josh and Gwen argued that the problem was made more challenging by regulatory bodies that were not organized to deal with the subsurface gas basin as a whole, rather than as disparate agency portions. The Four Corners region was a jurisdictional nightmare. The interests of four state legislatures intersected with those of tribal areas with their own regulations. Federal lands with Bureau of Land Management, three Environmental Protection Agency (EPA) regions, and U.S. Forest Service jurisdictions created additional complications. Overstretched state agencies were unable to deal with the massive scale of gas development. There were only nine well inspectors at the time for the whole state of Colorado, which had more than 37,000 active wells (Sumi 2012). Fewer than 25% of the active wells in Colorado were inspected in 2008 (Sumi 2012).

SJCA's citizen well-monitoring system responded to this lack of regulatory oversight by training volunteers to go to wells with a checklist to make sure facilities complied with local regulation. Amy from SJCA pointed out the benefits of this program: "We are trying to give people a little more power over what is going on and that starts with giving them the ability to look and say this is not right, this is not what they were supposed to be doing. There are so many people out there who have just succumbed to it." A neighbor from her rural community had responded to a noisy compressor station built nearby by sleeping with his TV blaring to drown out the noise. "That is what he had succumbed to," she exclaimed. He had not complained to the state because he knew nothing would be done. Gwen agreed that residents were so frustrated with state regulators they called OGAP or the alliance first. They knew "no one [from the state] is going to come out and look at the well." Since the state and counties were so unresponsive, Chris questioned how gathering more complaints would serve residents' interests. Josh, who had been a county commissioner before joining SJCA, felt that it was important for people to maintain hope and stay engaged: "I'm hoping that through this citizen monitoring program maybe we get a little more hope in the community.... They aren't going to stop and if we stop they win, so we can't stop," he said passionately.

Chris said a system for locally monitoring company behavior could "build a global mirror to these companies." Gwen smiled. "I kind of like the idea of going after companies," she said. "They are so sensitive about their image and negative PR." "Depends on the company," Josh added dubiously. Gwen agreed that "the bigger they are, the 'greener' they are." She explained that her advocacy had shifted its focus to lobbying for better regulation back in 1988 when Amoco (later bought by BP) planned to drill 1,000 wells in La Plata County but was confronted with aggressive responses from the community. Residents, who found they could light their tap water after drilling began, were passionately opposed to drilling, but there was equally committed support for drilling from mineral-rights owners, who would gain from royalties during an economic recession. "People were literally lined up across the aisle" at public meetings, Gwen said. "It was intense." She had even received a death threat.

OGAP then decided to lobby the state government rather than being directly adversarial to companies, hoping state legislators would do the right thing. Projections like TXTMob promised new ways to challenge and "trace" companies without the polarized anger and passions fanned by public forums. Our para-ethnographic conversations through projections such as those described here yielded a sense of the cultural, political, and economic critiques already in practice. OGAP, SJCA, and the Fitzgeralds raised the following concerns:

- The conflicting interests of surface owners versus mineral-rights owners
- A complaint-driven regulatory system that neglected to gather or respond to complaints
- So many unmonitored wells, which required citizens to try to make up for the state's shortfalls

— Gas development needed to be viewed in aggregate
— Many companies of all different sizes filled different niches in the industry
— The industry's strength and economic importance created deep conflicts between community members
— As the first point of contact between extraction companies and our interlocutors, landmen were a frequent source of division and misinformation
— Landowners were confused over which regulatory agency was responsible for monitoring the different aspects of well development and maintenance

CULTURAL CRITIQUE 4: INDUSTRIAL ECOSYSTEMS—A SPACE OF TACTICS

Hopelessness, intimidation, and frustration at persistent misinformation, mismanagement, and lack of oversight reemerged as common themes when we met Jack and Sug, Steve Kaiser's friends who owned land on the gas patch in Aztec, New Mexico. As two of only six or seven active SJCA members in what they described as "an industry county," Jack and Sug said that they had taken plenty of flack for their involvement from other residents. Sug joked that she had been accused of being funded by East Coast liberal obstructionists for years. She was glad to have finally met us.

Our gas-patch tour began behind a Wendy's restaurant in the center of Aztec. Sug told us that the well had tested positive for hydrogen sulfide, but the company merely questioned the validity of the test. Hydrogen sulfide is a rotten-egg smelling gas that can be produced with natural gas or by bacteria in wells. Sug told us how she had been exposed to the gas when walking to her mailbox and had "almost passed out." Hydrogen sulfide is neurotoxic and can be lethal in high doses (Kilburn 1995/2003; Skrtic 2006; Kilburn, Thrasher, and Gray 2010). The Occupational Safety and Health Administration's ambient exposure limit for construction for other outdoor operations is 10 parts per million (ppm). Sug said company sensors placed around her property had showed hydrogen sulfide concentrations over 100 ppm. The National Institute for Occupational Safety and Health considers a level of 100 ppm to be an immediate threat to life and health.[13] In response the company added a hydrogen sulfide treatment system to the well and provided informational leaflets that described hydrogen sulfide gas as naturally produced and safe in low doses.

Jack described how people had become sick and how paint started peeling off the sides of trailer homes near a drilling liquid-waste disposal and injection facility. The company involved eventually agreed to monitor the wells, but "forgot" to calibrate the monitors. After the monitors were calibrated, test records were "lost" on their way back to town. Finally, after conclusive formal tests and community legal action, the entire community except one family was bought out by an out-of-court settlement. Jack was disgusted. He said the case result had "allowed the waste disposal facility to continue operating as they had been and effectively shut up the opposition and concern." SJCA heard residents' concerns over air pollution and liquid waste, so it joined forces with Global Community Monitor (GCM), a California-based nonprofit, to teach the landowners how to use low-cost buckets to take their own air samples (GCM 2011).[14] Their 2011 study found that

> a total of *22 toxic chemicals* were detected in the nine air samples, including *four known carcinogens*, toxins known to damage the nervous system, and respiratory irritants. The levels detected were in many cases significantly higher than what is considered safe by state and federal agencies. The levels of chemicals, including benzene and acrylonitrile, ranged from three to 3,000 times higher than levels established to estimate increased risk of serious health effects and cancer based on long-term exposure. . . . At least two cancer-causing chemicals, acrylonitrile and methylene chloride, were detected at high levels near natural gas operations. Neither chemical is associated with natural gas or oil deposits, but both seem to be associated with the use of hydraulic fracturing (fracking) products. Resins acrylonitrile, 1, 3 butadiene and styrene (ABS) are suspected to be present in fracking additives. (2011: 2; original emphasis)

The local elementary school we visited next had a gas well at the end of its playing field. Sug told us that the school nurse had been told by school officials not to report to the state the chemical smells wafting into classrooms.[15] As our tour of Aztec continued, we gave up trying to photograph all the wells we saw. There were simply too many. Meanwhile Jack and Sug told us what it was like living in Farmington, a company town where gas was by far the biggest industry. People had complaints but were often too scared to make them. Those who did complain were silenced with enticements like brand-new pickup trucks.

Outside Farmington, we visited the 700 acres of cattle land that once belonged to Sug's family. Her family sold the land during the gas boom because adjacent land as far as the eye could see had been turned into a massive supply yard for gasfield equipment. At a nearby trailer-park community that was home to migrant laborers, one of the wells at the park's center had a condensate tank that used to leak fluid down the hill. Sug told us that it was only cleaned after New Mexico state officials on an OGAP- and SJCA-sponsored tour had watched kids playing in the oily, muddy run-off puddle. The day we visited, the top of the tank was left open.

A landfarm visible from the hill behind the supply yard spread toxic dust on homes downwind. Landfarms are named for their practice of spreading out, or "landfarming," contaminated soils and wastes from drilling pits so chemicals can evaporate. Residents had complained to Sug that closing windows against the thick dust during the summer had not prevented "sickening odors" permeating their homes. When Sug contacted state authorities about this, they had told her to stay out of other people's business. Why should she care if it did not affect her directly? they said. Although regulations forbade gas workers from bringing drilling fluids to the landfarm, we watched a "water" truck pull in. Jack said he and Sug had received insider tips that fluids were being dumped at the site. We saw another condensate tanker inside the facility as we drove away. The landfarm sign read "Industrial Ecosystems," a fitting description. We drove into a second trailer-park community alongside the hulk of an old refinery that was being rehabilitated as a Superfund site. A well erected directly up against a trailer was in blatant violation of set-back rules. A family, who had cautiously agreed to be taped by Chris, covered their license plate (the husband worked for the gas industry). They complained that they smelled rotten eggs all the time and were moving because they were concerned about their daughter's health.

Regulatory violations were commonplace in this industrial ecosystem reliant on its gas industry economy. Complaints were perfunctorily dismissed by authorities. The ExtrAct team stood out in this landscape as unwelcome outsiders, and we were pulled over by a San Juan County deputy sheriff as we drove away from the supply yard. The officer asked us why we were there and told us to move along, which we did. Was he simply investigating our odd behavior, or our interest in the gas patch? Marked as outsiders, we wondered how we could design something effective for the residents of a county in which basic regulatory violations could go unchecked, and where people seemed intimidated and opted to move or to close their windows rather than to contest the industry.

CULTURAL CRITIQUE 5: MEDIA RELATIONS—THREE ACTS OF PERFORMING THE GAS PATCH

Outsiders in Farmington, we blended quite well into the crowd of reporters at Garfield County's gas patch in Colorado. Our Farmington tour had focused on the social dynamics of an industry town, but this aerial ecotour over Garfield County directed by a pilot interested in wilderness issues was focused on oil and gas's impact on public lands. The mainstream media was far more interested in wilderness issues than a small trailer-park community near the Farmington landfarm. Although gas-patch issues were not a national media focus in 2008, we realized that media and performance were omnipresent during our trip.

Act 1: "This Rugged Wilderness Is Being Transformed". At 10 AM, the small regional airport was filled with reporters and their entourages and equipment. It turned out to be Earth Day, a day in the news cycle that demanded environmental coverage. A company specializing in ecological flight would take two planeloads of journalists (and us) over the Roan Plateau to see how gas development was affecting Colorado's wilderness. The pilot's conservation background came through in the topics and places he liked to cover. Conscious that any MIT study can generate media hype, we tried to keep a low profile. We did not want news articles drawing industry attention or raising any hopes about our project before it had even started. Unfortunately, we were not successful. A middle-aged male reporter with graying brown hair joined us with his lunch and started asking who we were. As soon as he heard we were from MIT, he started pressing for details about our work. I became profoundly uncomfortable in this first experience of being potential interviewee rather than interviewer. My awkward insistences that we were not quite ready to talk about our project only made him more interested. Thankfully, just as all the journalists began to circle, the pilot showed up. As we walked outside, the first reporter followed, repeatedly asking my name. Fortunately we had to be quiet on the runway so a film crew could complete their report. The auburn-haired TV reporter stumbled through five takes, trying to feign the right seriousness and steady-eyed expression to bemoan how the "rugged wilderness is being transformed" as wind disturbed her carefully coiffed hair. Chris delightedly filmed the whole scene. Wildlife, not human life, was the assignment brief for Earth Day.

Finally aboard the plane and free of the tenacious reporters, we asked to travel over Rifle and Silt. But the pilot had his own agenda—and the helm.

He wanted to show us the beginning of drilling on the Roan Plateau and even directed what pictures we should take. "Take a shot over there where the drilling hasn't started," he exhorted. It was shocking to see drill rigs in ravines on each side of the still snow-covered plateau, chains of them strung into the far reaches of each sharp-walled rift. The wilderness focus was not new to me, but I wondered why the gas-patch residents of Rifle and Silt had no news value for these reporters.

Act 2: "Oil and Gas Feeds My Family". The subject resurfaced later that evening. "This issue of wilderness protections tends to become divisive between groups," said a Grand Valley Citizens Alliance (GVCA) activist, Carol Baxter. We were sitting around on Carol's couches with her and a couple of GVCA members as Carol explained how environmental issues often played out along class lines: wealthier fishermen, tourists, and hunters on one side, versus working or middle-class trailer-park or subdivision residents on the other. Carol described how GVCA had organized a group of landowners to attend and testify at a local rule-making hearing where they were lobbying for surface owner rights legislation (OGAP 2007a). She had met one of her former students at the hearing, now a gas industry worker, who had explained how he and his colleagues had all been given the day off work to attend the hearing. All the gas industry employees dutifully sported "Oil and Gas Feeds My Family" buttons on their work shirts. The small group from GVCA was overwhelmed. Outside the meeting during a quiet conversation, Carol's ex-student urged her to keep going, not to give up. Rig work was making him ill, dizzy, nauseated. He had thought about quitting. When he had told his employers about these health issues, they moved him to a new job selling afterburners to reduce his exposure to emissions.

Another GVCA member recalled how workers disrupted a county commissioner's meeting the group was attending. An industry conference at the same hotel had closed its open bar just as the GVCA meeting started, in what appeared to be a coordinated attempt to disrupt it. GVCA felt pressure from the industry that was able to organize its workers as a collective and in greater numbers and force than the community group. GVCA members described being targets of personal pressure and direct harassment.

These two examples of gas-patch media play share a certain "staging." The airstrip TV reporter's multiple takes simulated exactly the right concerned looks and phrases. Her crew had filmed a plane taking off and landing, to be added to the piece at the studio for filmic believability. Similarly, the industry stages its workers to overwhelm a small community group like GVCA, which is no match for a wall of men in ready-made buttons, or for revelers full of company largesse from an open bar. These moments are performances. What will it take for communities to be similarly center stage? How can residents have a say in industry development untainted by mainstream media's momentary demands for novel entertainment, the industry's job-provider spin, and academics' prescriptions?

Act 3: "Local Access". During our potluck dinner with GVCA members, we revisited the need for a tool that might help communities respond proactively to industry development. After landmen appeared in his subdivision, Jim, a portly, friendly man in a Hawaiian shirt, made watching the Colorado Oil and Gas Conservation Commission website into a ritual to see when a drill permit would be submitted for his street. He was certain the industry would choose a time when there were limited chances for residents to contest the permit. Jim waited and watched. As he suspected, the company submitted its permit right after the county commission went on a scheduled two-week break. The next commission meeting would take place after the 14-day objection period had expired. Jim went from door to door in his neighborhood and condemned the tactic on local radio. The company promptly withdrew its permit requests in that neighborhood. Since not every neighborhood has a Jim, this case supported the benefits of the proactive organizing tool that Chris suggested during our conversation with the Fitzgeralds: an automatic alert to residents if a drilling permit is submitted for their area.

Another GVCA member, Dee Hoffmeister, had also made savvy use of low-cost media to document condensate tank explosions near her property. She and her husband, Harold, had bought their dream retirement home, a small house settled among the mountains outside of Silt, Colorado. After Dee's health declined sharply when gas development began, the couple was forced to leave their home for month-long stretches. The problems began when odors started creeping into the house and Dee became dizzy and lightheaded. The dogs refused to leave the house to play outside when the smell was in the air. Four of her children and grandchildren developed asthma. Dee felt lethargic and like her feet were on fire.[16] Her decline hit a nadir one night in 2007, when condensate tanks at the vacant lot next door exploded in a fireball. Dee was overwhelmed with fumes and could not get out of bed. She had to be carried to safety by her husband and was in hospital for two days afterward. They temporarily moved away from Silt and although Dee's health improved, she still walks with the aid of a cane.

The fire next to Dee's home became big news after she posted her daughter-in-law's video footage on YouTube. BBC World News ran the clip in its piece on western gas development (Willis 2008). It was later featured in the documentary *Gasland.* Although the spectacular event garnered significant media attention, there had been no news value in pursuing the follow-up story. Months later, the site had still not been remediated and the condensate tanks lay blackened and ruined, exposing molasses-like sludge to the open air. Dee joked that the company had come by to paint the side of the tank facing the road green. Dee became distraught as she described her inability to scientifically link her illness to gas development. The relationship was quite clear to her and her husband. We asked if her neighbors also had health problems and she said yes, many of them were ill, and one neighbor, Chris Mobaldi, had died. She wished there was a way to record and track all the health complaints in her neighborhood.

These different experiences of media participation in the gas patch created questions for us about what kind of system we could design to increase locals' ability to organize proactively. How could stronger links be forged between shocking events and the more mundane realities of dizziness and asthma? Jim and Dee's intended media audiences were not thrill-seeking prime-time viewers, but rather the companies themselves. They hoped to shame those companies into acting differently.

CULTURAL CRITIQUE 6: WORKERS?

None of our discussions with citizens' alliances had focused on the roles of workers. They appeared in the stories we heard as landmen, roughnecks, or anonymous trailer owners in New Mexico and who were organized only to attend county commission meetings. Workers had figured conspicuously in our discussions about health and safety, but we had not talked to them. How could we understand their role, and did we need to? Over dinner with the group, Theo Colborn told us about her encounter with gas workers on her flight back from a conference. She had asked two men sitting beside her if they were headed to Grand Junction, the nearest city to Paonia. They said they had been hired from Canada to work two week "fly-in, fly-out" shifts from a 400-person "man camp" outside the Junction. Theo then described her work with fracking chemicals and their health effects, listing the common symptoms of exposure—numb fingers and toes, burning eyes, upper respiratory infections, memory loss, erectile dysfunction. She indicated this last symptom with a discreetly drooping finger. The men laughed awkwardly.

“That’s me,” one of the workers, a 20-year-old, responded. Theo remembered how they had cried together right there on the plane.

The next day we met with Teresa Coons, whose epidemiological study of Garfield County (Coons and Walker 2005) appeared to exclude migratory gas-patch workers. Teresa explained that her two-part project, an environmental risk assessment and a health assessment, had been unable, because of lack of funding, to take environmental samples. She had to rely on available air-quality data, particularly a Colorado Department of Public Health study identifying EPA-recognized toxicants that combined continuous monitoring at county sites with grab samples made in response to public complaints. Her “point in time” health assessment of Garfield County residents sought to establish a health baseline against which any subsequent study could be compared. It did not seek out direct correlations between exposures and disease and it did not investigate pathways of exposure. The survey supplemented qualitative assessment and statistical data about births, deaths, reportable conditions, and hospital discharges with a household study restricted to residents of Garfield County who had landline phone numbers and addresses. It did not target the most highly exposed population, gas-patch workers, since they were migratory. The workers did not have landlines because they lived in hotels, temporary housing, or man camps. Also, reportedly, companies can require their workers to go through their own medical services, so workers’ illnesses could be missing from hospital records. Teresa suspected that industry workers were reticent about describing their employment. For example, they would say, “I drive a truck,” which could mean many things and further limited the survey’s ability to track workers’ health concerns.

As a result of these conversations, we began to discuss the possibility of generating a whistle-blowing website particularly for workers.

CULTURAL CRITIQUE 7: NEIGHBORHOOD WELL WATCH

The lawyer Lance Astrella was the first person we met to whom we did not have to pitch our ideas. Astrella argued that the industry was far too distributed for effective regulatory oversight and that a platform for public monitoring could alter the gas giants’ behavior by promising to produce legal action rather than actual litigation. Providing landowners with contact details for experts able to carry out testing would establish baseline standards for air, water, and health quality. For an industry that relied on self-reporting, “publishing complaints would hold the oil and gas industry and regulators to a higher standard,” Astrella said. He told us how small-scale complaints about everyday industry practices such as building roads that were supposed to be 20 feet wide but ended up being much wider were "very inefficient to handle in the context of a lawsuit. If you can aggregate a number of similar cases, administrative or judicial relief may be affordable." "More importantly," Astrella added, "is the fact that, if the industry knows you have that data at your disposal, they are going to change their practices."

Foundational information for epidemiological studies was "horribly lacking," Astrella said. "If you don't have the science, you can't go to court." He understood the value of designing a platform for such information, and he thought that online tools for generating oversight would force the industry to behave better, because of the threat of legal action. He encouraged us to develop a tool for whistle-blowing: "You don't want a rash of litigation, what you want to have is the threat of litigation that causes a change of practice. So a whistle-blowing mechanism will do that." Whistle-blowing would be particularly effective in the case of inadequate water-quality testing performed by the companies employed by the oil and gas industry. Sampling errors occurred if samples were not immediately placed in airtight containers, because volatile organic compounds evaporate. Bore-hole tests on contamination concentration from leaking tanks or pipelines only sampled groundwater, neglecting to test the soil beneath the water table. While flowing groundwater could temporarily flush contaminants, contamination could continue after water levels subside. Unless landowners employed their own environmental experts to contest the methodology used by the industry testers, government regulatory agencies were inclined to approve industry activity based on the industry's inadequate testing, Astrella said. He believed that whistle-blowing from inside such companies could really make a difference; he was interested in its "deterrent effect": "There are always unhappy people out there, people that have been fired. If the bad actors in industry know that this whistle-blowing site is out there on the Internet, they will shape up. Those following proper procedures will have nothing to worry about."

Our meeting with Astrella was one of the most productive on the trip, in part because our ideas were concretizing into a prioritized program. "If you get done what you're planning to get done and you establish credibility, the industry is going to change," Astrella said. "Industry executives don't want their companies to face mass liability. If credible data is generated and published on the Internet, it will be hard to ignore. In addition, the younger people in industry tend to be more open-minded when it comes to looking at potential environmental hazards and they are more in sync with the electronic media. Through them, change can be encouraged in a positive way."

Astrella helped us see that the end game for our tools might not be industry reform through actual legislation or litigation, but rather amplifying oversight by raising the threat of both. Creating new channels of information flow and novel collaborations between citizens and experts could achieve potentially heightened industry vigilance.

Return to MIT . . . What Now?

After our eye-opening, warp-speed field experience, we had to compile what we had learned, formalize those ideas, and formulate our first plan of action. We spent our summer in a gray-carpeted, corporate boardroom that was a far cry from homesteads and mountains and tourist towns, designing tools for those now-distant places and people. We needed tools that would work for a diverse audience, without us being perceived as interfering outsiders.

Astrella's idea for protecting whistle-blowers' identities inspired us to think of building a "toolkit" from which our collaborators could choose, rather than a single website to handle many different issues. Different tools would address different audiences and needs. With rhizomes in mind, we decided to produce a group of separately operating sites that shared a common back-end data structure. "Back-end" is the term for the database and code that the user never encounters, while the front-end is commonly referred to as the user interface (UI). We decided that the site UIs would be different and appear separate for each tool.

We excitedly discussed how developing a different UI for each site mirrored the corporate tactic of separate branding. Chris added that individual branding would also save the rest of the tools if one site failed. It would also allow us to be more experimental with our designs. He suggested that individually branded, apparently unconnected sites could take advantage of the popular, but erroneous misconception that technology is apolitical. He compared "technology" to Teflon, because technology, like Teflon, has a sheen that prevents labels from sticking to it and that lets it slip in as apolitical.

Our para-ethnographic field-trip discussions stressed that natural gas development had created a terrain of structural information asymmetries. We listed some of the significant information gaps between landowners and companies:

1. Surface owners are isolated from each other and unable to aggregate their complaints. Furthermore, there are many different and unresponsive regulatory agencies and the complaints process is time-consuming. We thought of

a website that would primarily be used by surface owners to improve the process of gathering and networking landowner stories and complaints. We imagined gathering complaints about many issues, including health, noise, property damage, and contamination. These complaints would then be filed with the state when possible, as well as with companies and boards of health (where applicable). This site, which came to be called WellWatch, "community monitoring of the oil and gas industry," will be discussed in chapter 8.

2. Companies have better internal systems for organizing information and sharing it than gas-patch communities and their advocates. Landmen have comprehensive information on the terms being offered by their companies. We wanted to help develop internalized information structures for citizen alliances that would allow them similar distributed coordination.

3. Confusing regulatory agencies and problems with aggregating and analyzing complaints made company behavior hard to trace. Our third site idea, Corpmap, was a corporation map that would show well ownership, including information about subcontracts, so that landowners and community organizations could have a way to understand, reflect on, and track the changing companies acting in this industry. Such a map might usefully indicate the tight relationship between water-testing companies and the gas industry, for example.

4. Communities where gas extraction is just beginning have little knowledge of the industry. The pattern of gas development is established very early when the first leases are signed. It is only later that landowners realize the pitfalls of these binding contracts. A landman review website where landowners could share information about landmen as well as surface and mineral contracts would help improve community-to-community education at the beginning of drilling. This landman review site became Landman Report Card, or LRC, which is discussed in chapter 7.

5. Gas industry workers afraid of sharing information, about everything from spills to faulty testing methods, or lacking the means to do so, prompted the need for the anonymous whistle-blowing site discussed with Lance Astrella. This site would be addressed to the needs of both workers in the financial sector and workers in the field.

6. Landowners who are unaware of regulations or of what constitutes a good lease or a dangerous practice suggested the need for a sixth site aimed at facilitating and strengthening communication between specialists and landowners. This site would gather information about relevant laws to help people make complaints or research the best lease deals.

7. Our final idea was for a "proactive" site that would alert residents to the potential for drilling permits to be filed in their area so that they could become organized and manage development collectively rather than be taken by surprise. Information for this site could be garnered from state sites, providing locations where drill permit applications had been submitted.

Dan gave us his perspective on the back-end design issues raised by these outlined tools, discussing which were feasible and how long they might take to develop. He ranked the ideas from easy to harder to build in the following order.

LRC was the easiest to build because it could be modeled on an existing site like Yelp, a website where users review local businesses. The most challenging aspect of this site would be striking a balance between user anonymity and verification that contributors were actually residents rather than anonymous landmen giving themselves favorable reports. We needed to determine a threshold, such as having an address in an area or providing a name, before allowing someone to submit a review.

The whistle-blower site, the next easiest, raised the same anonymity and verification issues as well as the question of the nature of the site interface. Would anyone be able to see remarks left on the site or would they only be accessible to specific groups such as OGAP or a network of lawyers?

The proactive site took third place, as it was easy to know where new leases and permits had been granted, but also difficult to find addresses for where to send that information once we had it.

Aggregating complaints and mapping stories seemed like the largest site that might require some type of web mapping platform. We wanted to develop a TurboTax-type complaint framework so we could standardize complaint formats and automatically determine if they qualified under the states' categories. We would need to mine data from many databases, each with a different format. Where and how to route complaints raised a further challenge.

The information structure for distributed communication and solidarity presented a few problems. A phone-tree tool required Dan and our other programmers to become familiar with the vast, poorly organized world of telephone programming languages and to build telephone interfaces that route through a server using a programming language called Asterisk.

Corpmap came in last because we did not know how we would gather the data.

Dan and Chris would lead building these tools. The terrain of gas development can be bewildering, but Dan's and Chris's back-end development discussions flew by me in a flurry of incomprehensible terms—Django, Postgres, SQL, Asterisk, edge-node, open-layers, Ubuntu, API. Languages intersect, communities battle over terms, and learning involves constant adaptation to continual change. For me, staying abreast of software development was like trying to catch a high-speed train that left 10 minutes ago. We wrapped up the meeting with a sense of what we would build, a remarkable accomplishment considering the range of potential options. We also had a tentative order beginning with the mapping and complaint site and the whistle-blowing tool. However, the most pressing goal for us was developing demonstrations of our ideas so that we could have something concrete for the first annual civic media conference at MIT. As we were the only project yet to get off the ground, we had to put on a good show.

Landman Report Card

Developing Web Tools for Socially Contentious Issues

Landman Report Card, the tool that we imagined was the low-hanging fruit, turned out to be the most culturally and legally complicated of our proposed "toolkit." We chose the name Landman Report Card, or LRC, because we hoped landowners would be empowered by its suggestion that they could "grade" landmen. The idea was inspired by websites like RateMyCop, which allowed the public to rate police officers, and Angie's List, a service offering user reviews of plumbers and repairmen. Our field research showed that gas development frequently caught new landowners off guard. They were isolated and divided during the leasing process. We wanted LRC to encourage community-to-community education so people could visit the site to find out more about the person at their door.

Unlike plumbers and repairmen, landmen arrived with industry strength behind them. They were not public servants like police officers, but rather oil and gas industry representatives. Landowners were very aware that reviewing landmen put them in a precarious position. Landowners we had interviewed found speaking about their experiences with landmen hard, but committing those opinions to writing and posting them online turned out to be much harder. The popular assumption that no one knows who you are on the Internet has proved inaccurate as regulation, control, and Internet traffic monitoring have grown (Lessig 2006). The web's open architecture creates radical accessibility and fluidity between each individual's personal computer and those of others on the network (Lessig 1999, 2006; Benkler 2006; Zittrain 2006). While this fluidity creates new possibilities for redressing structural imbalances, such as those between surface owners and extractive industries, new laws and the necessities of web design practices make it difficult to develop web services that protect users' identities.

With LRC, we learned that creating a strategic networking space for landowners to gain more control over the leasing process not only provided new tactical opportunities, but also carried significant risks for both designers and users. These risks extended to the ExtrAct team as website developers and hosts because, in legal terms, we became an Internet service provider (ISP). We would be constrained by two kinds of code: one legal, and the other in our software programming (Lessig 1999, 2006). The next section explores how ExtrAct negotiated the challenges in these codes to both protect site users' identities and weed out corruptive users. This balance between user anonymity and credibility and our concern to protect ourselves from liability as website providers took place in a social context fraught with explosive issues of identity and infiltration, as I will illustrate.

Fort Worth, Texas, February 2009

Our second LRC training and field-testing trip was to Fort Worth, Texas. The first had been in Cleveland, Ohio, where neighborhoods were enraged by suburban drilling, some within 100 feet of their homes (Johnston 2009). ExtrAct planned to pair Cleveland with Fort Worth, where urban drilling had also started recently (Lavandera 2008). We hoped that people in the two cities would have greater Internet access than most rural areas and might more readily use our website. Jen Goldman, in training to replace Lisa as the Oil and Gas Accountability Project (OGAP) liaison to ExtrAct, accompanied Lisa Sumi, Dan Ring, and me for the training.

Lisa was concerned that, since word about our training session had been spread through the 400-member email list of Tom York, a local activist, it was likely that industry would have heard about our meeting.[1] We discussed at length how to deal with potential disruption. Lisa was in favor of just asking industry representatives to leave: "This isn't a public forum after all." As the meeting's goal was to speak with people who had direct experience with landmen, we decided to split the group, to create a space where people would be able to talk freely. Only those with direct experiences with landmen would be allowed to stay in the room to try out the LRC software. Everyone else would go to another room to have a discussion with Jen and Dan about WellWatch, another website we were developing.

The meeting began with Lisa's introduction about OGAP and ExtrAct and my brief description about the goals of the session and the broader ExtrAct project. Lisa then invited everyone to introduce themselves and speak about why they were attending the meeting. This would help those with issues about the industry to get them off their chests. Sharing experiences with those in similar positions builds trust. But there was a more tactical reason for the introductions; we would be able to separate those who had not yet interacted with landmen from those who had.

The introductory stories we heard, about people's experiences with the gas industry, ranged from mildly troubling to totally outrageous. A long-term resident of a poor inner-city African American neighborhood, Jimmy Johnston described how his neighborhood had been particularly badly treated during leasing. He knew of neighbors, drug addicts and old ladies, who had been pressured into signing leases way below market price. Renters were pressured into signing documents just so there was an authorizing name. The contract documents inevitably failed to fulfill the promises made verbally to people, but the terms of the written agreements were binding.

Few of these tales raised red flags. The storytellers seemed heartfelt and genuinely aggrieved. One woman, in her late 20s or maybe early 30s, wearing "business causal" style clothes, stood out. Neither a landowner nor a reporter, she said she was "just interested in the issues." Hmmm. As she spoke, furtive, meaningful looks were exchanged around the room. You could see people stiffen. She so clearly was not part of the middle-aged, angry-homeowner demographic. But she was not a cowboy-hat-wearing landman type either. And she seemed a little bit afraid, particularly when someone asked her how she had heard about the meeting. It was a difficult moment, and I was glad that we had a plan of action to deal with it. The introductions concerning people's life stories took time, then were followed by a short spiel from me omitting any details about LRC. Those without direct experience with landmen were then asked to leave with Jen and Dan.

The testing and training session with the 10 remaining people passed uneventfully, interrupted only by the usual problems with site navigation, overly complicated password protection, and getting people to answer a brief paper questionnaire about their Internet experience. At times I needed to be in three places at once, as we were called here and there for questions: "Logging in isn't working!" "What does this mean?" "How do I start a report card?" A white-haired lady, possibly in her 60s, joined us in the middle of the session. She was known by activists in the room to have a devastating story about how she had been treated by the gas industry, but she was averse to using computers. Thankfully I had a voice recorder and she and Lisa were able to work in a small, side office to record her story.

With about 30 minutes left, we concluded with a discussion about ways to improve the site. The remaining people who had been talking with Jen and Dan filtered in and word spread through the group that someone from the industry was in the room. Although I was caught up in a discussion with Jimmy, I could see that Tom, who had advertised the meeting, was restless. I received a note that read, "Announce that industry is in the room." I felt awkward creating a scapegoat in this way, because the only evidence that the smartly dressed woman was industry affiliated was that she looked different and had no experiences with landmen. My reticence became irrelevant when Tom stood up and stentoriously announced: "I just want to let everyone know that someone from industry is in the room." "What?" exploded the woman who had refused to use the computers. "I'm not staying here if industry is listening in. I've been fighting the industry for the last six years. . . . Who is it?" she demanded. "Who here is with industry?" She was out of her chair and angrily surveying the crowd. She focused on Jimmy, who was at the front of the room with me. "Is it you?" Jimmy tried to reassure her with an unequivocal "No, ma'am," but she was already moving along the aisle to leave. "I'm not staying here. Who are you with? You've taken everything from me—my property, my money, my peace of mind. You've put me and my family at risk." She did not want to hear Tom's and Jimmy's desperate reassurances.

In the meantime, the face of the smartly dressed woman reddened as people turned to look at her. "I'm not from industry," she called out. "I swear I'm not." The woman who had been leaving the room redirected her ire. "Are you?" She demanded. "Who are you with?" The woman again protested her innocence, looking more and more bewildered. I felt sorry for her and a little for myself. The situation had spiraled out of control and my responses were slow. I was aghast at how quickly we had lost control of just 15 people. Many were victims of very poor treatment by the gas industry. Their feelings were powerful and running over. I tried to restore some kind of order: "Everyone just calm down for a moment," I said over the din. Sometimes British accents help. "No, Jimmy isn't from industry and perhaps this woman isn't either. Let's not jump to conclusions. It's been a really great session and I'd just like to see if we can wrap things up. We're almost finished."

The conversation was not quite the same after that. There was no way of settling down the angry woman, so Jen left with her. Everyone else calmed and we wrapped things up with a semblance of civility. As people were leaving, the scapegoated woman (who was still blushing awkwardly) came over to me and apologized. She truly was not from industry, she said. I apologized and said I felt terrible if that was the case. Honestly, I had no idea what to make of her or the situation.

We tried to discover the truth as we drove as fast as possible to our next training session in Denton. Jen and Dan used their Blackberries to track the email address the woman had added to our sign-in sheet. It belonged to a small company that a brief Google analysis revealed was probably a public relations firm. "That would explain it," Lisa said. "She can truly say she's not with an oil and gas company if she was hired by a contracted PR firm." But who knows? I still don't.

Mock-ups and Demos: Learning Legal and Social Codes through Design

MOCK-UPS

This ethnographic moment illustrates the chapter's central theme: the problem with mapping the polarized field of gas development onto our web platform. Polarization is a tactic employed to reduce the complexity of an issue and create a target for action (Alinsky 1971). Within social movements resisting gas development, it is a common practice. "Industry" is a catch-all term for advocacy targets. In a polarized environment, getting inside the other's territory becomes a good tactic for either side.[2] People, events, and organizations are frequently captured or are the targets of capture by the other side (J. Scott 1985). An extreme case of this occurred in Texas, where industry funded a grassroots citizen organization and organized rallies in a process known as astroturfing (Greenpeace 2009; Sheppard 2009). Spotting industry infiltration is a game in activist circles. Just as creating a script for managing industry infiltration was vital to organizing our testing session, so too was creating scripts and protocols to manage the relationship between ExtrAct, users, and industry in the design and development of LRC's online meeting place.

The user interface for LRC became a new terrain for establishing such scripts, as it was the place for socially and legally establishing relationships between the website and its users (Fuller 2008). It was here that we had to establish our responsibilities and those of our users through the site's visual presentation, terms of use, and various legal disclaimers (Berkman Center for Internet and Society 2010: 255). As we tested site ideas and gathered stories and experiences around which to model the site design, landowners told us how concerned they were about the risk that posting their stories online could make them the targets for lawsuits. Would they experience legal retribution? We met Susan Watson through San Juan Citizens Alliance (SJCA). She was a landowner who was very enthusiastic about our site for reviewing landmen after suffering repeated problems with a landman because of her absentee surface ownership. She lived in California but owned land in San Juan County.

Our mock-up paper version of what a landowner might encounter on our review site included the landman's name, company name, and the time and location of a visit.[3] We knew that people needed to fully express their stories, so the description of the problem or incident would be central. However, there were technical drawbacks to free-form text. It was hard to parse bits of information from a description for later use and it was difficult to flexibly reorganize the data (Berners-Lee and Fischetti 1999). If it was well "tagged" (organized by some searchable conventions that could be flexibly associated with other pieces of data), the text would be much more helpful. So we asked our contacts to come up with such tags, landman "tactics" that they had experienced. This potentially created folksonomies, generative taxonomies based on common knowledge that users could borrow from and add to (Mathes 2004).

LRC attempted to turn tagging into a tactic for social analysis. The tags would not only organize report cards but also make users aware of potentially hazardous landmen behaviors and practices. It turned LRC into a research tool through which users could generate a classificatory analytic so landowners, landmen, academics, lawyers, or reporters could study and track the structural dynamics of the industry, rather than merely focus on the "bad apples" in the bunch of landmen.

The following are Susan's responses to the mock-up interface:

— V.
— User: SWatson
— Landman: [name redacted] For Company P
— Company: Company P
— Honesty: F
— Knowledge: C

— Courtesy: D
— Reliability: F
— Time of interaction: April through July, 2006
— Type of Interaction: certified letter, e-mails, telephone calls, on-site meeting
— Tactics:
— Unresponsive to phone calls and other attempts to communicate
— Lies and subsequent apologies
— Description:

I am a nonresident surface owner maintaining property in San Juan County, New Mexico, but living in California. Last May, I received a letter from an assistant to a landman for Company P. The letter said that I had to respond within 5 days if I wanted to be present at a staking for a well location on my own property in New Mexico.

I tried to get in contact with the assistant in several different ways. The email printed on her letterhead was invalid, and she never picked up her phone. When I called the receptionist at her office, the receptionist refused to take a message for me or tell me if she was in. Eventually, I figured out that she had spelled her email wrong on the Company P letterhead and heard she had taken a 3-week long vacation.

Eventually after getting in touch with her boss, I thought I managed to be present at the staking rescheduled in July when I would be in the area. He neglected to call me until after the staking was done.

— VI.
— User: SWatson
— Landman: [name redacted] Landman for Company P
— Company: Company P
— Time of Interaction: August, 2006
— Type of Interaction: phone calls, on-site meeting
— Honesty: F
— Knowledge: F
— Courtesy: F
— Reliability: F
— Tactics:
— Last minute changing of plans
— Preventing landowner involvement

— Lies and subsequent apologies
— Description:

> After being given the runaround by Company P to be present for a staking of a well site on my property a month earlier, I thought everything was settled for me to be present at a time that would be convenient for me as a nonresident surface owner. I was visiting New Mexico for my mother's memorial service and I specifically arranged to meet with Company P when I was there. Monday morning after the service on Saturday, the landman called and said "I heard that there was a memorial service for your mother. I'm sorry to hear it. I apologize, but we went ahead and did the staking without you."
>
> As you might imagine I was very upset in this period of grief and felt the company had taken advantage of me and they definitely lied to me. Adding insult to injury they have staked on exactly the part of my land that I had told the landman I did not want them to drill a well on—an area next to the county road (frontage) which is the most valuable portion of that property. When pushed about why they done the staking, he claimed that they didn't understand that I had wanted to be present for it!
>
> So far they haven't drilled the well yet, but it has been a very frustrating experience. The landman keeps telling me to just tell him where I want the well, as if I really want a well on my property.
>
> I am now working with the same landman on a different well on another piece of property, too.

These situations are similar to those described by the Fitzgeralds (see chapter 6). The landman structured "miscommunications" to benefit himself and shape outcomes in his favor. He manipulated timelines by rushing Susan with a five-day window and the runaround of vacations, missed calls, and failures to respond, exploiting opportunities when the landowner's ability to respond was limited. Susan's and the Fitzgeralds' parallel cases raised questions about whether these unethical practices were widespread. What other communicative media or technologies would increase landowners' abilities to recognize and effectively respond to these tactics? Susan had few formal channels of response once the well was staked. She had neither the time nor money nor sufficient legal standing to sue the company over her interactions with the landman.

This structural power imbalance was illustrated in our email conversation about filling out this form, in which Susan expressed discomfort with naming the landman because he was currently in negotiations with her. She was worried about the potential liability created by participating in the project: "I know how hard it is to fight big oil and I don't want to have a slander suit filed against me or against you. Has this legal issue been researched?" Her caution gave us significant pause. Could LRC become a new terrain through which users might experience industry backlash, and should we anonymize the service to protect them? An anonymized site might be flooded with fictitious positive reviews. Chris had drawn an inverted scale representing a conceptual balance for each tool between anonymity and verification in which the more anonymity there was, the harder verification became. Balancing anonymity and verification is a common design issue for websites working within the open structure of the Internet (Lessig 2006; Zittrain 2006, 2008). We discussed possible technical solutions to this problem. Would supplying a local area address be an appropriate verification threshold or, alternatively, a name that was stored in the back-end database but not visible to other users? As the project progressed, we realized that a technical solution for risk management would be neither easy to come by nor foolproof.

DEMOS: MOMENTS OF DISCOVERY

Our first LRC demonstration for the Center for Future Civic Media conference unwittingly illustrated many of these design problems. The classic cowboy image on the site's banner revealed our assumption that we were designing for western audiences such as those we had researched in Colorado and New Mexico (see figure 7.1).

The demo added landowner reports compiled using the paper survey and divided according to company (marked by smokestack icons) and the landman's name. Report quotes highlighted common experiences such as "We realized then that the landmen were attempting to turn us against each other." We hoped these clips would quickly convey a colloquial perspective, micro-moments of recognition among those facing similar problems. A mountain landscape in the background would be inviting for rural landowners, while minimalist, simple graphics and underlined links catered to slow Internet connections. A Wall of Shame held the worst landman reports and we included a Top Landmen section, so users could find people they might prefer

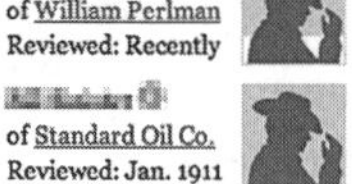

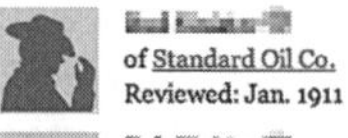

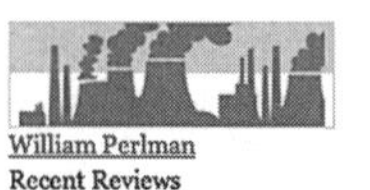

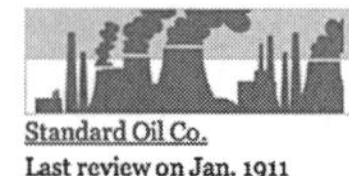

FIGURE 7.1. Landman Report Card, first demo, summer 2008 (names within the image have been redacted).

to work with, and so landmen might be encouraged to compete to improve their ratings. The demo draft was an unreflexive attempt to design a useful site for western landowners that would reflect western landowners' experiences. It was not real, but it felt real if you clicked the right buttons. Matt Hockenberry had created a set of webpages that were interconnected using links and the buttons appeared to work because they took you to a different webpage. For instance, clicking a landman's name on the home page brought up a review for him drawn from our stock of stories that was actually another hard-coded page, as the back end did not exist yet. It looked good, if you took the prewritten path. Turn the wrong corner or search for a page that had not been hard-coded, and the illusion crumbled. It was like an old street scene in a Hollywood Western, a pure façade. Walking down such a street one could think the houses real, until looking back to find them nothing more than painted plywood held up by frames.

Although constrained, the demo was an opportunity for ethnographic reflection, a moment for preparing a story about why the site was worthwhile and bringing ideas to a variety of audiences. Discussions with our consultant on the ground, Lisa Sumi, and the lawyers at the Berkman Center's Citizen Media Law Project (CMLP) were particularly helpful and transformed our understanding of the relationship we could build between our users and ourselves as website hosts. They changed our concepts of both the audience and LRC's development timeline, and foregrounded difficulties with overlapping the physical and online sociopolitical geographies that the site conjoined.

A New Frontier

In 2008, four states and thousands of individual surface owners were suddenly united by a new gas "play" within the Eastern Appalachian region that took on the name the Marcellus Shale play.[4] The Marcellus Shale spans about 600 miles from southern New York, through the west of Pennsylvania and into eastern Ohio and West Virginia. Newspapers in West Virginia began reporting in early 2008 that record offices were filled with landmen researching minerals and surface ownership (Snedegar 2008).[5] OGAP was flooded with calls from the east, an area with which it had not previously been involved, and Lisa Sumi was contracted to put together an informational package about the Marcellus Shale and potential concerns for landowners in the region (Sumi 2008). Lisa reviewed the LRC demo while she was researching this publication and pushed us to develop the site quickly so it could be used in the Marcellus Shale region. Suddenly our audience was no longer western, but western and eastern, and our project was being influenced by the frontiers of gas development.[6] The project also became more personal for the ExtrAct group; my family lived in Ohio, as did Christina's, and Matt grew up in rural Pennsylvania. We organized a trip to visit groups in the region through OGAP. During that week Christina, Dan, Lisa, and I, as well as our two undergraduate research assistants, David and Jian, met with both landowners' organizations in West Virginia, Ohio, and New York and the Pennsylvania State Extension Service. As discussed below, these vastly different groups faced similar but differently refracted problems and approaches to gas development and posed new design and implementation challenges for us.

Three years previously I had not imagined that my gas development research would bring me to a suburban street in Cleveland, Ohio, familiar to

FIGURE 7.2. A Cleveland street, visited by ExtrAct in August 2008.

me in the way that all suburban streets are when you grew up on one (see figure 7.2). "A house with children" was my first thought as we filed into the small, split-level suburban home to meet the founding members of the Northeast Ohio Gas Accountability Project, or NEOGAP.[7] The homeowner, Mark, shook hands with us at the screen door and waved us straight through to a classic American family kitchen complete with kids' drawings adorning its cabinets. A veggie platter, brownies, and soda were laid out for us on the small dining table and we settled in around it exchanging pleasantries while Mark called his neighbors Harry and Mary. They arrived soon after, Mary laden with a crockpot and a bag of hot-dog buns for Sloppy Joes. The meeting began after Kari, a NEOGAP cofounder, arrived, having traveled 41 miles to meet with us. A tall, long-blonde-haired software consultant, Kari looked more like a realtor or lawyer than an oil and gas activist.

We moved into the den, squeezing onto a ring of mismatched sofas surrounding a round coffee table on one of those kids' rugs with streets and railroads on it, which teaches children to imagine systems from a bird's-eye view (Stilgoe 1985). Our NEOGAP meetings all had this provisional quality, because NEOGAP lacked funding, offices, and regular meeting spaces. Mark explained NEOGAP's origin in this way. In 2007, he received what he perceived as a threatening letter from a gas company saying it was planning to drill on his block. He could either sign up with the company or it would "force pool" his minerals into the drilling block:[8]

> Because our leasing efforts have been unsuccessful in the past, [because Mark was not interested in leasing his minerals] we must now inform you of our plans to initiate a mandatory pooling action, with the Ohio Department of Natural Resources Division of Mineral Resources Management pursuant to 1509.27 of the Ohio Revised Code Section, requesting that the above-described property [Mark's property] be mandatorily pooled in the drilling and spacing unit. . . . Should you elect to lease your mineral interest in lieu of the mandatory pooling, you will be dismissed from this action. (OVE 2007)

Mark described his impression of the letter on the LRC site: "This sure sounded to me like OVE [Ohio Valley Energy] was attempting to coerce my family into leasing our mineral rights to them."

Before it could drill a well in Ohio, the gas industry needed 20 acres of minerals rights around each well to be assigned as part of that well's block. Unlike those in the West, most eastern homeowners also owned subsurface mineral rights and were entitled to a share in the well block's royalties. While these rights sometimes put them in stronger negotiating positions, mandatory pooling meant that they could be forced into a well's pool. If a company had close to 20 acres of mineral access secured, the remaining three acres could be "force pooled" into that well block (as had happened to a local daycare center). Importantly, the subsections composing the 20-acre block only needed to be contiguous, that is, one abutting another at only one corner. Unlikely geographies of neighbors linked tenuously by an adjoining corner or a sidewalk were being created to compile the necessary 20 acres.

Negotiating the 20-acre blocks brought neighbors together in new ways, but it also tore apart the fabric of the community. "They use the same technique and I'll tell you what it is," Harry said. "They say, 'We're going to drill a well on Mr. Smith's land here. You ought to become a part of it because you're going to make some money out of this. Oh and by the way, if you don't decide to do it you're going to lose out because we'll just get your neighbor to go along with it.' Or, a little more underhanded, say you guys all own property rights where you are sitting, and I go to David and say, 'We're going to drill a well. We've got the OK from Dan here,' and he [David] says, 'Nah, I'm really not interested.' So I go over to Lisa and say, 'You know we're thinking about drilling a well over on Dan's property and David has already agreed. He's already in on it, he wants it.' And they just lie, and, the way America's become, people don't talk to their neighbors that much. They find out later—'I thought you were in on it' and it's too late, the leases are already signed."

Harry said this tactic had been used by landmen to deceive elderly and other vulnerable community members into signing leases that completed the 20-acre blocks. Unfortunately, contracts signed under duress and as a result of fraud or misrepresentation can only be rescinded if there is sufficient evidence of these illegalities. Unethical industry representatives easily exploited landowners who rely on verbal promises without recording meetings or having independent witnesses present during negotiations.

As events like these were disrupting his neighborhood, Mark saw a letter to the editor in a local newspaper written by Kari in 2006 about a hydrogen sulfide leak from a well in her area, which hospitalized a neighborhood kid. The child's home had to be evacuated and decontaminated. She was disturbed that the company and the Ohio Department of Natural Resources (ODNR) were still refusing to recognize that hydrogen sulfide was the cause of the child's illness, despite reports from a doctor and the local Department of Health, as well as an independent environmental assessment that confirmed the gas was present (Kuns 2006; Lele 2006; ODNR 2007). The company's continued refusal to acknowledge that the problem was due to hydrogen sulfide meant there was no confirmed complaint on record with the ODNR. Kari was stunned that emergency services were unaware of the potential hazard of hydrogen sulfide in this area, despite her research showing that Ohio is considered a major hydrogen sulfide zone. She had brought a map of "sour gas" regions in Ohio to show us.[9] Additionally, there had been no warning of the leak to any of the neighbors and Kari had woken with a headache that night.

After reading Kari's letter to the editor, Mark looked her up online and called her. NEOGAP was formed out of that conversation. Like the other small grassroots groups we had encountered, NEOGAP had around 10 very active members and more than 100 others on its mailing list. It had also become a mini research organization. Kari carried oil- and gas-relevant news articles and pictures of local wells around in binders. She said that many people were completely unaware of local news events, even alarming incidents such as well explosions and hydrogen sulfide leaks, because they received only fleeting coverage in small newspapers and were then forgotten (Rusek 2009).

NEOGAP's research into local ordinances revealed a law called House Bill 278, for which the oil and gas industry had lobbied in 2004, which gave the ODNR total control over zoning regulations that might once have constrained drilling. The ODNR now had "exclusive authority to regulate the permitting, location, and spacing of oil and gas wells in the state, and to revise the laws governing the drilling of oil and gas" (OGA 2004).[10] The appointed division chief had broad powers to "adopt, rescind, and amend, rules for the administration, implementation, and enforcement dealing with oil and gas law" (OGA 2004). Local zoning regulations for oil and gas extraction throughout Ohio were soon negated, allowing drilling to occur within 100 feet of homes and right across the street from Mark's home (Johnston 2009). Notwithstanding municipalities' concerns that this rule was tantamount to giving the oil and gas industry a "free pass around local planning and zoning" (Cave 2004), the bill enabled the Marcellus Shale boom in Ohio by opening the state to residential and urban drilling (Stewart 2005, 2007).

NEOGAP argued that there had been little resistance to the bill in 2004 because landowners had yet to be affected by drilling and residents were largely unaware of what it entailed for their communities. Debate over this rule change took place at the State Capitol in Columbus, where no drilling would occur. About 60% of the state, including the Capitol, was not located on the Marcellus Shale. Republican congressperson Tom Niehaus, who proposed HB 278, notably came from an area that would not be affected by oil and gas drilling.[11] Energy and natural resource companies and representatives were the second biggest industrial contributors to his 2003 and 2004 election campaigns, contributing $35,740.[12] Following his congressional service Niehaus went on to register as a lobbyist for the Ohio Oil and Gas Association and BP America (Rowland and Siegel 2015).

Unlike landowners, state regulators were well aware of the possibilities of a gas boom throughout the Marcellus Shale region. The natural gas industry's systematic, regionwide awareness campaign in Pennsylvania, New York, Ohio, West Virginia, Kentucky, and Maryland aggressively promoted "opportunities" in the Appalachian region. A brochure authored by the Interstate Oil and Gas Compact Commission (IOGCC) along with individuals from state agencies such as the Departments of Environment and Energy, which "intended to serve as a reference source for government and industry decision makers," attested to the tight interrelations between the oil and gas industry and state regulatory agencies (IOGCC 2005: 3).[13] Laden with promissory words, the brochure stressed opportunities not to be missed in the region's "most drilled, but least explored basins" (2). Gas was referred to as "stranded" seven times, framing it as a resource that needed "recovering," as if it were unnaturally trapped and needed returning to its rightful place. The brochure's images of urban gas wells were picture-perfect, fenced in and barely noticeable.

FIGURE 7.3. A flaring pipe within a waste pit during well drilling in Cleveland. Taken by a NEOGAP member.

The brochure's quaint image, allegedly of a completed well, was not only tiny compared with the urban wells I had seen, it also entirely neglected to show the impacts of gas-well installation—the pad that had to be cleared and the roadways, surrounding pipes, and electrical lines required.

NEOGAP's photograph of well drilling directly beside a Cleveland apartment complex offers a more realist perspective (see figure 7.3). In contrast, the IOGCC images showed gas wells and jacks nestled among cornfields, inviting viewers to make a visual parallel between gas and food as vital resources to cultivate.

This is an unstable analogy because, unlike corn, gas is a finite resource. Once used, the industry and its jobs are gone. Unsurprisingly, the brochure did not address the boom-and-bust cycle it helped to begin. It merely estimated the added tax revenue each state could expect and the number of new and secondary jobs politicians could promise (IOGCC 2005: 12). It omitted the fact that since the industry's labor is migratory, these new gas jobs (except for those produced in service industries like hotels and restaurants) would be unlikely to employ local residents, nor would they be permanent. The brochure elided the differences between the natural gas industry and the state, the inherent structural misalignment of interests between short-term corporate profits and political elections, and permanent, irrevocable environmental and social damage.

This report and Ohio's regulatory changes showed that the Marcellus Shale frontier did not magically appear. It was brought into being by promises between tightly coupled industry and bureaucracy. With industry and regulators' aligned interests, there was never a question whether gas development would occur, only about its speed of progress: "Supportive policies are needed to provide access to resources on public lands in an environmentally sound manner, to resolve mineral rights conflicts, and to address unique access issues in urbanized areas" (IOGCC 2005: 26).

Moving regulatory control to the ODNR exemplified a basinwide strategy that addressed urban "unique access issues" by erasing local legal uniqueness. The ODNR dealt with the question of minimum setbacks of wells from houses through its Mineral Resources Board, a group of eight people, six of whom were or had been industry employees. The remaining two members represented royalty owners and the public (O'Brien 2009). This board composition ensured that the vast majority of rulings on force pooling landowners' mineral rights into drilling blocks went in favor of well development: "Landowners can challenge a company's demand for pooling, but of the 56 applications decided in 2007 and 2008 . . . only one was denied" (Johnston 2009).

The environmental and human impacts of the 2004 ruling and of long-running industry campaigns became apparent in 2008, in cases of hydrogen sulfide poisoning, house explosions, unprepared emergency services, and fragmented neighborhoods (Kuns 2006; Lele 2006; Demirjian 2009, 2010; O'Brien 2009). Industry and government cooperation had left landowners out of the loop until they were surprised by the landmen showing up at their doors. These surprising moments for landowners had been manufactured by a slow, deliberate effort by industry to open the frontier. Their experience of accelerated social change was produced by the technical (discussed in chapter 5) and political (discussed in chapter 1) infrastructure of this industry. The ability to control those inhabiting this frontier's experience of time, to make time accelerate, emerged from the industry's infrastructural, strategic position, which allowed for the development and execution of long-term plans (Tsing 2004). Once its conditions of possibility were established, gas development occurred with great speed, leaving residents with the hopeless feeling that "resistance is futile." The prevalent feeling of the group was that the battle over gas development in the region had already been lost.

Other community organizations emerging in the east did not share this feeling. We next met with New York's Damascus Citizens Alliance (DCA), a group of wealthy environmentalists, who were well connected in the city and eager to battle the industry.[14] DCA already had a large mailing list of 400 concerned people and was actively working with The Endocrine Disruption Exchange and lawmakers to try to prevent drilling in the New York City watershed.[15] Gas development's potential threats to the water supply for this urban area of 50 million people raised alarm bells and community resistance, particularly with respect to pollution. Here, we sat on miniature chairs and tables in the children's section of a pretty, well-maintained red-brick public library to watch a short video produced by one of the alliance members, an homage to the author's relationship with the Hudson River that the alliance sought to protect, that showed him bathing at the river's edge. NEOGAP members had stressed that they were not against drilling per se, just its urban manifestation in their backyards since "living, breathing humans have less protection from drilling than endangered species or wilderness areas." By stark contrast, DCA members were proud to call themselves environmentalists.

We heard a different rural fighting perspective at a well-attended community meeting held by the West Virginia Surface Owner Rights Organization (WVSORO, pronounced "West Virginia sorrow") in the local town-meeting barn.[16] After a solemn rendition of "God Bless America," WVSORO introduced us by talking about the Knight Foundation and journalism because this audience might not appreciate environmentalism. West Virginia's historic coalfields had left many landowners in split-estate situations, in which surface owners with no mineral rights could not prevent drilling on their land. WVSORO was pushing for laws similar to those in New Mexico to improve surface owners' rights, particularly to negotiate fair lease terms. As the law stood, drillers could do "whatever is fairly necessary in the contemplation of the parties"[17] to access subsurface minerals. Surface owners had only a 15-day window in which to lodge a complaint when industry submitted a permit to drill on their land. Once that window closed, they lost the ability to directly resist the well. This situation led to minimal consultation with landowners. One local landowner had simply found a gas company staking a well on his property one day. The WVSORO presenter provided tips on how to prevent or delay these situations that included enforcing trespassing law that could prevent companies from entering a surface owner's property without written permission. The meeting ended with WVSORO outlining the landowner bill of rights it was proposing. The bill's first article required companies to meet with landowners before any gas development began.

These disparate groups had very diverse legal problems, goals, and identities, but they had considerable commonality in their experiences of how landmen behaved. They also shared the opinion that our Internet site was urgently needed in the Marcellus Shale region. LRC would need to be technology that everyone could use for their own ends and that would be of most use before the tide of leasing receded. Like these organizations, ExtrAct was drawn into the emergence of this new frontier, and LRC became the tool we sought to develop first and fastest.

Back to Development: On Designing Teflon

The LRC's design had an inherent problem, one of passing potential risk on to its users. Chris's earlier analogy equating depoliticized technology with Teflon turned out to be an appropriately dark metaphor for the troubles we had designing LRC. Like Teflon, which creates unpredictable health risks and is now found in the bodies of most nonstick-pan users, LRC similarly passed risks on to its users, which we found impossible to evaluate ahead of time.[18] Discussions with our pro bono legal advisors from the CMLP at Harvard's Berkman Center for Internet and Society and Harvard Law School's Cyberlaw Clinic revealed how little we knew about rapidly evolving Internet law.[19]

Section 230 of the Communications Decency Act (CDA), passed by Congress in 1996 in an attempt to control Internet pornography, dramatically shifted legal responsibility away from online publishers whose sites were used to host prohibited materials. Whereas newspapers and other traditional print media were regarded by law as the publishers or speakers of any material printed on their pages, the website hosts, or ISPs, could not be held liable for defamatory or other prohibited material that was published and uploaded by others (Berkman Center for Internet and Society 2010: 255). Defamatory speech (slander) or writing (libel) is the communication of false or private information that might reasonably cause material harm to a person's reputation or their mental or emotional well-being. Statements of opinion that cannot be proven true or false are not defamatory.[20] A person who claims that he or she has been defamed must prove both that statements of fact are false and that he or she has suffered material damage as a result (255).[21] Case-law precedents interpreting Section 230 have held users liable for defamatory statements where a host ISP has proven itself a mere distributor of the material posted on its site.[22] However, ISPs who act as publishers by exercising editorial control over their sites, such as posting guidelines or screening software, have been sued successfully for defamation.[23] As the CMLP pointed out, "The perverse upshot of the CompuServe and Stratton [precedent] decisions was that any effort by an online information provider to restrict or edit user-submitted content on its site faced a much higher risk of liability if it failed to eliminate all defamatory material than if it simply didn't try to control or edit the content of third parties at all."[24]

Circulating negative reports about named landmen made LRC and its users potential defamation-suit targets. The site might also provide material that could be used for Strategic Lawsuit against Public Participation (SLAPP) suits. Sociologists coined the term "SLAPP" to describe the use of frivolous lawsuits against small NGOs or activists to dissuade them from continuing their activism (Pring and Canan 1996). OGAP documented the effects of a 1998 SLAPP case targeting a small grassroots group from Colorado's Las Animas region:

> As part of the lawsuit, the company asked the judge to require SoCURE to provide the company with the names of its members. This was seen as a deliberate attempt to frighten people out of supporting the group. And it worked. According to one SoCURE member, people in the area became so worried about being named in a SLAPP suit that they would only support SoCURE through anonymous contributions. (Sumi 2005: III.34)

SLAPP suits have also been used in an online context as a means to identify anonymous speakers.[25]

Limiting our liability as an ISP and protecting our users from defamation suits would require a delicate balance in our design for the LRC site. The passive role of content distributer made it hard to both inform our users about the dangers of defamation and protect them from SLAPP suits. It also arguably reduced the site's utility and transparency for users and researchers. Prior to our legal discussions, the tag line under the LRC's banner had read: "We realized then that they were attempting to turn us against each other." Afterward, we were concerned that this highlighted quote potentially editorialized or made a comment on what kind of content we valued.[26] We replaced it with a legal disclaimer in our next version of the site: "Please note that these reviews are generated by users and do not reflect the opinion of the

ExtrAct group." Protecting ourselves as the ISP entailed creating strong divisions between ourselves and our users. The site had to be experienced like Teflon—no particular politics would stick; we just provided the pan. While we provided the space for discussion, the dialogue and tone had to be created by our users.

We redesigned the site to encourage users to add both positive and negative reviews rather than only their "gripes." When we removed the "Wall of Shame" because it highlighted and potentially encouraged bad reports, we realized that landowner users and the site would lose a potential avenue for placing social pressure on the industry. We were also concerned that the site's use of the term "tactics" could be construed pejoratively so we replaced it with the softer, more value-neutral "practices," but we were then unsure how users would interpret it and even whether "practices" and "tactics" were synonymous. Additionally, the site's prepopulated list of possible "tactics," taken from the landman accounts we had gathered, again risked editorializing. The CMLP had been closely following a California case related to the website Roommates.com.[27] This website was sued for violating the Fair Housing Act, which seeks to prevent housing discrimination. Drop-down menus forcing users to characterize themselves by gender, sexual orientation, and whether or not they had children violated antidiscrimination legislation and Roomates.com lost its Section 230 immunity by shaping and limiting users' content. Because we were unsure if our prepopulated drop-down menu constituted editorializing, we replaced it with a user-generated tagging approach in which users could fill in practices or leave the section blank without any input from us.[28] This drew attention to another more immediate concern in that we had added a couple of sample reviews to the system based on our interviews and feedback on the paper forms. These were hastily removed because they made us users and publishers and therefore ineligible for Section 230 immunity.[29]

Our legal advice on the CDA's Section 230 made it clear that we could not explicitly warn our users about defamation risks without losing our immunity as platform providers because it might be construed as advising our users on how to work around defamation laws. While newspapers and their lawyers had decades to learn this issue and develop practices to work around it, Internet users are often ignorant of legal consequences. Many novice online users wrongly believe their identities are protected because their faces are hidden. In fact, the presence, location, and duration of a visit to a website can be tracked via IP addresses. Even as we were trying to develop a system that required as little private information as possible to start an account on LRC—a username and password—users occasionally denied themselves any privacy by using their real names as usernames.

In the end, we decided to include information on defamation on our "Frequently Asked Questions" website page, under legal questions, rather than on the site. A link for "risks associated with publication" took users to defamation information provided by the CMLP.[30] Our terms of service made it clear that users were responsible for their own report cards. From a user's perspective, these methods protected ExtrAct as the ISP rather than the user. We knew that users rarely read or understood the implications of websites' terms of service and that the need to navigate their way to defamation information made it less likely it would be seen. Indeed, the site itself set up the expectation that it was okay to make a report card, or why else would it be there?

Of course, it was absolutely legal to make a report card as long as any factual allegation in it could be proven to be true or it consisted entirely of subjective impressions that were clearly opinions. The form on LRC where users describe their experiences with landmen was now labeled with "Please use this box to describe your own personal [read subjective] experience."

The more objective front page that was released is shown in figure 7.4. The ominous landman silhouette that carried western connotations and smokestack company images had been removed and two legal disclaimers stated that all site material was user-produced. An algorithm randomly selected the opening line of any site report posted and another algorithm (rather than a site administrator) calculated report cards' grades and rankings. Here the apolitical politics of counting helped the site keep its nonstick shine.[31]

Beyond defamation liability, more direct fears of reprisal or harassment in "industry counties" were legitimate concerns for site users. We developed a series of technical fixes to manage these problems and protect the identities of users who sought privacy. These fixes walked a fine line between flexible privacy and user identity verification to prevent false reports. On the one hand, requiring and maintaining minimal data, a username and password to login, rather than a full name or email address, shielded users from liability and harassment. However, the absence of an email address or a private message system through which to request one meant that users were unable to contact each other and forgotten passwords would remain that way, because we could not send reminders. Lack of direct accountability also increased the likelihood of false reports. We required those signing up to the site to disclose any relationship to an oil and gas company in an attempt to guard

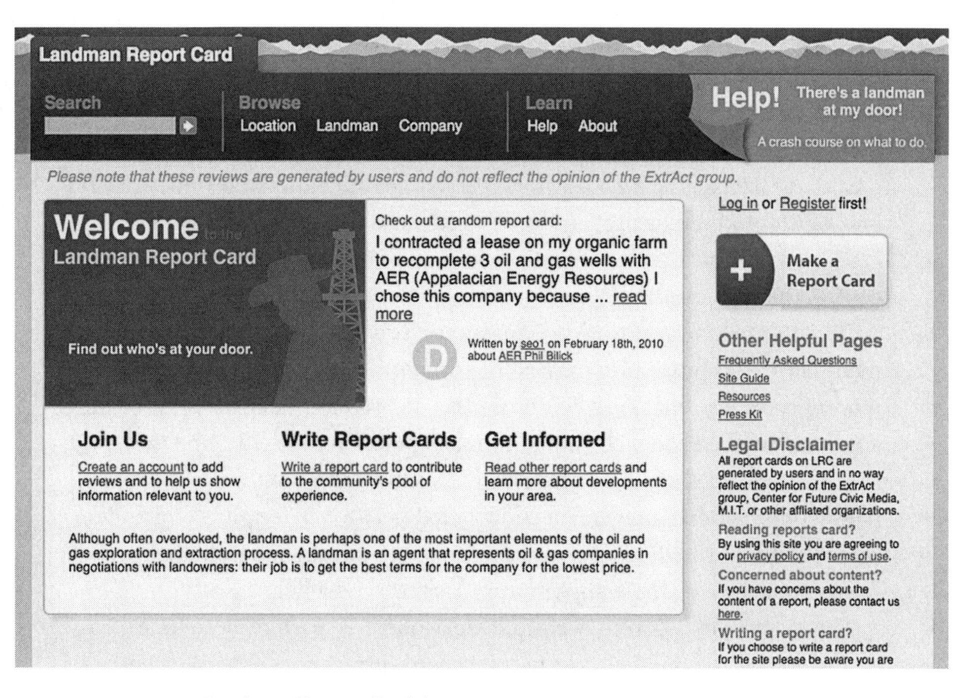

FIGURE 7.4. Landman Report Card, home page, 2010.

against this. These disclosures were intended to help report-card readers make more informed decisions about biased reports. False statements would not only violate our terms of service but also potentially attract state and federal legal penalties.

Arbitration in the form of a moderating system offered another way to manage misuse. The site's eventual hosts, OGAP, would recommend community organizations that would then elect a moderator to review reports flagged by users. Users had to be LRC community members before they could flag a report. If moderators decided a report had been validly flagged, it would be moved to a publicly accessible graveyard where the moderator's editing decision could be reviewed or contested by users. Developing this transparent system took much negotiation and thought, as discussed further in the next chapter, which deals with the question of how to make the ExtrAct tools self-sustaining.

Users were also given options to "fuzzify" reports' locations and times to prevent those familiar with their location or the time of a landman's visit from identifying them. Reports could be located by street address, county, or state and with a specific date, date range, or no date at all. Although these choice levels were intended to help protect users from reprisals, it was difficult to make their purpose explicit, short of including a note saying, "Hey, this is how you hide your identity." Would they encourage people to think through the possible implications of making a report? There was a mysteriousness to developing a website like this. We could neither know nor predict how our tool would be used. Indeed the purpose of a website driven by user-generated content was that it had no cardinal point. The developer could not control the site's direction.

The idea behind choosing openness over verifiability was to give users as many choices as possible in making their reports. They could even keep their reports private so that they could use the site as a journal. But most of the angry landowners we met wanted to make public reports. "I hope they sue me," said one landowner infuriated by a bombardment of threatening company letters. His encouragement contributed to ExtrAct deciding to include a means through which landowners could upload documents from their dealings with landmen. These also improved reports' reliability.

Should research-driven websites, developed as part of academic programs, be able to offer their tool users the same protections as those accorded to research subjects, particularly if they are vulnerable to intimidation, like the landowners we met in Farmington, New Mexico; Cleveland; and Fort Worth? CMLP may have misrecognized LRC by viewing it as a web service, like Yelp or Angie's List, that could inform individuals' choices. Although this is arguably a sort of public service, LRC and ExtrAct's other tools are more than *Consumer Reports*.[32] They provide a platform for collaborative grassroots analysis in order to reconstruct systematic knowledge about the leasing process. As such, their online stories are less like published or broadcasted news and more like neighborly conversations, sharing personal experiences or popular epidemiology.[33]

LRC was designed to facilitate the same kinds of interactions that brought NEOGAP together, frank exchanges between residents in a small Cleveland suburb about how they were intimidated into leasing their minerals. Its scholarly aim was to bring together landowners who had been systematically isolated by industry tactics to make opaque leasing practices more transparent to public scrutiny and analysis. Site users were not complaining to their local coffeeshop about bad service. They were enjoining industrial development to take notice of its impacts on their economic, physical, biological, and social well-being. But the parties to industrial development are unevenly weighted. Landmen on the doorstep command comprehensive knowledge about leases and land parcels, whereas landowners have little or none. While landmen are the vanguard of changing economic and social circumstances, landowners have limited experience of oil and gas development and what it will mean for them. The leasing process aggravates these asymmetries. Stories from elderly and vulnerable people in Ohio showed that even when landowners have been exploited by criminal misrepresentation, they could be too frightened of legal repercussions or too short of solid evidence to break their leases. ExtrAct was concerned that the new legally compliant version of LRC would not adequately fulfill our goals to resolve this fundamental asymmetry and offer landowners protection for speaking freely about their personal experiences. Worse still and despite all our user-anonymizing measures, ExtrAct could be required to disclose our users' IP addresses if subpoenaed for them through a defamation, SLAPP suit, or other action under CDA Section 230.

The National Institutes of Health developed certificates of confidentiality to manage research with vulnerable populations, protecting researchers from "compulsory legal demands, such as court orders and subpoenas, for identifying information or identifying characteristics of a research participant." Such protection can be sought for research when identity disclosure could potentially damage a participant's "financial standing, employability, or reputation within the community" through "information that might lead to social stigmatization or discrimination if it were disclosed."[34] Certificates of confidentiality were initially developed to protect socially vulnerable people from the stigma associated with research on sexual preferences or sex work. Intimidated landowners silenced by threats such as SLAPP suits and social alienation in "industry" towns should arguably qualify for such protections, if sites like LRC were viewed not as consumer choice services but rather as research tools for investigating questions relevant to the public good, questions like whether low-income areas such as those described by Jimmy from Fort Worth are being exploited because of their lack of social and economic capital.[35]

LRC aimed to offer a new arena for both researchers and landowners to collaboratively study this industry through "science and technology studies (STS) in practice." It offered a novel way for landowners to simultaneously provide each other with social support as well as learn how to analyze common leasing patterns. I describe this matching of individual assistance with research for the social good as civic social science, which, as described by Kim and Michael Fortun, allows communities to question the "state of things" (2005). The site aims to shift power dynamics present in leasing by enabling landowners to see themselves as part of a larger social, economic, and political dynamic that is shaping their experience and giving them the means to act on that recognition by sharing stories, contacting other landowners, or finding well-reviewed landmen.

This mode of social science research reduces the lag between a research study and actionable data for its subjects (Minkler and Wallerstein 2008).[36] Unfortunately, cautious interpretation of defamation law and current Internet law precedents, in which websites and ISPs must be mere platform providers, work against this combined approach by failing to adequately protect already vulnerable populations and researchers.[37] Although many of the LRC site revisions were positive, such as algorithmic report-card selection and the more balanced presentation encouraging positive as well as negative report cards, I was ambivalent about some of the other changes, particularly in handing risk over to landowners: for example, the removal of plain language that warned users about defamation and showed them how they could increase their anonymity. I feared these revisions created additional unnecessary risks for those we had most hoped to serve, socially and economically vulnerable landowners. They also harmed the site's research goals by failing to create a safe communication space for such landowners and by preventing us as website providers from acting like researchers. Researchers, while supportive of user tagging, would likely have also actively organized content through a consistent coding of landmen practices, so that related reports were kept together.

There were no answers about how these risks would play out until LRC was released and used. As developers, we hoped that users would flock to the site so our experiment would produce a new space from which to pressure industry into changing its negotiating tactics, but we feared users could be vulnerable to SLAPP or defamation suits. Whatever the consequences, the site opened new possibilities for managing these issues. While this chapter focused on negotiating the site's front end, the role of ExtrAct as an ISP, and framing the homepage of LRC, the next two chapters explore the concept and practice of "civic social science" to address how and why a more participatory and recursive mode of governance should be evolved for the gas industry (Kelty 2008).

From LRC to WellWatch

Designing Infrastructure for Participatory and Recursive Publics

Landman Report Card

ExtrAct released a public version of the Landman Report Card (LRC) in January 2010. This chapter reflects on what we learned from our users and it discusses how LRC's reception influenced WellWatch, our map and database tool for community gas and oil well monitoring. Amy Mall, an energy organizer for the Natural Resources Defense Council (NRDC), released a press statement on the Earthworks/Oil and Gas Accountability Project (OGAP) listserv discussing gas extraction issues for communities and nonprofits, praising LRC as "a resource to support citizens negotiating with big energy." The media release said LRC was an "easy-to-use, web-based resource [that] helps landowners educate and assist each other as they negotiate drilling rights with oil and gas company representatives." It described LRC as a rich resource for landowners in fracking regions that enabled them to share their experiences and learn about industry practices, citing Chris Csikszentmihályi on how "the impacts of natural gas mining are profound and lasting": "A decision to sign a lease—or not—can change a family's life and their community, so the more people know about landmen and their business practices the better prepared they can be to negotiate."[1]

LRC was used by many websites involved in debates over natural gas extraction. It was featured as a "landowner advocate organization" by PAgas Lease, a website for landowners to compare stories on lease pricing. PAgas Lease is certainly not against leasing. It is simply interested in ensuring that landowners obtain fair terms. LRC was also linked to sites providing services to landmen, environmentalist groups like NRDC, and smaller landowner organizations such as the West Virginia Surface Owners Rights Organization. LRC's appeal across this broad range of interests suggested that it had achieved its aim of providing a balanced, nonpartisan tone about landmen and leasing. An industry blog article about the site said, "I'd like to hear from both landmen and property owners who have worked with them. Is this a useful tool or just another way to harp on the energy industry?" (Fowler 2010).

The following comments discuss the politics behind the tool. None assert that LRC expresses any particular partisan politics. A landman called "J Dub" argues that the report-card concept is great, but that extreme reports will bias readers:

> As a young but seasoned landman (7 years in industry; both contract and "in-house") I have seen some of the issues that can be difficult for both landowners and the contract landmen to navigate. The concept of a report card is great in that it can help police the "unscrupulous oilmen" from the public; my question would be how to make it universal. Just because I didn't pay someone what they wanted, would I get a lower score? If I am a tougher negotiator, will that make me seem "unreasonable"? We mustn't forget that a property owner has to be responsible for his/her own knowledge and actions. What would you do if someone randomly approached you to buy your house for $10,000? 100,000? 1,000,000? A responsible property owner would investigate the fair market by hiring a professional (a realtor or an appraiser) to see if they were getting a good deal. If you have questions about what rights you are giving in exchange for the money you are receiving, go to an attorney or hire a Certified Professional Landman (usually $450–$700 per day) to analyze the deal on your behalf. Remember that unless there is specific language in your lease protecting you, you are not entitled to a higher rate because "that is what the neighbor got paid." If you don't think you are getting enough money, don't do the deal. If it is enough to convince you to sign the contract as is, don't complain that you didn't get top dollar and claim ignorance down the road. Ask all the Barnett Shale homeowner associations in Fort Worth about that;

they thought $25,000 per acre was too low. When oil fell to ~$40 a barrel, they couldn't lease it for anywhere near that price. That is the game an owner plays in timing the value of their property. In Texas we have a line from Pat Green that applies to this: "You got no one to blame but your own damn self."[2]

LRC also stimulated a long discussion thread on the PAgas Lease site. Comments about it stretched over four pages and had been read 2,615 times as of July 31, 2010. The discussion speaks less about the site than about people's experiences with landmen. It generally confirms the need to be cautious of landmen because so many landowners are unfamiliar with the process of gas development. The overall response to the site was good, confirming our sense that LRC was useful at the beginning of the Marcellus Shale boom.

One user named Fossil posted the following on February 19, 2010: "Wow! I thought it was just me. I've experienced unethical, rushed, misinformation and have seen them bring a notary to hurry the process. Landmen are pushy flatlanders. The sad thing is that some landowners don't see it through all the dollar signs. It must be okay with the companies the landman indirectly work for." Fossil added in a later post: "People have far less knowledge about the gas leases. That's why this site is so very important. I wish my neighbors would have tried to learn more before they jumped into a lease. I know several people that wish they didn't sign." Joining this discussion on the following day about how unfortunate it is that many landowners sign the first deal they are offered, rfscala added:

> It is appalling to me how many people signed the first lease offer that came to their home. I will never understand the thought process that caused it. With any luck those same people will at least have a second chance to do it correctly when the first lease runs out. IF THEY ARE PAYING ATTENTION. I don't know how many different people wanted to know why we haven't signed yet, especially when they know how many offers I have gone through. It is rather scary to contemplate. And the fact that I have to explain . . . do they sign EVERY LEGAL DOCUMENT without knowing what it is??

On February 21, Duffy raised access and equity issues but was optimistic about LRC's potential to educate those new to gas development:

> Come on robin, you have been here long enough to know the answer to the first part of your question. This web site wasn't even on the web, and then who knew it was available, if people signed the early leases that

was their business. After the site was on the web, if people signed, that was their business, not everybody has a computer or knows how to visit this site. And yes people sign things they don't understand with relation to oil and gas, it's been done for years and nothing ever happened, another reason to tell people about this site.

On the same day, Saxonprincess affirmed the site's benefits as a research tool: "We learned the hard way the first go around. Of course, this site wasn't available (as far as I know) 2007 and anyway, was basically blindsided. Won't do that again. This time, I wised up, have taken my time, researched, found this site, etc. This has all been a very good learning experience."

rfscala's response to Duffy on February 21 blamed unethical landmen for hastily signed leases early in the gas boom before LRC was available:

> True, this site was not available when many people were first approached. But my father was approached years ago too, and did not sign because too many questions could not be answered.
>
> I have to put the majority of the blame on landmen and their tactics because even in the case of my father, they never told him they wanted a lease . . . they said "we just want to see if there is any gas on your land, okay?"

Most of those who posted comments about LRC to the PAgas Lease thread, even industry representatives, were enthusiastic about its helpfulness as a research tool. However, despite the large number of viewings and cross-posted links to LRC, the site itself has not been very well trafficked. As of August 2013 only 33 report cards had been submitted from eight different states. Of these, 64% were negative or mixed reviews and 30% positive reviews (6% were not reviews of landmen but general news or, in one case, a review of a landlord). Nonetheless, LRC produced some interesting findings. For example, rfscala, presumably the same rfscala who posted on PAgas Lease, added a report on February 4, 2010, describing how a landman had treated his sister:

> I was present at my sister's request. She was being coerced into signing a lease she did not want.
>
> Two [company name redacted] landmen came over around 8:30PM, and introduced themselves as David and Randy. They insisted the DEAL ENDS RIGHT AWAY as usual, it was imperative that the lease be signed NOW (it was all filled out and ready) because the lease bonus amounts will be MUCH LOWER after this. When the landmen were questioned and proven wrong they became insulting to my sister and

myself. They told her she didn't understand anything (she's been a title searcher since 1985) and they told me I knew nothing and argued with them since they stepped in the door. (An audio tape proves otherwise). When it was clear that I would not let my sister be coerced into anything, they got mad and the older guy flipped me the bird as he was leaving![3]

Hopefully, advice from OGAP and the LRC encouraged their use of a tape-recorder. On January 19, 2010, another user from Ohio, Director, wrote angrily about the threats his family received after they resisted a landman's offers:

> ON LABOR DAY WEEK END 2009 [JOHN] CAME TO OUR DOOR, WITHOUT A PERMIT TO SOLICIT AT, HE APPROACHED MY FAMILY MEMBER WITH THE STORY OF HOW HE WAS GOING TO MAKE US ALOT OF MONEY FOR OUR NEIGHBOR IS PUTTING IN A GAS WELL AND WE TOO CAN BE BENEFACTORS OF THIS BY SIGNING A LEASE. MY FAMILY MEMBER DID NOT OPEN THE SCREEN DOOR SHE ASKED HIM SEVERAL TIMES TO JUST PUT HIS CARD IN THE MAIL BOX AND SHE WOULD SEE THAT THE OWNERS OF THE PROPERTY WOULD GET IT. THE NEXT DAY MY HUSBAND CALLED THIS [JOHN] AND HE WAS AROGANT [*SIC*], ILL INFORMED, INSULTING, UNPLEASANT AND JUST PLAIN RUDE. HE CAME TO OUR PROPERTY AGAIN TO OFFER US MORE MONEY AND I WAS HOME, HE SAID HE HAD INFORMATION AND A LEASE FOR US TO SIGN AND THAT HE WOULD LIKE TO TALK TO US. I EXPLAINED TO HIM THAT WE WERE NOT INTERESTED IN A GAS WELL ON OUR STREET SO COUNT US OUT. I PUT FORTH AWARENESS TO MY NEIGHBORS AND WE DECIDED TO ADDRESS OUR MAYOR AND COUNCIL AT A CITY COUNCIL MEETING. JUST BEFORE WE WERE LEAVING FOR THIS MEETING. [JOHN] CALLED US AT HOME AND THREATENED US. SAYING THAT WE BETTER BE CAREFUL WHAT WE SAY AT THE MEETING FOR WE WERE BEING VIDEO TAPED. SINCE THEN THIS MAN AND HIS COMPANY HAVE BEEN PUSHING THEIR WAY THROUGH OUR CITY WITH THEIR BAD MANNERS AND UNPLEASANT TACTICS.[4]

Director used LRC's upload function to add a list of violations from this particular landman company that he had been given by the Ohio Department of Natural Resources (ODNR). Other users have also uploaded files in order to share information, such as notices of violations (also from ODNR) for Beck Energy, a company trying to lease land from a neighboring church, after "[David] [of Beck Energy] assured [them] that they had drilled wells without incident. In this matter, it depends how you define incidents."

The following report (by melvinn on February 4, 2010) draws attention to a case of lease flipping in which a landman used an alias to gain a lease and then disappeared:

> An individual calling himself Mr. Rush had several individuals sign leases. He promised them $2k per acre and said he represented Chesapeake. After obtaining the signatures, he became increasingly hard to locate. He did not return calls and upon further examination, it appears he was using an alias and his real name is [█████ Chesapeake] advised they don't have anyone by either name working for them. This individual was apparently misrepresenting himself and trying to flip these leases and make a profit.[5]

Although relatively few cases were uploaded on LRC, the reports were high quality and not overly inflammatory, revealing interesting new practices and providing readers with thorough supporting information. The most frequent user-generated tag was "unethical business practices," reported in respect to 12 landmen and nine different companies. Unfortunately this tag became a catch-all term covering many different situations, including companies that have failed to live up to verbal or written promises,[6] and high-pressure "sales" tactics pushing landowners to immediately sign lease deals,[7] as well as landmen who have hidden conflicts of interest[8] or who misinformed landowners about royalty rates.[9] Tagging categories would be more nuanced if LRC researchers were able to suggest classificatory layers or tags. Misinformation (documented on 18 report cards) and rushing (documented on nine report cards) were the second most frequently reported negative practices, experiences that resembled the ethnographic stories collected by ExtrAct and that showed that landowners developed the types of tags we had anticipated. Users developed the novel negative tag "bait-and-switch" to describe two cases in which landowners were misled into signing contracts.[10] "Knowledgeable and forthright about development plans" was one user's tag for a positive experience with a landman, which was then adopted without variation in the 10 positive report cards. There was some evidence of negative report cards generating positive "replies." In two cases, negative reports were closely followed with positive accounts of the same landman,[11] and the tone and content of positive reports could be eerily similar, as in this report on "Mike": "HE HAS AN HONEST STRAIGHT FOWARD [*SIC*] APPROACH,

AND HIS INFORMATION IS ACCURATE. HE IS VERY PATIENT AND TAKES TIME TO GO OVER ALL DETAILS AND ANSWER ALL QUESTIONS. WE RECOMMEND HIM HIGHLY WITHOUT RESERVATION." The previous user's summary of "Mike" is echoed in this subsequent description of "Andrew" from East Resources: "FROM THE START, AN HONEST STRAIGHT FOWARD [*SIC*] APPROACH, NO FALSE INFORMATION. HE IS VERY THOROUGH & STEADY TAKING TIME TO GO OVER ALL DETAILS AND ANSWER ALL QUESTIONS; WE DO RECOMMEND HIM HIGHLY." Positive and negative reports, as well as accounts from landowners who had leased their property but had complaints about particular landmen, suggested that the site had achieved its goal of impartiality.[12]

LRC also generated further inquiry into larger questions concerning landmen's behavior. The *New York Times*' front-page story on December 1, 2011, "Learning Too Late of the Perils in Gas Well Leases," drew on the site to describe how "millions of Americans" had been duped by gas company landmens' sales pitches into signing unfavorable leases (Urbina and McGinty 2011). The investigative piece analyzed the terms of 3,200 drilling leases, linked to the LRC website and it quoted Chris: "When it comes to negotiation skills and understanding of lease terms, there is a gaping inequality between the average landman and the average citizen sitting across the table" (2011). The *New York Times* also developed a piece they put online and analyzed the language of more than 111,000 oil and gas leases.[13] LRC was discussed in a *Columbia Journalism Review* article and received favorable coverage on public interest websites and radio programs (Santo 2011).[14] *Inside Climate* reported on LRC (Song 2011) and Reuters used LRC's entire database of stories for an investigative piece in the fall of 2011. Since then, it created a series of reports on the problems of landmen (Schneyer and Grow 2011; Reuters 2012). Landmens' leasing ethics were even the subject of a major Hollywood movie, *Promised Land*, in which Matt Damon plays a beleaguered landman who learns the hard way just how far his employers will go to swing public perception in their favor and secure mineral leases (Van Sant 2012).

OGAP: ORGANIZATIONAL RESISTANCE

An unexpected source of resistance to LRC came from within the OGAP itself, one of the groups with which we had most closely collaborated in developing ExtrAct tools. OGAP was unsure how to support and promote LRC to gas-patch communities because the site was not directly relevant to the group's focus on political advocacy for stronger protective legislation. Rather than using LRC as the basis of a call for more regulatory oversight of gas leasing, OGAP appeared to think of the site as only ancillary to its goals. Furthermore, the group did not have the staff skills or funding for someone to maintain the LRC site. LRC was a useful service but not something to which they could meaningfully contribute or own.

The ExtrAct team identified three design issues that contributed to this problem: (1) the back-end database was designed so that a skilled programmer was required to modify or adapt the system; (2) there was no "community around the site" because the front-end design enabled people to participate in developing the project only by submitting their stories; and (3) it was hard for the site's first users to benefit from it because it was not prepopulated with stories.

Investigating the Limitations of LRC's Design

ExtrAct's programmer, Dan Ring, built LRC as quickly as possible so that it would be available to help inform communities in the Marcellus Shale gas development region. He used an open-source version of a very common digital database language called Structured Query Language, SQL (pronounced "sequel"), that can be used to call or query data stored in a table. This table contained one column for each of the elements in a report card, such as "landman," "company," "username," "location," and "description." Each of the table's rows had a unique ID number (called a "primary key"), representing a new report added to LRC. All data entered into the columns were associated with its row's primary key so when a user searched for a particular report the web service queried the database "asking" for all the data associated with that report's ID number. The data were then gathered and organized into a webpage generated from the database in response to the query. The inverse of this process occurs when a user initiates a new report. Information entered in the website's various boxes was remembered by the database as strings (sequences of characters and letters) and associated with the tag "company" for the "report card" with ID number *X* (Mackenzie 2012).

SQL allows programmers to write relatively simple commands for searching, analyzing, and making calculations with the data stored in a database. For instance, the command "Select" begins a database query and "From" will indicate where in the database (from which column) the program should gather information. Selected data can then be compared with other data by using the command "Where." This command establishes a statement for where something is true, such as "SELECT ID FROM reportcards WHERE company='BP,'" and the query would display all the report cards filed about BP. Although "Select," "From," and "Where" have colloquial English meanings, in SQL they stand for further functions that the program executes to produce particular behaviors.

SQL is a relational database management system. The "Select" command is the most complex query because it allows the user to create "relationships" between items within a table and across different database tables. There was no stored table organizing all of the report cards that mentioned BP. The user initiated a relationship with a search term query that brought them together as a temporary subset of the database. Relational database structures

> engineered a split between the physical storage of data in the computer and the representation of that data. . . . Each record might be stored in the system by its record number—the order in which it was entered. However at any point, the user of the system could specify a set of relationships that it was interested in . . . and produce a "virtual" archive that reflected that set of relationships. (Bowker 2005: 131)

Relational databases are very effective for data analysis because of the flexible internal relationships that search queries generate. However, adding a column to an existing table could be problematic for anyone other than an expert programmer. During our second round of LRC development we realized that we needed to distinguish oil and gas companies from the subcontracted landman companies they employ. Landowners field-testing our first report cards had been unsure which company to enter in the site's "company" section so we needed to either add a column to the SQL table for "landman company" or allow multiple companies to be associated with a report card. While this problem was trivial to fix on the front-end user interface by adding another field to the web form, only a programmer familiar with SQL could rewrite the database's structure and add another column to its organizing tables. This entailed rewriting all prior report cards to include a space, blank or otherwise, for information about landman firms, a time-consuming process that made it clear to us that maintaining the site would always require a skilled programmer. LRC users were unable to reshape the site's structure. They could only participate in the project by adding report cards, reading site contents, contacting one another, or flagging unusual reports. Although we had collaborated with many community groups to develop the tool, it felt as if ExtrAct had done all the "development" and our collaborators were limited to using the site rather than owning it.

During ExtrAct's organizational meeting for LRC in January 2009 we discussed site maintenance and how we could design for recursivity and sustainability as well as participation. Recursivity is a concept developed by Chris Kelty, an anthropologist who studies free open-source software (FOSS) programmers, whom he calls "recursive publics." Kelty's "recursive publics" are groups of people who are aware of and work to refine and remodel the underlying structures that bring them together (2008). For instance, FOSS communities redesigning the LINUX operating system are brought together by the code that structures and is continually and self-consciously refined through their work. Similarly, ExtrAct tools needed to be both participatory and recursive so that their user communities could add to, maintain, and reshape them to their purposes.

Recursivity was a central concern during the design process for WellWatch, our mapping tool for monitoring oil and gas wells. We started with a whiteboard list of all of the functions we hoped our next site would fulfill. The bracketed explanations below expand on our shorthand terms:

— Share forms/flyers/org tools [WellWatch would be a place to share forms and materials from various states and organizations]
— Alternative interfaces [phone and fax interfaces would make its information accessible to non-Internet users]
— Mapping and scraping [the site would scrape gas facility locations and information from state databases and map this along with user-generated events]
— Person <—> person [users would be able to connect with each other so people could find others with similar issues even if they were across the other side of the country]
— Person <—> expert [the site would help connect experts and community members]
— Expert <—> data [experts could present their data on the site and because it would also compile user information, it would be a useful research tool for such experts]
— Community <—> community [communities dealing with gas development could interconnect through the site and share resources]
— Person <—> data [nonexpert users would also be able to research and gather data about the industry]
— History [the site would help establish historical track records for the wells and the companies drilling them]

— Civic groups <—> to infinity! [this was a joke that spoke of the project's scale and our hopes that it would be helpful to the widest variety of users we could imagine]

Interrupting the list both literally and conceptually were two quotes, the first from Christina, "The community needs to become sentient!," and the second from me, "They [the community] already are, we need to become relevant!"

These quotes neatly express the problem with cultivating a site's community. Christina argued that people isolated and impacted by landmen needed to become sentient as a community through using the site. I argued that Citizens Alliance communities were already sentient, but our tool needed to become relevant to their current work practices. LRC also suffered from a problem common to participatory databases; it could only be useful if people added and kept adding data to it. The first users had added data without the benefit of seeing others' stories. This created a chicken-and-egg problem of how to achieve a critical mass of users to make the database valuable. Wikipedia solved a similar issue by paying users to add entries and it was many years before everyday web users began contributing significantly to the overall number of entries (Bruns 2008; Niederer and van Dijck 2010). We could not pursue such a solution with LRC. Although we collected many landman reports prior to the site's public release, we could not post them to the site without confusing the boundary between developer and user, as discussed in chapter 7. We could ask people who worked with us to add their stories, but only they could enter their individual accounts.

Cultivating communities that not only used the tools but also owned, maintained, grew, and adapted them led Chris to observe the following: "People can get involved in both the project and the data." Both front-end and back-end involvement were key to building a sustainable online community. If users participated in building and maintaining the back-end infrastructure of the websites, the ExtrAct tools could be recursive because users would be conscious of and able to adapt the structure that brought them together.

THE INTERNET AND PARTICIPATORY MAPS AND DATABASES

ExtrAct's participatory mapping tools were inspired by grassroots online mapping projects, such as Ushahidi, that revolutionized the Internet during the first decade of the twenty-first century (Szott 2006; Goldstein and Rotich 2008; Sui 2008; E. Zuckerman 2008; Crampton 2009; Farman 2010; Graham 2010).[15] The Swahili word for testimony, Ushahidi was devised by a

U.S.-trained Kenyan blogger to circumvent the media censorship of violent events after the 2007 Kenyan elections. The service received text-message reports of regional violence and added them to an online map that tracked the election's violent aftermath in real time (Goldstein and Rotich 2008; Mäkinen and Kuira 2008; E. Zuckerman 2008).[16] The site's success motivated a new field of crisis mapping that has since enabled distributed grassroots coverage of crises in Afghanistan, Haiti, and the Gulf of Mexico.[17] Ushahidi leveraged the online availability of satellite imaging and map data along with data layering made possible in 2006 by Google Maps (Farman 2010). Open-source versions of online mapping software, such as OpenLayers, a free library of code for displaying map data in any webpage, were also released in 2006. Many participatory mapping programs have been produced using OpenLayers, including OpenStreetMap, which creates new street maps from user-submitted GPS data (Graham 2010).[18] These online participatory databases and mapping tools have been described variously as neogeography, social media, Web 2.0, convergence culture, and crowdsourcing (Surowiecki 2004; Jenkins 2006; Szott 2006; Howe 2008; Shirky 2008; Graham 2010).

ExtrAct was part of this broad movement but differed from other contemporary social media projects in crucial ways. Rather than tracking crises, ExtrAct's sites were devoted to monitoring a large-scale industry structurally organized to exceed and avoid government oversight. ExtrAct offered specific rather than general knowledge (Wikipedia and OpenStreetMap) that was not primarily concerned with recording neighborhood events (EveryBlock) or enhancing consumer choices (Angie's List). It built on sites that encourage public monitoring of bureaucratic and government systems, like RateMyCop or social media pothole mapping, by designing searchable tools that could be the basis of academic and public research.[19]

Rather than social media, crowdsourcing, or Web 2.0, terms like "participatory" and "recursive" more accurately describe exciting new methods for the mundane activities that ground all of these new social forms of building databases and maps. These two ways of organizing, remembering, and representing knowledge pervade and constitute the social and technical practices of bureaucracies, companies, countries, and computing (Soja 1989; Edney 1990; Mitchell 1991; Winichakul 1994; Burnett 2000). Novel map-making and databasing methods are socially and technically important because they enable new ways of imagining, studying, and organizing the political, social, economic, and physical world (Bowker 2005).[20]

Widespread Internet access enabled participatory mapping and databasing by nonexpert, nongovernmental, or corporate communities, and the Internet itself can be comprehended as a massive shared database. Its standardized, radically open architecture, over which post hoc accessibility control measures have been only unevenly added (Galloway 2004; Palfrey et al. 2008; Zittrain 2008; Deiber et al. 2010), requires very little knowledge about participating computers except that they use a protocol for sending and receiving information, a process called TCP/IP, or Transmission Content Protocol/ Internet Protocol.[21] The web further opened this structure by providing modes for Internet publishing.[22] Stable web pages, which are easily distributed and interpreted by other computers on the network, can broadcast information rather than restricting it to peer-to-peer communications through emails or bulletin boards. Nonprogrammers can create content with prepacked page-building tools such as Blogger and WordPress, while users can share alternative forms of information on hosting sites like Craigslist and YouTube.

Making and contesting maps and using the web to comment on geographical distinctions have become common forms of participatory politics online (Halavais 2000; Enteen 2006; Bholer 2008; Graham 2010). Map making in various online forms has become a tool that further enhances the participatory, egalitarian, antiauthorial character of online culture (Bholer 2008; Farman 2010), particularly because maps, as inherently rhetorical representations, are easier to contest and debate than databases in which value preferences are so often hidden (Bowker and Star 2000). Nevertheless, databases such as WikiLeaks illustrate the ability of online databases to challenge embedded power dynamics.[23] ExtrAct similarly contested the "authorial nature" of natural gas extraction maps that habitually exclude surface residents, discussed in chapter 5. Internet theorists have begun to question the increasing control that nondemocratic, for-profit corporate entities have over the flow of data (Hindman 2009; Vaidhyanathan 2011). Despite their participatory aims, many web services are fundamentally nondemocratic and structurally opaque yet increasingly gather vast amounts of data that their users often consider private (Berry 2008; Vaidhyanathan 2011; Mackenzie 2012). A vast warehouse of every single one of its users' data, Google exemplifies this problem. Its individually optimized search protocols give all unwitting users alternative search algorithms, comparing their results and search terms to improve and evaluate the search algorithms' effectiveness (Levy 2010). Google users are effectively test subjects whose behavioral patterns are analyzed by the corporation to improve its service and profits. Google illustrates how the combined web and Internet can create massive new possibilities for field science. Google users are mostly unaware of the benefits their use of its search engine provides to the corporation. They are not given the option of informed consent and they are not paid for the value they add. Google's participatory infrastructure is opaque to its users and parasitically develops itself from the data it gathers through tracking its users' behavior in aggregate. After our experience with LRC, ExtrAct sought to enhance and democratize the participatory possibilities of web and Internet databases through its next tool, WellWatch, so that users would be aware of and able to shape the development and representation of a data structure they could collectively own. The group took three complementary paths for realizing this ideal without turning users into programmers. First, a tool called News Positioning System would prepopulate WellWatch with geographically tagged industry-related news data (We hoped this tool could become an everyday part of work for our collaborators at the Citizens Alliance).[24] Second, we decided to build tools that nonprogrammers could use to participate in building and maintaining the database. We called the first of these Syncscraper. Third, a semantic wiki framework supported by SQL would replace our purely SQL database, allowing participants to create a larger variety of content. By translating between participatory maps and participatory databases, each of these tools aimed to encourage recursivity within the technical and social structure of the ExtrAct tools.

Designing for Recursivity

To build recursivity into ExtrAct's tools, we aimed to exploit the notion that websites are front-end maps in respect to databases' hidden structures: "In general, creating a work in new media can be understood as the construction of an interface to a database. In the simplest case, the interface simply provides the access to the underlying database" (Manovich 2002). Interface design differentiates between computer users who access and add to a computer's databases and the computer programmers who shape and extend computers' capabilities. Although graphical user interfaces, or GUIs (pronounced "gooeys"), such as Mac and Windows "desktops" make it easy for nonprogrammers to use websites and computers, they help render those users functionally illiterate and incapable of generating new operations beyond those already encoded into representational objects. As Matthew Fuller states, "Pictorial symbols simplify control languages through predefined objects and operations, but make it more difficult to link them through a grammar and express custom operations" (2008: 172). Comparing my level of computer use with that of ExtrAct's programmer, Dan, may help to explain these differences.

Dan's coding work convinced me that I would need many more years of fieldwork to comprehend the programming of a site as deceptively straightforward as LRC. Dan preferred to work at home from his small room on the top floor of his co-op. Periodically I traveled up three flights of well-worn stairs to sit with Dan and learn more about his work. Two screens being insufficient, his working desktop was split into four windows, parallel black panels displaying command-line terminals busy with arcane commands highlighted in various colors meaningful only to Dan. These four windows gave him instantaneous control over the processes being executed on his PC.

Although we use the same hardware, keyboard, and screen, programmers do not interact with their computers in the same way as a computer user—even at the most basic level. Dan's screens displayed no folders, family pictures, or friendly icons, metaphorical hints to help users. He was aware of the space in a different way, one that dispenses with spatially arranged metaphors in favor of the speed and computational transparency of command-line access. "How should I make sure my kids are truly familiar with computers?" I asked him. "Make them use the command line," he replied. The command line's emptiness belies a computer's vast contents and capacities for processing data. Open the terminal on your machine and you are greeted by what looks like a blank page headed by a line containing a colon and right arrow awaiting your command. Through this simple interface you can perform anything possible on the desktop: opening files or programs, reading emails . . . but to do so you need to know the right commands. Working with the command line involves interacting with the computer as a database rather than as a map.

Desktops encourage users to think of their computers spatially by displaying a layout of saved information organized as a map, and it invites them to treat desktop items as material objects that they can physically manipulate (Fuller 2008: 31). Though filled with representations of real-life objects such as buttons, file folders, icons, and trash cans, the metaphorical desktop is controlled and defined by code. Commands are enacted by clicking on or moving these icons rather than by typing a command. From the computer's perspective, either mode of opening a folder—through the command line or a clicked icon—achieves the same computational result. But from a human perspective they are quite different. Using the desktop's spatially arranged icons makes the computer easily accessible by reference to material things such as folders but it prevents users from expressing "custom operations" (Fuller 2008: 171–72). Furthermore, icons and spatial desktop displays tend to obfuscate computing processes. The user's spatial activity on a desktop—moving the cursor, clicking and dragging—requires icons ordered in a spatially coordinated system.

The command line reminds users that the spatial display is only a conventional representation of the computer's underlying data structure. It reinforces the fact that computational interactions are produced by linear, processional commands rather than spatial activities. Rather than simply clicking to open a folder, a list query must first search for it among all the folders available and then ask to list that folder's contents. By stripping away the desktop's physical metaphors, the command line allows one to see that the same command "list" is used to display folders and the objects in them. The command line shows that folders and the desktop are one way of viewing a nested data structure and it reemphasizes the computation process: an extremely well-ordered data structure for the input, recall, mathematical manipulation, and representation of data. GUIs and web environments may hearken back to our mechanical past with their metaphorical allusions to newspapers and buttons, but they are vastly different media, built on recursion, artful classification, and layered inter-referential data structures with nested hierarchies and shared conventions.

Spatial illusions and metaphoric GUI design also mislead users about the practical requirements of how they interact with a digital environment. Consider the differences between a digital button and a button you might use to call an elevator. Anyone can press the elevator button, leaving no record of the action needed to make it work. Clicking a digital button is entirely different. A website automatically knows that, at a certain time, a particular IP address used a specific browser to call the web address initiated by that button. This information is gathered because it is required to produce the result the user is seeking, whatever action the button is meant to complete. Any online action—say, opening a web page—leaves a trace that must be instantaneously recorded for the user to access the medium at all. Gathering this trace metadata is a structural requirement for computing. A website host can review how many people accessed the site that day, how long they stayed, in what order they accessed the pages, and the IP addresses of the computer or wireless hubs used to access the web. The identity of the actual computer through which the data were accessed can often be traced. These traces can be deleted by erasing data, but this is an action that must be consciously chosen.

Because collecting metadata is an inherent requirement of Internet functions, computing enables data analysis and visualization not offered by any other media. The web alters the terrain of what it means to be present. It produces all manner of new social forms to hide, erase, and reveal those traces.

Yet the metaphors we build into our data representations actively discourage users from thinking about the traceability of their online actions, by encouraging them to think "button" very much in the sense of "elevator button." The description above calls attention to the inherently maplike structure of any web page. It is a map not of any physical terrain, but rather of the terrain of a database that literally helps the user navigate that data structure. Unlike maps of old, this map includes within it the process through which it was rendered.

ExtrAct leveraged these two key properties of digital media to build WellWatch for recursivity: using websites as maps to underlying databases that are designed to be interpreted, and collecting metadata as a necessity of this process. These two properties allowed us to conceptualize a tool allowing nonprogrammers to add not only new entries to the underlying SQL database but also whole new tables that they could then relate to others.

Designing Recursive Publics: From LRC to WellWatch

SYNCSCRAPER: MAKE YOUR OWN ROBOT

A strange outcome of humanity's archival pursuits through the Internet has been the formation of a massive distributed database too large for any human to make sense of, yet designed for only humans to read. The web was not made in such a way that programs can access and amalgamate the data sets served and collected in its pages. Internet development was geared to making information accessible to human readers rather than ensuring its machine readability. We ran into this problem of building websites that humans rather than machines can read when we tried to build a meta-database of all oil and gas development wells within the five states we had researched, one with the ability to add in other states and nations as necessary.[25] Inspecting various state databases for information was the first stage of developing our map. Our first research assistants[26] were initially tasked with determining how to gather information from the states' databases or, in programming parlance, how to "scrape" them.

Although they managed to collect data from Colorado, Wyoming, New York, and Ohio, our researchers found that each state maintained its database differently. Some, like Colorado, had web-accessible records of each of the many pieces of data produced when a well was drilled, including the permits, records of bond agreements with surface owners, and records of well-production data. Colorado also had a mapping program. Indeed, while

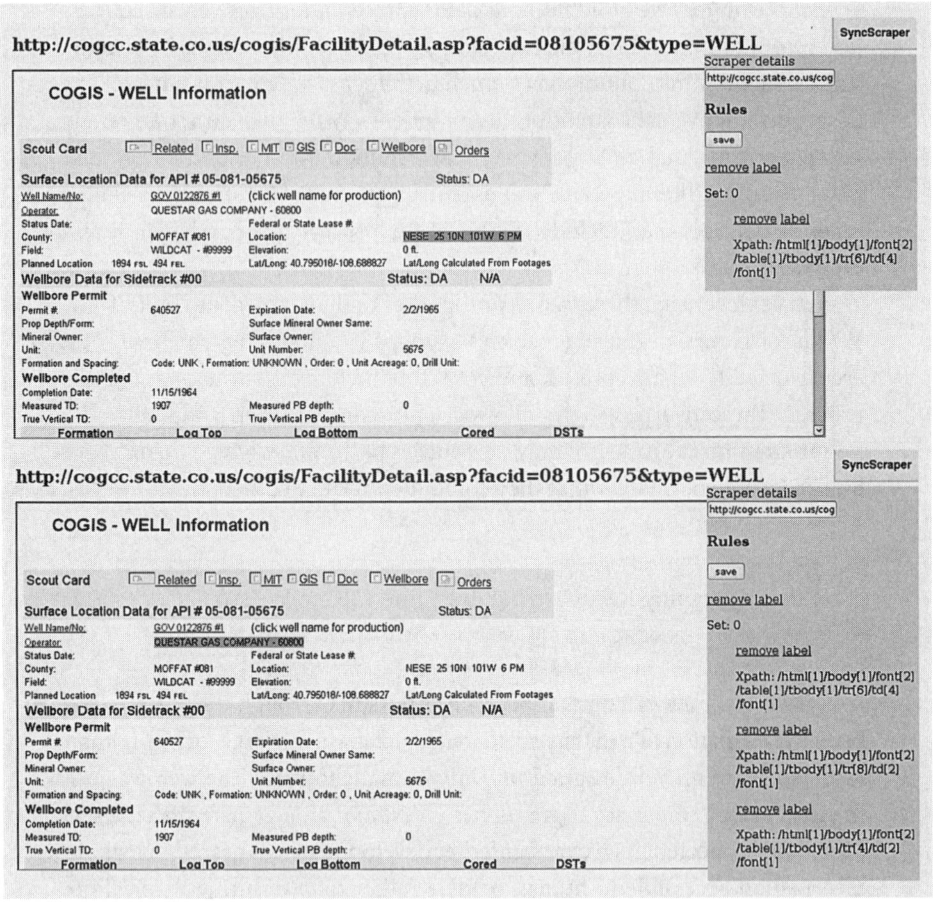

SyncScraper

http://cogcc.state.co.us/cogis/FacilityDetail.asp?facid=08105675&type=WELL

COGIS - WELL Information

Scout Card Related Insp. MIT GIS Doc Wellbore Orders

Surface Location Data for API # 05-081-05675 Status: DA

Well Name/No: GOV 0122876 #1 (click well name for production)
Operator: QUESTAR GAS COMPANY - 60800
Status Date: Federal or State Lease #:
County: MOFFAT #081 Location: NESE 25 10N 101W 6 PM
Field: WILDCAT - #99999 Elevation: 0 ft.
Planned Location 1894 FSL 494 FEL Lat/Long: 40.795018/-108.688827 Lat/Long Calculated From Footages

Wellbore Data for Sidetrack #00 Status: DA N/A

Wellbore Permit
Permit #: 640527 Expiration Date: 2/2/1965
Prop Depth/Form: Surface Mineral Owner Same:
Mineral Owner: Surface Owner:
Unit: Unit Number: 5675
Formation and Spacing: Code: UNK , Formation: UNKNOWN , Order: 0 , Unit Acreage: 0, Drill Unit:

Wellbore Completed
Completion Date: 11/15/1964
Measured TD: 1907 Measured PB depth: 0
True Vertical TD: 0 True Vertical PB depth:

Formation Log Top Log Bottom Cored DSTs

Scraper details
http://cogcc.state.co.us/cog
Rules
save
remove label
Set: 0
remove label
Xpath: /html[1]/body[1]/font[2]/table[1]/tbody[1]/tr[6]/td[4]/font[1]

SyncScraper

http://cogcc.state.co.us/cogis/FacilityDetail.asp?facid=08105675&type=WELL

COGIS - WELL Information

Scout Card Related Insp. MIT GIS Doc Wellbore Orders

Surface Location Data for API # 05-081-05675 Status: DA

Well Name/No: GOV 0122876 #1 (click well name for production)
Operator: QUESTAR GAS COMPANY - 60800
Status Date: Federal or State Lease #:
County: MOFFAT #081 Location: NESE 25 10N 101W 6 PM
Field: WILDCAT - #99999 Elevation: 0 ft.
Planned Location 1894 FSL 494 FEL Lat/Long: 40.795018/-108.688827 Lat/Long Calculated From Footages

Wellbore Data for Sidetrack #00 Status: DA N/A

Wellbore Permit
Permit #: 640527 Expiration Date: 2/2/1965
Prop Depth/Form: Surface Mineral Owner Same:
Mineral Owner: Surface Owner:
Unit: Unit Number: 5675
Formation and Spacing: Code: UNK , Formation: UNKNOWN , Order: 0 , Unit Acreage: 0, Drill Unit:

Wellbore Completed
Completion Date: 11/15/1964
Measured TD: 1907 Measured PB depth: 0
True Vertical TD: 0 True Vertical PB depth:

Formation Log Top Log Bottom Cored DSTs

Scraper details
http://cogcc.state.co.us/cog
Rules
save
remove label
Set: 0
remove label
Xpath: /html[1]/body[1]/font[2]/table[1]/tbody[1]/tr[6]/td[4]/font[1]
remove label
Xpath: /html[1]/body[1]/font[2]/table[1]/tbody[1]/tr[8]/td[2]/font[1]
remove label
Xpath: /html[1]/body[1]/font[2]/table[1]/tbody[1]/tr[4]/td[2]/font[1]

FIGURE 8.1. A mockup design for the SyncScraper interface.

working with The Endocrine Disruption Exchange (TEDX), I had struggled to load the Colorado map produced by the Colorado Oil and Gas Conservation Commission (COGCC). It required me to install numerous new plugins and when the data eventually emerged, as a gray map of Colorado covered with red dots, there was no information about well ownership or attendant complaints. It was possible to zoom in, painfully slowly, to examine a specific well. There was no way to easily filter the map's data, for example, to identify wells owned by a particular company. In fact, the map loaded so slowly that it had no practical use for TEDX.

Unfortunately, Colorado's web interface was sophisticated compared with those of other states. New Mexico had no map interface at all, and its database was impossible for a machine to read because it contained scanned paper documents, rather than well data in an electronic format. This is a quintessential example of the problem with making web pages that are human-readable rather than machine-readable. The form was perfectly legible for a human, but for a machine, currently incapable of interpreting handwriting, it was functionally empty.

This heterogeneity of states' databases, like that between web pages and databases generally, meant that computational speed and strengths were hard to apply. A computer's linear logic and storage capacity ensured that it could easily make quick calculations across web pages, compared to the time and labor such work would take for a person. It was analogous to teaching a child the alphabet, so she can organize your library, but then neglecting to teach her to read. She could find any book you possessed but would not be able to make any use of the material inside it. GUI interfaces might encourage functional illiteracy in their users but computers are functionally illiterate in terms of content.

Using the web to compare data from New Mexico and Colorado was almost impossible. Selecting all of the wells owned by BP across two states would take hundreds of hours, and although our research assistants David and Jian made a gargantuan effort to scrape data from each of these states, we realized that database maintenance would require routine re-scraping and reprogramming. Individual state databases were also limited in how they displayed gas-well data. Colorado's complaints database was only searchable by predetermined parameters, such as Inspection and Spill/Release, rather than by key words. It was also difficult to see a complete picture about a particular well, because permits to drill, inspection forms, and any complaints were disaggregated onto different pages that were not interrelated. The data were theoretically relatable because the pages used American Petroleum Institute (API) numbers to uniquely identify individual wells.[27] Each API number could serve the same function as a report ID number in LRC and could aggregate all the data about a particular well.

As well as being heterogeneous, records presented by the state databases we analyzed were also incomplete. Most offered no way to follow a well's history, such as tracking the owners and subcontractors who had worked on it. For instance, the well map for Texas was not updated when a well changed hands or was sold, so its ownership information could be many years out of date. Most states had some way to map wells, but there was no way to look at complaints geographically to see whether particular wells or facilities had problematic histories. Landowners would be hard-pressed to follow well development even on their own properties because there was no means of receiving automatic updates about changes in the wells around them. Although overall these databases presented information, they did not provide it to those who needed it most. They complied with the letter of legal requirements to provide public information, but neglected the spirit of those laws in failing to inform the public of potential impacts. We wondered how public participation in monitoring could help to correct or check inaccuracies in these databases.

The fact that the well data could be read by people but were practically impossible to bring together online is favorable for industry. Because it is impossible to compare data across states, the industry has an odd geography that appears to be organized by state. But the only way in which the gas industry is state-organized is in terms of its regulatory compliance and agreements. Industry operations are actually organized by the geography of geological formations bearing natural gas. Communities across many states and in different formations would be much better served by having access to shale-wide data or by a company's national holdings. Data presented in a shale-wide fashion would allow people to compare royalty rates and production quantities across a formation. Companies operate across state lines and manage their businesses transcontinentally. Thanks to supercomputing, satellite uplinks, and smartphones, Schlumberger is aware of its workforce's activities across the country and the world. Communities across the country (and world) dealing with the same companies do not have any shared information infrastructure. Dividing this data by state makes the gas industry's countrywide actions almost impossible to study for everyone but the companies involved. ExtrAct understood that Colorado landowners with water contamination "coincident with" fracking in their gas patch could be helped by finding people in Ohio dealing with the same problem. An online platform integrating that data would be of great help to communities and regulators.

To build an interactive map of gas extraction across the country, we needed to not only gather the data from each state database but also be able to scrape it reliably and continually to keep it updated. There are two competing web-development approaches to overcoming web architecture that prevents machine-readability. The semantic web approach attempts to standardize tagging and naming for web content so that it is machine-readable. This does not solve a program's inherent blindness to content. It merely adds more forms so that content can be more minutely categorized.[28] Applying such conventions in everyday practice is problematic because humans and computers will always "read" differently. A variety of different tags can be applied to the same content that can in turn be interpreted in many ways. Achieving consensus on how content and ideas should be classified is inherently arbitrary. While computers have no problem embracing the arbitrary, people name objects in political, cultural, and other contextual ways so that no singular classificatory structure fits the physical and social world (Bowker and Star 2000).

The second approach to increase the accessibility of data from heterogeneous sources involves gathering together or "scraping" data as it is presented online rather than attempting at the outset to systemize information input. We chose this path for WellWatch because state agencies would be unlikely to change their websites for us or make available their back-end databases. Our scrapers would use the web pages' spatialization to generate maps that led back to the data we sought to collect. The idea behind our scraper tool, which Matt Hockenberry named Syncscraper, was that any web page from any of the state databases could be used as a window into the back-end database itself. Essentially, Syncscraper would allow a user to teach a program how to read the web page like the user, so someone familiar with the COGCC database could click on each relevant bit of information on a well page that she wanted the program to gather. The link between the two-dimensional web-page representation and the place in the code containing the actual content is called an "xpath." It is a route generated by the translation in the digital environment between maps and databases. Once an xpath is found for an item on one page, it could be found easily on any other page that has the same form but different content. The user could then add the label used for the relevant piece of information in our database (e.g., "well number") and the scraper could move through every possible page like it on the client side in order to collect the data held on the "server side" and reorganize it in our database. Once completed for every piece of information on any web page, this operation could regenerate the back-end database as if a human had drawn up and looked at each page.

Scraping enables a human to move between databases and to draw together similar data. Colorado's reports on spills and releases associated with oil and gas production, which were labeled "spill/release," were found under inspection/incident inquiries on its database. Similar reports in Pennsylvania could be found in "compliance reports" from the Department of Environmental Protection.[29] A computer would be challenged to recognize the similarities between these two naming conventions. However, a person using

Syncscraper could recognize that the terms were essentially the same and merge them in our back-end database without Pennsylvania or Colorado changing their naming conventions. The disadvantage of scrapers is that they take time to set up and someone has to go through each kind of page in a database and tell the program which bits of information to collect. They rely on the human ability to recognize similar content in dissimilar form, something that is fundamentally perplexing for computers. More important, they do not force the content producer into standard content classifications. In WellWatch's back end, the states do not even have to have exactly the same data structure and no single über database is required to put all of the well information for each state into the same table. States' data can be compared across a map because WellWatch adds data together from these data sets at the moment of representation. To actually compare data across the states, however, there also do not need to be shared categories of information across the databases. Rather, comparing across states can be achieved by generating equivalencies; for example, "inspection/incident" in the Colorado database could be set as an equivalent to "compliance report" in the Pennsylvania database. It is much easier to undo or modify such equivalencies if a mistake is made than to try and convert all data into the same classification system.

Syncscraper aimed to allow nonprogrammers with no prior knowledge of SQL to generate a new table in the back-end database. Citizens Alliance members whose daily business is to study and collect information from state databases could employ their knowledge to build and maintain WellWatch without knowing any programming. Using websites as maps of databases enabled not only participatory databasing but also recursive participation because users could potentially build and modify the underlying architecture of WellWatch. Syncscraper had the additional benefit of drawing everyday users' attention to the fact that humans are not the only actors when it comes to maintaining and producing participatory databases and maps (Niederer and van Dijck 2010: 1384). Often hailed as the most successful social media tool, Wikipedia is effectively coproduced by "bots" like those that Syncscraper generates. When a researcher in 2004 began intentionally making editorial mistakes on wiki pages, he was surprised that his 13 erroneous edits in 13 different articles were resolved within a matter of hours. While some researchers credited the watchful eye of human Wikipedians for fixing these mistakes, later analysis showed that many of the edits were in fact picked up by a bot that was able to identify all of the edits made by his IP address and reverse them, as a result of the traceability of "presence" embedded in online tools (Niederer and van Dijck 2010: 1377). Dan used Syncscraper to automate data scraping from Colorado, New York, Ohio, Pennsylvania, and Texas sites in order to populate the back-end database of WellWatch. Unfortunately, although the scraper code was successfully written and functional, the interface necessary for a nonprogrammer to use the tool was stalled at the mockup stage due to the project's untimely end, as described in chapter 9.

WIKIS AND WELLWATCH

Syncscraper was our answer to designing WellWatch so it could be sustained and extended recursively by nonprogrammers. A second problem for Well-Watch was that people had many different kinds of complaints: about water contamination, site reclamation, noise, human health problems, wildlife and animal health problems, sickening odors, company behavior, and royalty payments.[30] For a database of such varied complaints to be useful, content needed to be categorizable using many different parameters. We were reluctant to create a database that locked in any complaint parameters because we were unable to predict what future categories of complaints might occur. We also wanted to gather as much structured data as possible so that the complaints would be as machine-readable as possible while making sure that the data structure was modifiable. SQL-like data-structure and tagging made sense for LRC, but they would prove too limited for WellWatch.

Initially, we hoped that WellWatch would be a TurboTax-like tool (an online service for filling out tax returns) that would "take" people's complaints and enter them into the database systematically by moving the user through a tree-like question structure. We imagined users would begin to enter their complaints by choosing whether they were issues with a company, health or environmental problem, or property damage. Depending on the choice made, the system would then ask a series of questions. We realized that a complaint about a well explosion after a nearby frack job could simultaneously be complaints about the environment, property damage, the company, and health, so we attempted to make the data structure heavily redundant so that information would be collected gradually on all aspects of the complaint, irrespective of the initial category choice. There were many relationships between the questions and sections in the database so the user could get to the same questions by many different paths, depending on what seemed most natural or obvious. We spent hours building nested question trees and filling them out with real complaints we had heard, a fraught process because there was no "correct" classificatory structure. In trying to get as much structured data as possible, we designed a form that took an age to fill out and would probably be both frustrating and confusing to use, since people would not know exactly how to make their complaint fit into the system. We eventually abandoned this structure completely in favor of a much more free-form data entry system by using Mediawiki, the software that generates Wikipedia.[31]

Wikipedia has a much simpler structure than SQL table-based databases. It is a traditional database in the sense that it has units of information such as a page, but that page is essentially the only "type" of data in its structure, the basic unit of classification. Interrelationships formed between pages make this data structure more complex (Zimmer 2009). Any user can add, name, and relate Wikipedia pages with others. This data structure is somewhat like the web's, where the basic classificatory unit is the URL, or page.

Wikipedia differs from the web by allowing users to collaboratively edit information on its pages. Anyone can add or delete content from a wiki page. Content can be collaboratively refined and transformed, and the page itself can expand as a database containing all of the modifications added to it over time. Unlike previous databases, where there was only one fundamental database to which units were added or subtracted, Wikipedia is internally plural, in that each page holds its own history of changes that can be restored at a later date (Niederer and van Dijck 2010). LRC's singular database is always in the present, like a library that has exactly the same number of books available on the day you get there, but Wikipedia multiplies archiving possibilities by holding previous states for each page. It is disaggregated so that restoring a page to its older version does not affect any other pages, except those that are linked to the changed page. Nondigital-memory devices like books and libraries obscure the ways in which wikis have fundamentally changed our archiving and remembering practices. A recent art piece foregrounded these differences by printing the entire history of revisions for Wikipedia's entry on the Iraq war and binding it within 11 hardback volumes (Brindle 2010). Saving older versions creates a memory system with unprecedented historiographical depth. As soon as the volumes were printed, the information they contained went "out of date," as modifications to the post that expressed changing attitudes to the past that it represented continued to be made online.[32]

WellWatch combined this archival historical depth with a map interface so that users could enter an address to see gas wells in their area (see figure 8.2). After locating a well or facility on the map, a user could then select it with a click and make a wiki page about the location (see figure 8.3). Users could add a note or complaint to wiki pages that automatically contained all of the data scraped from the state website about that well (see figure 8.4).

These notes could be photos, videos, news articles, or LRC report cards associated with that well (see figure 8.5). The wiki page for each well does not exist until a user selects a particular well and chooses to add information to it.

Because WellWatch is semantic, it is searchable across pages that can also be interrelated to other pages in the database. We also used the wiki to develop other kinds of pages for community organizations describing their work, and for counties and states, which included any information relevant to oil and gas extraction. WellWatch was not structured like a TurboTax form, but it could be easily expanded for new uses, and users could easily maintain its contents. Chapter 9 investigates how WellWatch was used and evaluates the outcomes of our projects.

Tracing and Transforming Industry

Users' online traceability makes the Internet a surveillance tool of unprecedented power. States and companies alike have dimensionally scaled up the pervasiveness and quantity of monitoring, data gathering, and analysis to which they subject citizens and consumers. For companies, structurally enabled online tracking provides new ways of manipulating consumer habits with targeted advertisements. The Internet's structures offer new test subjects for corporations seeking to optimize their products, services, and profits, as well as new ways for contracted private security firms to investigate activist organizers.[33] For nation-states leveraging corporate data systems, traceability facilitates tools like PRISM,[34] a massive project shared by the United States and four allies that uses data from online companies to aggregate and analyze information on individuals in order to identify potential terrorists (Gellman and Poitras 2013; Greenwald and MacAskill 2013). States and corporations both use field sciences' databasing and mapping tools to maintain their hegemonic power; consumers and citizens are the subjects of intense research at an individual level without access to or conscious influence over the process. Contemporary democracies may be recursive publics to some extent, because the public participates in rule making and elections. However, at a technical and institutional level, many bureaucratic systems do not enable public participation in data gathering, interpretation, and agenda setting. In the gas patch, communities do not have access to the kind of information required to understand and follow the structural shifts around them. In order to be a part of a recursive public, an individual needs to be able to attend to and understand the processes that make her public a possibility. The essential

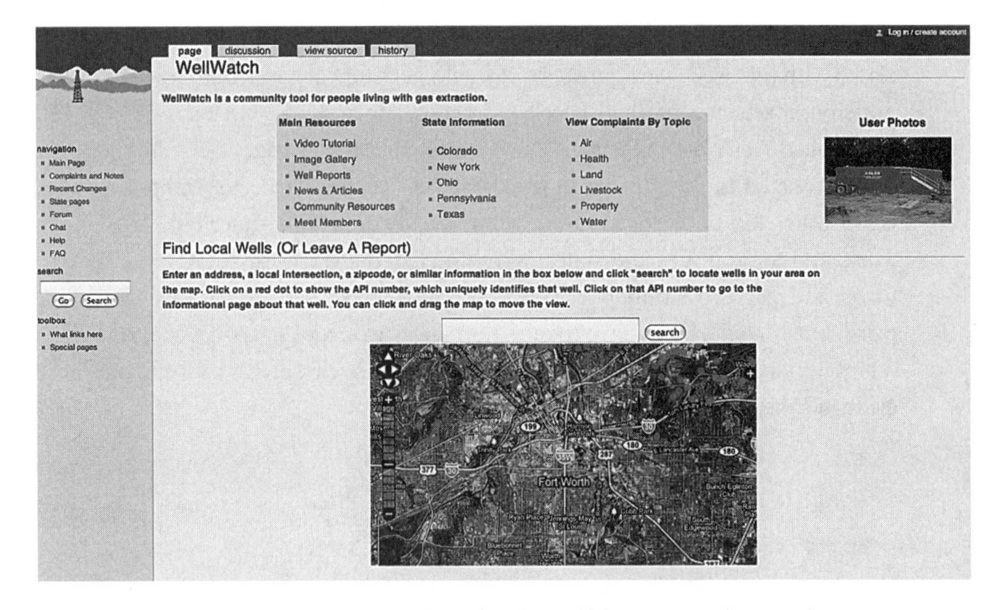

FIGURE 8.2. An annotated screenshot of WellWatch's home page, showing the website's features and functions.

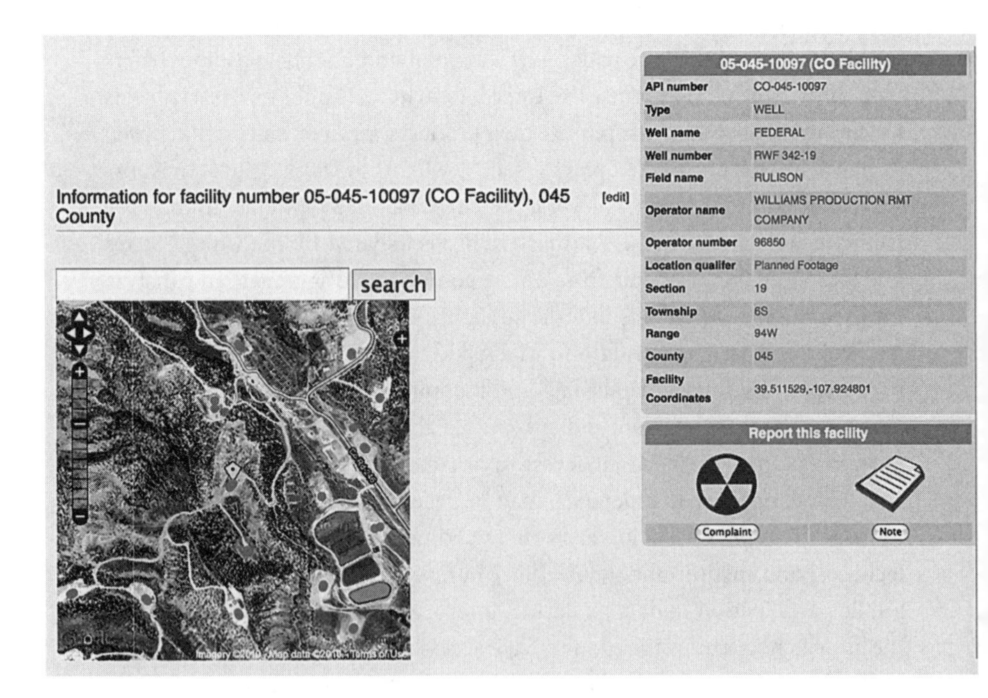

FIGURE 8.3. A screenshot of the "Information for Facility" page.

Add data

Warning: You are not logged in. Your IP address will be recorded in this page's edit history.

Complaint Form

Submitted by: 174.147.90.120

API number: 42-503-37710

Complaint headline:

What caused the complaint? facility / employee / subcontractor / company conduct / unknown / other

What has the complaint impacted (separate different things by commas, i.e. air, water, livestock)?

Start date:

End date:

Is this a recurring complaint (Yes or No): Yes

Companies involved (Separate with commas):

Was there a specific employee/s involved? (Separate with commas):

Detailed complaint explanation:

Upload supporting documents (pictures, pdfs, etc.): Upload file

Watch this page

Save page | Show preview | Show changes | Cancel

FIGURE 8.4. A screenshot of the "Add Data" page.

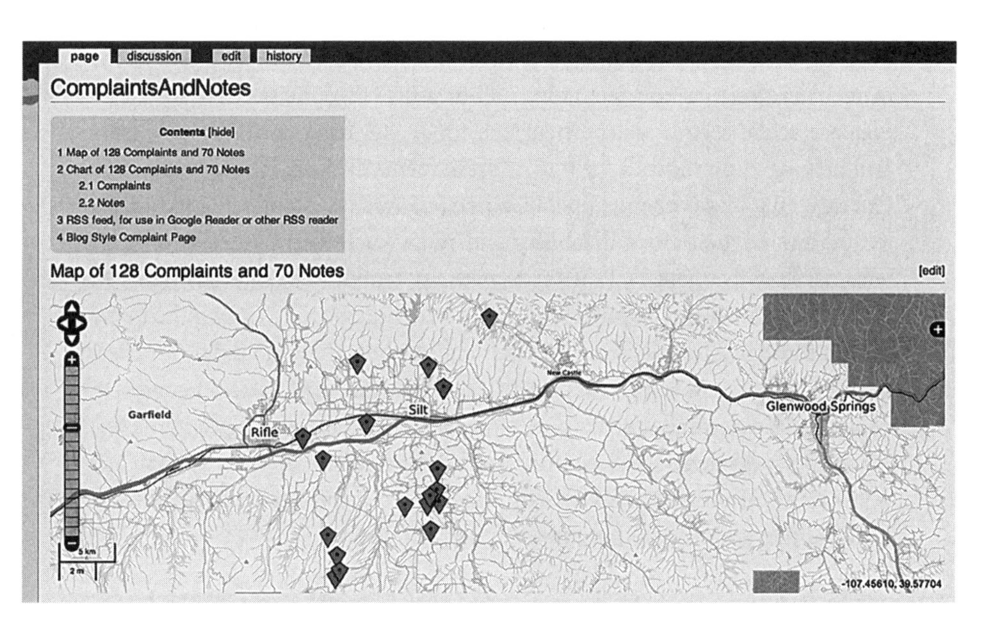

FIGURE 8.5. A screenshot of the "Complaints and Notes" page.

aim of both WellWatch and Syncscraper was to make gas-patch communities more like recursive publics, such as the FOSS community that is engaged in maintaining and shaping the structures that make them possible.

Open-source software and the FOSS community have evolved a much more distributed structure for making knowledge, in which individual users consciously contribute to the development of a larger project (Chan 2004; Coleman 2004, 2009, 2010, 2012; Feller et al. 2005; Kelty 2008). WellWatch was part of this lineage, attempting to create recursive and participatory processes so individuals could investigate shared structural issues. WellWatch was designed to aggregate grassroots experiences to enable collaborative study of shared problems. Designing with and for communities, ExtrAct experimented with a system that could be supported and maintained by its users so that they could understand their shared issues with natural gas extraction. Individuals would no longer be the isolated subjects of study.

Generating a participatory and recursive infrastructure for field science opens the doorway to reconceptualizing how both social scientific and scientific fieldwork about environmental and social issues can be performed. Digital translatability between maps and databases and the possibilities of developing participatory, distributed databases and maps change the process of field research. They change scholarship as well, by allowing "research subjects" to actively participate and contest researchers' findings. Those kinds of maps are science and technology studies (STS) in practice, because they create a social-technical infrastructure for recognizing common problems and helping communities form to address them. In the same way that Theo Colborn's databases changed people's perspectives on chemicals used in gas extraction, participatory databases and maps such as the ones developed with user involvement by ExtrAct have the potential to change gas industry practices, by making the process of gas production visible through the many eyes of those in gas-patch communities and the bots they could develop. Perhaps such tools could narrow the distance between experts and laymen, to challenge the dominant structures of regulatory agencies and academic engineering and science that support and maintain contemporary capitalism's petrochemical-industrial ecosystem. But could our infrastructure deliver on the haunting promise of these ideas?

WellWatch

Reflections on Designing Digital Media for Multisited Para-ethnography of Industrial Systems

> I don't open my windows in the summer, it is like a flume of toxins blow in from the South West, and smell is very thick here. . . . I have plastic on the window. I can't risk a garden, I can't sit on the deck, I drink bottled water. The fumes were really bad a month ago. I called Company X and complained, and one of their employees, the liaison, told me that a lot of the fumes around where I lived were coming from Company Y's wells. The liaison said that Company X "maybe" would put afterburners in on their future wells. The neighbors down the road had high-end horses, and there were reported problems with the births—the people moved out. The neighbor across the road gets royalties and moved to Florida. When the fumes were really bad a few years ago you could see a funnel cone shape of fumes from the deck. I am very ill—I have been threatened with eviction if I "talk and make trouble." I don't know what is going to happen. My former Doc. who worked at the Rifle hospital told me there were lots of alarming, unaccounted-for illness around here—and he advised me "to get out of Dodge" . . . he is gone now from the hospital . . . away from gas development areas.
> —WELLWATCH REPORT FOR COLORADO (February 2011)

Human experiences, like those described above on WellWatch, are the front-line indicators of changing technical, economic, and social conditions. This chapter reflects on how WellWatch functioned as a space for individuals and communities to express their concerns and build networks of knowledge. It evaluates the extent to which the site met ExtrAct's development goal to provide a new approach for community-based participatory research (CBPR) that both empowers communities and improves research. The chapter also investigates the technical and sociopolitical obstacles that prevented WellWatch from moving beyond a pilot project.

WellWatch for Individual Landowners and Communities

WELLWATCH RELEASE

After pilot testing in Colorado in the fall of 2010, WellWatch was advertised more broadly during January 2011 and remained online until the fall of 2012. The site prototype received more than 200 reports from our three trial states—Colorado, New York, and Pennsylvania—and garnered 110,000 page views between March 10 and April 16, 2011 (Song 2011). Two law firms also used its data for research. WellWatch integrated data from Ohio, Texas, Pennsylvania, Colorado, and New York into a searchable open-layer map that served data on nearly one million wells. Mapping well data allowed users to evaluate state records against their experiences. For instance, one Colorado landowner noted a newly drilled well had not been recorded on WellWatch's map.[1] Additionally, users monitored industry activity, such as reporting on fracking operations that appeared to be occurring during a moratorium.[2] As the site was a wiki, users added stories, notes, and complaints about particular facilities and were able to generate their own pages containing uploaded evidence of gas extraction activities, including documents and photographs (MIT LabCAST 2009; Palfrey and Bracy 2011). The site enabled landowners in Texas and Pennsylvania to find people in Colorado who had suffered similar health issues after the gas extraction industry moved into their neighborhoods. Residents used WellWatch as a diary of their illnesses and exposures, documenting skin problems in their children as well as toxic spills and their aftermath. *Inside Climate*, Reuters, and local newspapers praised the site's utility for gas-patch residents (Colson 2010a; Song 2011). In the WellWatch contributor and organizer Tara Meixsell's words, WellWatch was "much easier to navigate than the official state databases" and allowed users "to compare state-by state trends like never before" (Song 2011).

THE IMPORTANCE OF WEBSITE ADVOCATES FOR FOSTERING PARTICIPATION

Engaged local advocates who gathered reports from friends and neighbors were vital to the site's success. As a platform for popular epidemiology, WellWatch was most effective when there was a personally invested individual who acted as a project advocate. This pattern is common in the toxics movement and CBPR (Allen 2003; J. Brown 2007; Minkler and Wallerstein 2008). A Newcastle, Colorado, resident, Meixsell became such an advocate in Garfield County, Colorado. She was drawn into oil and gas issues by the story of a coworker who was diagnosed with very rare neurological illnesses after gas development in her neighborhood. After her health declined precipitously, Tara's friend and coworker asked her to write up her story, as Tara had previously written historical fiction about their region. This led Tara deep into the lives of landowners in her area whose health and environment had been impacted by the shale gas boom (Meixsell 2010). She worked tirelessly to document their stories and encouraged landowners whose computer experience extended to email to share their experiences through WellWatch. This effort to capture and reflect the lived experiences of those in her social network was a mode of nonprofessional ethnography or oral history that helped create a physical and online network of landowners experiencing issues with gas extraction. Like community health workers who have contributed to CBPR, Tara and other advocates played a pivotal role in building bridges, trust, and literacy between community and academic partners.

BENEFITS FOR INDIVIDUALS WHO PARTICIPATED IN WELLWATCH

In the time that it was online, WellWatch united those who had been isolated and divided by the gas extraction industry by enabling them to share their experiences. One community member described her first encounter with the site as "'jawdropping.' She had recognized the need for this sort of access to information but was unaware that the data existed or that it could be made available in this way" (Palfrey and Bracy 2011: 19). Users said they were empowered by learning about other people in similar circumstances:

> Thanks to a series of social networking tools called ExtrAct developed by the Massachusetts Institute of Technology, [Jill] Wiener has connected online with citizens who have lived for decades among the gas

fields. A small but growing group of users shares personal stories of drilling impacts such as contaminated water wells and health problems. "It sure is nice to be able to see what people in other places have experienced," says Wiener. (Song 2011)

WellWatch helped participants to recognize themselves as part of a larger community of affected landowners. It gave residents a glimpse of the power that multinational gas and oil giants accrue from their strategic, state-spanning operations.

WellWatch contributors also used the site to catalog their daily experiences, creating a public record of their health experiences and events. Odor logs and diaries are common in environmental justice and community-based environmental health research (J. Brown 1997; Steinzor, Subra, and Sumi 2012, 2013; Saberi 2013). Personal narratives are increasingly being recognized as vital to medical practice (Charon and Montello 2002; Campo 2006; Charon 2006). Recording such narratives online could add to admissible evidence in law courts by documenting changes in health and living conditions, and these stories offered windows into the lives of those who were suffering. They illustrated that fracking is not simply the execution of a technical process to break a shale formation but is a lived experience in which landscapes, bodies, and social and political relations are reconfigured (Cartwright 2013):

1/8/2011

— Adult female, no sleep due to horrible pain in top of shoulder that will not leave

— 8:03 AM took picture of smoke coming out of pipe over the well pad by the driveway

— 8:16 AM took another picture of white fumes still coming out of pipe/stack on well pad by driveway

— adult female, experiencing terrible pain in shoulders, chest, breast, and left arm. Got State to do water test on now stinking foul house water today. Headache tonight.[3]

By combining symptom logs with mapping features, WellWatch allowed for symptom and exposure mapping as well as longitudinal diaries of experience. For example, using the geographic locating feature, one well in Colorado had 12 complaints associated with it from four different users.[4] Such mapping processes are essential to CBPR because they link experiences with physical places that can be further investigated or become focus points for advocacy efforts (Allen 2003; Corburn 2005).

Patterns of problems that emerged on WellWatch enabled contributors to recognize emerging trends without the delays of academic or formal scientific research or state reporting. Personal accounts of human and animal health deterioration "coincident with" gas extraction made up the majority of air-quality complaints and were particularly affecting (Bamberger and Oswald 2012, 2013). Of the 42 complaints made by May 8, 2011, 29 described health issues experienced by people and their pets, with 27 of these centered on people. A wide range of symptoms were reported, including nausea, sores, rashes, dizziness, headaches, respiratory problems including asthma attacks, internal bleeding, issues with childbirth such as miscarriages, gastrointestinal issues, lack of appetite, muscle spasms, swelling, and neurological problems such as memory loss. These symptoms aligned with academic and community studies on shale gas extraction's health impacts (Bamberger and Oswald 2012; Steinzor et al. 2012, 2013).

Uncontrollable nosebleeds similar to those recorded for other community studies (Steinzor et al. 2013) were documented on WellWatch by residents of Colorado and Pennsylvania. A Pennsylvania landowner wrote the following report on February 20, 2011:

> I had to take my daughter to the hospital for nosebleeds that would not stop—they were torrential nosebleeds—the seat belts in the car are wrecked from them. . . . After the nosebleeds she would then be nauseous and have bad headaches. . . . The nosebleeds were very frequent, the school would call me when they happened at school. I had to send extra shirts for her to change into at school. When we stopped using and bathing in the house water about 2½ months ago the nosebleeds subsided.[5]

A Colorado landowner living a half mile from a well pad told a similar story. Her odor log beginning June 2010 documented a variety of petroleum, diesel, and organic odors:[6]

> Feb 9th, 2011—had first nosebleed (had never had nosebleeds before, maybe once or twice in my life—)

> Feb 10th, 2011—had second nosebleed, it went on for over half an hour, I called the EMT's—they couldn't stop it, and then I was taken to the ER in Rifle. They put a balloon in my nose. That same night I went to the ER in Grand Junction with issues from the balloon in my nose. They

wouldn't take the balloon out of my nose because they were afraid the bleeding wouldn't stop if they did.

Feb 14th 2011—I went back to the hospital and they took the balloon out of my nose.[7]

Another Colorado homeowner also described her son's experience of nosebleeds: "I am uploading more pictures of the welty rashes my son has been experiencing since Company Z started flaring two mammoth towers of flames, that stink to high heaven. His bloody nose bleeds are getting so bad, that we've been using tampons to stop the bleeding."[8] This family sought the advice of a physician who eventually treated many people in the region. The physician wrote a letter to the Colorado Oil and Gas Conservation Commission (COGCC) detailing the family's symptoms and describing his efforts to get oil and gas companies operating in the area to release information on chemicals being emitted there. The physician permitted users to share the letter on WellWatch, thus providing useful information for others experiencing similar symptoms and for other physicians whose patients reported similar problems. WellWatch participants were for the first time able to recognize and position themselves in patterns rather than waiting for an outside authority to impose a schema on them. Tagging by company allowed geographically dispersed users to also recognize patterns of company behavior and their impacts. For example, one user's comparative research revealed that company safety policies could be more lax in rural areas:

> In 2003 the landowner asked Employee from the Company X office why his property's well, and other nearby wells had no afterburners/secondary burners (or reclaimers). Employee told the landowner that "he'd look into it." . . . When the landowner asked some of the subcontractors the same question, he was told that due to the "sparse population" of the area in which these wells were located, afterburners/secondary burners (or reclaimers) were not necessary. The landowner feels that regardless of the population density, secondary burners or reclaimers are necessary for proper health and environmental reasons. Why should a more rural landowner get weaker regulations for protective infrastructure on the wells at his or her home than a landowner in a more densely populated area? Are their lives and health and well being less important that the health of those in more densely populated areas?[9]

Densely populated areas are more expensive to compensate should legal claims arise. Guiding environmental protections via cost-benefit analysis can create systematic environmental injustice where those whose lives are "worth" less receive fewer protections (Faber 2008).

Users were also able to advise each other on how to navigate the complex bureaucracies associated with gas development by recounting their successes and failures in resolving problems. Here, a user relayed his experience of navigating the regulatory structure at the Pennsylvania Department of Environmental Protection (DEP), potentially pointing others towards seeking out the DEP's help:

> July, 2006
>
> When the pond became contaminated I called the DEP. Official A was Official B's replacement while Official B was recovering from a heart attack. I walked around the pond with Official A and told him how disgusted I was about my property being destroyed and not getting any relief. I met with Official A several times on the farm after the pond incident. He told me he couldn't determine where the contamination was coming from. He did tell me that my farm "was a textbook case of what not to do." He said that he had an album full of photos of my farm of which the DEP reviews in their meetings. Official A also said that my name has been brought up numerous times during DEP meetings. He agreed with me that the roads and locations were poorly constructed. Every time it rained locations #3 and #6 slid down the hill and into the field causing extreme erosion.[10]

PARTICIPATION IN WELLWATCH AND IMPROVED RECURSIVE EFFICACY

WellWatch publicity produced tangible benefits for at least two previously unaddressed cases. Rick Roles from Colorado reported the COGCC's lack of response to his repeated reports from 2008 to 2011 of a failed wastewater pit liner. The pit was finally repaired after Rick posted a picture of the liner floating uselessly in the middle of the wastewater on WellWatch in 2012. Rick said WellWatch participation gave him confidence to pressure the COGCC to address a frack spill that had occurred on his land six years earlier, and his increased communication with the commission led to three additional visits as well as air-quality tests by the Colorado Department of Public Health and Environment (DPHE). The DPHE also reviewed the lack of afterburners on his property's well sites. Tara Meixsell, who had trained Rick to use the website, commented that

> if we had not done the work on Wellwatch I seriously doubt any of this would have happened—the COGCC links/contacts were all a result of the work on ExtrACT. This is just for the ExtrACT team to know—but I wanted you all to know that in a back door way some efforts toward better situations for landowners seem to be happening. He is very happy. . . . He's coming over Sunday to do a report on the last 3 COGCC visits—he's been wanting afterburners on his wells for 3 years. (Personal communication, June 4, 2011)

That WellWatch increased Rick's self-efficacy suggests a similar but larger-scale online tool could bring concrete positive changes to people's lives by giving them a voice in gas development on and around their property and supporting them to recursively shape the extraction process.

COMMUNITY BENEFITS OF PARTICIPATING IN WELLWATCH

Communities in Colorado, New York, Texas, and Pennsylvania added contact information for governmental agencies to the site's wiki pages for their states and counties. These detailed entries described how to make complaints and additionally provided links to Environmental Protection Agency (EPA) regions and Poison Center networks for reports about industry waste dumping or spills.[11] Created by regional community organizers, these vital resources showed how WellWatch's wiki structure enabled nonprogrammers to adapt the site to meet their needs. A participant reported an incident in which WellWatch Colorado's emergency contact section helped gas workers experiencing acute health problems:

> On the night of 2/18/2011 the employee working the disbursement and pump station at the site of the contamination became ill (she was the replacement employee after the other employee got ill with similar symptoms a day or so before). The new employee suffered burning nose and sinuses, swelling in the tongue, burning in the mouth, that then pushes down in their lungs, "they couldn't barely breathe"—lost voices, unable to speak, similar to laryngitis—also lips went numb. Employee called a friend, was taken to Meeker Hospital, and physicians had no diagnosis.
>
> Dog was vomiting "his guts out" at the site, and after the humans left the hospital they took the dog to a vet clinic, where the dog was tested and his PH tested at 9—the vet was concerned. Today, 2-19-2011, calls went in for more emergency offices to call the numbers were given for the Rifle office of the COGCC which has an emergency #, plus num-

> bers were given for the Colorado Department of Public Health and Environment. These numbers were given through a second party to the afflicted workers. These numbers were pulled off the MIT Garfield County page, under the Emergency Contact information. These numbers are not in the local phone books to the best of our knowledge. The doctor at the hospital could not diagnose the illness of the patient brought in. The name of the disbursement used on that site by the Colorado river is "Micro Blast" according to name plates. The employees on the site say when their illness symptom were ongoing the fumes were coming out of a certain location—hole # one.[12]

Further WellWatch notes about this event showed the platform working as a system for collecting local knowledge and research. The participant who made these reports, Homeland, did further research into the chemical "Micro Blast" and found that it potentially contained 2-BE, or 2-Butoxyethanol:

> Preliminary investigations into the materials contained in the "Micro Blast" disbursement chemical mix seem to indicate that 2 companies make this type of product. One company uses a granular mix that reportedly contains the chemical "2BE"—this company is based in South Africa. Another company in the US makes a liquid disbursement mix. At this time it is unknown which type of disbursement was used at the gravel pit location, but the symptoms endured recently by the individual are similar to what occurs after exposure to 2BE (reportedly).[13]

This chemical was familiar to landowners in the area after Laura Amos's experiences (Amos and Amos 2005; Nijhuis 2006; Amos 2012). WellWatch captured and spread this research, educating those new to the effects of 2-BE and dealing with possibly dangerous exposures. Links to community organizations and organizers added to the Texas, Pennsylvania, and Colorado state pages also helped build knowledge and support networks that had the potential to unite communities across the country.

COMMUNITY BENEFITS OF CREATING CONNECTIONS

The above report illustrates WellWatch's promising ability to bring local communities and transient workers together. As discussed in chapters 5 and 6, these groups were frequently separated through the processes of resource extraction. The following worker's WellWatch report gave insights into workplace hazards, lax safety precautions, and the costs of whistle-blowing:

Hello: My name is [redacted], I am married to [redacted] and we live in Garfield County, Colorado. There are over 10,000 active $CH4$ (methane) gas wells inside Garfield County, Colorado.

I spent one full year working on these $CH4$ wells. I was employed by Company P, a Canadian company, and I worked on a "swabbing rig." We were not part of the exploration (drilling). We serviced existing and producing natural gas wells. A swabbing rig is a "workover-rig" and our purpose was to remove water from the well that had stopped the flow of natural gas in a producing well. It turns out that the water we were removing was residual "fracing" water and we were never informed of the benzene and other chemicals that exist in this "production water."

In March of 2008, my swabbing rig was working on a well site where the gas company had reused the same "fracing" fluids in an attempt to save money. The problem with that is the reusing of "fracing" fluids causes the manifestation of H_2S (Hydrogen Sulfide Gas) which, in doses above 50 ppm for 30 minutes is deadly. My co-worker died and I spent three months in workers-comp recovery. My employer and the billion dollar energy company basically lied and covered-up their gross negligence in ordering my co-worker and I to expose ourselves to the deadly H_2S without proper safety equipment, such as supplied-air-respirators. Federal OSHA [Occupational Safety and Health Administration] fined my employer. I blew the whistle to OSHA and was fired for doing so. My whistleblower case with OSHA is still pending and I have been blackballed from the industry.[14]

This powerful example had a number of ripple effects, particularly for my own research. Since meeting this WellWatch contributor in 2011, following his report on WellWatch, I have been working on a low-cost method to map H_2S by using photographic paper. The method has successfully mapped community exposures to H_2S in Wyoming.[15] Along with my collaborator, Deb Thomas, a community organizer from Wyoming, I presented the results of this work at the National Institute of Environmental Health Sciences conference in 2014, and I received funding to further develop the tool with communities in Texas (Wylie and Thomas 2014). Our project was inspired by this worker's report and those of landowners who also reported on WellWatch smelling the hallmark rotten-egg odor of neurotoxic H_2S:

Back in the first week of August, 2010, when we saw the paddock behind us being destroyed, we called [Company B] and asked them to help us protect our water, before drilling commenced. I spoke with [a Company B employee] and he told me in no uncertain terms, that they don't help people protect their water, and that's not the business they're in. They did, however have Olson and Associates test our water, the first week of August. We also had a company out of Oregon test our water the same week. Both tests came back saying that we had brilliant water. Well all of a sudden on Monday October 29th, we walk into our house, and it stinks of rotten eggs, badly! I wasn't at all thinking it was the water, I thought it was some sort of leak, but then we don't use any gas. So my husband turns on the cold faucet in the kitchen to start washing some Swiss Chard, OMG!! The stench was so strong, that I told him to turn it off, and get out of the kitchen. The hot water smelled just as bad. It was too late to call anyone, so I sent out an email, and a fellow [Community] Alliance friend told me to call [the county's oil and gas liaison], and the COGCC. Well, [the county's oil and gas liaison] was great, she actually called employee X at the COGCC, and they both dropped everything to come to our place and test the water. They were here for about four hours, answering our questions and testing the well water. The test will be back in about 4 weeks. Judy smelled the hot tap when it was running, and said she could smell the rotten egg smell.[16]

The community subsequently organized bucket testing in the area with Global Community Monitor (GCM), which found levels of H_2S in homes that were over 185 times the EPA's long-term exposure limit (GCM 2011: 14). WellWatch brought together both community and worker experiences of this issue to identify a shared problem and stimulated further research. In doing so it formed an effective space for para-ethnography where communities become self-reflexive about their shared conditions (Marcus 2000a; also see chapter 6). The worker's report also identified another occupational hazard that has since been confirmed by researchers:

That said, I do want to bring to your attention a very deadly practice that is taking place on every "fracing" job site. "Fracing" uses silica sand in the "fracing mix." The truck drivers, pulling "sand-cans" (box-car-size trailers) full of silica sand arrive at the well site and using high pressure pumps unload from the "sand-cans" the silica sand into the "fracing tanks." During this process there is created a silica sand dust cloud that is much more dangerous than asbestos. Just as cut glass will lacerate the flesh of your arm, this silica sand dust is an airborne particulate, that when breathed into the lungs will cause lung damage that is a quicker death than asbestos exposure and extremely painful for the victim. I informed Federal OSHA of this danger to Americans but nothing has happened from OSHA yet.

Therefore, I am informing you folks. The gas drilling industry has an expression: "WELL-FIELD-TRASH." The corporate officers of the natural gas industry considers all of their well workers to be "TRASH." The worker safety-protection measures on these gas well sites is non-existent.

Pass the word about the silica sand dust these Americans are breathing.

[The worker shared his full name, along with his email address.][17]

This worker's concerns about silicosis in workers has become an active area of both agency and academic research. The National Institute for Occupational Safety and Health released a report confirming the hazards of worker exposure to silica during fracking (Esswein et al. 2013).

Just as we had hoped, WellWatch created connections across communities. A county planner in Elbert County, Colorado, drew from model regulations shared by community members in Garfield County on WellWatch: "The Elbert County Planner spoke after [Tara Meixsell] (and very well) and said great things about how the WellWatch information and conversations with me and access to 'model regs' from other counties for gas and oil regulations helped him draft their proposed regulations" (personal correspondence from Tara Meixsell, June 8, 2011).

Lisa Song reported that a workplace injury attorney, Todd O'Malley, was "one of several lawyers helping to spread the word about ExtrAct." O'Malley said WellWatch was "an outstanding database" that "could help provide evidence in court cases over fracking regulations" and "help lawyers find witnesses for court cases" (2011). WellWatch's Community Resources page listed model regulations as well as contact information for legal services.[18] Song notes that the policymakers and FRAC Act[19] cosponsors, Representatives Diana DeGette and Jared Polis (both Democrats from Colorado), had expressed interest in the site for its ability to inform other policymakers (2011). Unfortunately, WellWatch and its server were brought down before it could be useful in such arenas.

WellWatch for Researchers

BROKEN PROMISES AND STRUCTURAL OBSTACLES

While the wiki made WellWatch easily modifiable by wiki-literate participants, it was also very easy to spam. The WellWatch programmer reported on July 29, 2011, that one landowner's story about her illness was modified multiple times by multiple IP addresses in the course of a single day, as was the page for a Texas well:

> http://wellwatch.org/wiki/42-059-32930 (TX_Facility) The page for this well is currently the target of an automated edit war. It has been edited (at least 18 times, by various IP addresses) to contain text and links related to buying prescription drugs, Sacramento locksmiths, porn, cameras, and wallpapers. I'm now locking down the page but I thought you all might be interested to know in case this turns out to be the work of a brand management company hurling Internet spam to cover up a page they don't like! (personal communication, July 29, 2011)

Spam attacks contributed to the site crashing in the fall of 2012 and eventually led to the irretrievable corruption of the server that housed both the WellWatch database and its backup. This loss was particularly keen, given the personal stories held by the site for landowners, making professional security protection vital to the sustainability of any successful WellWatch redesign.[20] Additionally, a second iteration for such a system would not host the website back-up on the same server as the site and would improve spam filtering, as well as limits on which pages can be edited and by whom.

The tension between academic research cycles and maintaining web projects of this scale is another important issue affecting site sustainability. When researchers move to another institution, their research funding and infrastructure may not travel with them. The Center for Future Civic Media became the Center for Civic Media in 2011 after Chris Csikszentmihályi left MIT (Yoquinto 2014). He was unable to take funding for the center or ExtrAct with him to his new position in Los Angeles, so ExtrAct was in administrative limbo. Moreover, in the summer of 2010 I had to stop directing the project in order to write my dissertation. Christina Xu ably took over the co-directorship, but with Chris and I both searching for academic positions, she lacked the personnel to thoroughly support the websites. Chapter 10 further investigates the structural challenges that projects like ExtrAct encounter within academia, and it recommends some ways in which these obstacles to civic science and science and technology studies (STS) in practice might be addressed.

With the benefit of hindsight, a project of this scale would be more sustainable if developed as an independent nonprofit. The proof of principle WellWatch supplied is similar to the traditional process for spinning off a startup company from an academic design or engineering project, but with an emphasis on the not-for-profit aspect rather than commercial applications.

WELLWATCH'S UTILITY FOR RESEARCHERS

Before WellWatch crashed, the site not only confirmed ethnographically identified problems with oil and gas development but also opened up new avenues for research. It proved an effective medium for gathering detailed, rich, and candid personal narratives that were accessible for systematic observation. For instance, the following stories were submitted by two landowners in Colorado who visited a well site as part of a *New York Times* article on the effects of gas extraction on their neighborhood:

> I went to the location and met with a group in front of a well pad by a resident's home, we were about 300 feet or less from the well pad that is just off the road. The fumes being emitted from the tank (and afterburner) were visible (wavy distortion lines going up into the air). The chimney was making a ping-ping rhythmic noise as the metal was heated and the air was cold. I have an extremely compromised sense of smell, and did not detect fume odors at that time although another chemically sensitive individual who has had over exposure to fumes did smell odors immediately after arriving at the location. When the *NY Times* interview ended about an hour later, I drove home (15 min. drive). After getting home I experienced an intense headache, and my sinuses completely plugged up. That night at 2 A.M. I woke up coughing and felt unable to breathe. I had an excruciating headache, I took Afrin to attempt to relieve the symptoms—when I blew my nose there was no colored or thick discharge, just a watery run off and blood on the tissue each time I blew. My respiration problems continued to be severe. I would lose my breath just walking from room to room in my house (15 ft. or so). Doing the smallest physically exerting chore (throwing half a bale of hay to the horses (40 lbs.) would leave me totally winded. . . . I had to sit and rest about 5 minutes before feeding the next horse. My coughing fits continued, every time I coughed I felt like my head was blowing off/ the fits lasted from 5 to 15 minutes. The first day I coughed about every 30 minutes. They were worst in the middle of the night when I was trying to sleep (in a chair—where I had to sleep due to severe body pain that prevents me from sleeping lying down—I've been this way for years . . .). The second day after the exposure the fits were not as frequent. By the 4th day (every 3 to 4 hrs), by today, the 5th day after the exposure I have had 2 coughing fits as of mid-day. When I have these fits it feels like I'm not getting any air in my lungs, and it is

much worse after any exertion. I also experienced body cramps all over after my exposure to the fumes, plus back pain, and Tuesday (the day after the exposure) my right foot swelled up, it went down by Wed A.M. My headache was severe for four days after the exposure, it is receding more on the fifth day (today) after the exposure.[21]

Here is the second WellWatch narrative:

> I went to the X property to meet with the *New York Times* reporter and many of the landowners who have been inundated by the gas wells and the fracking. Across the road were condensate tanks. The well pad and condensate tanks are roughly 350 feet from the X's front door, and we were standing closer than that to the pad by about 30 feet, we were in the front driveway close to the road. . . . The smell started immediately when I got out of the car. I tried to cover my face to not inhale the horrible caustic fumes. But it did not work. . . . (Intermittent emissions/wavy lines in the air effect were visible, and the black stack emitting the fumes was ticking—due to the heat of the metal stack reacting with the quite cold winter air that day.) I came home after an hour or so and was very ill. I had so much pain in my lungs and became very lightheaded. By the time I went to bed I couldn't breathe and the pain became more intense. I was also coughing continuously. . . . This morning I was still very weak and having problems breathing. . . . I will not be able to go to anymore meetings around the areas of gas company facilities as it is too dangerous for my health and it takes me, as my husband says, "for an hour or so of visit you are sick for three days after." He tells the truth.[22]

These reports parallel experiences by landowners who have developed chemical sensitivity, which they believe is related to air pollution from the gas development in their areas (Colborn et al. 2011, 2014; Macey et al. 2014). Not only do they detail similar symptoms—difficulty breathing, coughing, headaches, and a slow recovery process—they also reveal a shared issue; helping reporters document their experiences of nearby gas extraction required them to put themselves in environments that, in their experience, result in physical harm. The challenges of interesting the media in these issues are not reflected in news coverage. Advocates are often exhausted by their struggle to gain recognition and remediation for their problems and from having to perform their story for media outlets (Edelstein 2003). WellWatch provided these landowners with a forum in which they could speak about their experiences rather than being represented. This shared structural issue was brought to light by their reflections. CBPR breaks down the boundaries between experts and laypeople by refusing a structure of "researcher" and "research subject" (Minkler and Wallerstein 2008) and allowing people to speak for themselves. Rich, personal narratives challenge the reductive structures of surveys that are commonly used to capture and study communities (Brasier et al. 2011).

WELLWATCH SOCIOLOGICAL FINDING 1: IMPACTED LANDOWNERS FEEL ISOLATED AND POWERLESS

Earlier chapters ethnographically documented landowners' isolation and exhaustion over their lack of epidemiological or similar evidence to incontrovertibly prove the link between their illnesses and gas extraction. WellWatch reports supported this finding:

> We arrived from MN after our 50th wedding anniversary and there it was the most horrendous thing, the gas rig 700 feet from our home. When we drove up in our yard we noticed a plume of grey smoke hanging in our deck. We ran for the porch and tried to get in the door before the fumes got into our home. The smell of benzene and other chemicals were so strong. Little did we know that the fumes had gotten into our home already while we were gone. After being in our home for 15 minutes I passed out and my husband called my son-in-law to come and get me out of there. He took me to their home. While there I could not walk because of the disorientation. I had headaches and dizzy spells. . . . I kept trying to go home but I would immediately get sick or pass out when I arrived. . . . I called so many people with the government and gas company for help. The senator's office stated that the landowners had no rights and no laws were in place in Colorado. The federal government was in charge. I thought how can that be and I also did not know of others what kind of tests to perform. . . . Finally I did get hold of a gentlemen with [Company X] in Denver and he said that they would put us up in a motel or get a rental house. But I had to do the work. I stayed at my daughters for 8 months and tried to find living places but [the] motel was all full with workers of gas companies and rental houses were not available around our area and that is where we worked and went to school. . . . The man from Company X stated that they would let me know when the fracing was done and then they would clean our

home and vents so I could move back in to our home. They would not pay for the medical bills as their attorneys said they could not it would show guilt on their part. I felt like a person on an island, deserted at that, and no one really cared. I had no rights. . . . The sad part is that it took our lives away and we could not enjoy our lovely home and land anymore. Our dogs would not even go outside. . . . I know all the stories that the gas companies use like it does not cause any problems and any problems we have it is our burden of proof to prove that the illnesses and air and water problems had nothing to do with the gas companies.[23]

A Pennsylvania landowner described her fear and isolation after reporting on her family's health impacts following contamination of their water:

At this time, my future is uncertain—re what is going on at our home and the future health of my child . . .
— I'm not sleeping at night, and when I do sleep I have nightmares.
— I recently had a nightmare of me falling through my water well—the ground where my well was opened up and I fell in and couldn't get out—then I woke up in an anxiety attack.
— The stress of caring for someone dying from cancer for a prolonged time is equal to the stress I am dealing with now re this issue.[24]

Managing these uncertainties makes daily life intensely complex as homeowners struggled with recalcitrant bureaucracies and companies, on top of work life and new pressures to fill basic needs for safe, clean water:

Feb 20th 2011

I call my house "death row." I'm now getting water from other family members and hauling it. I can't keep this up—I go to the laundromat and don't use the house water for anything anymore.[25]

Landowners literally displaced from their physical and social environments described feeling alienated from their homes: "I am afraid to have friends over in fear of making them sick, plus I'm also embarrassed because the water from the toilets and sink stinks."[26] Another Colorado report read, "Landowner had a vegetable garden near the house and had eaten the produce for years, but concerns over increasing health problems prompted him to stop eating the vegetables. Also, he stopped drinking milk from his goat herd after they began experiencing stillbirths and more health problems within the herd."[27]

Weak regulatory structures and unresponsive investigators, combined with epidemiological frameworks for disease, condemned landowners to a lonely wait for "expert" confirmation of what caused their illnesses (Allen 2003; Steinzor et al. 2013).

WELLWATCH SOCIOLOGICAL FINDING 2: CITIZEN SCIENCE

WellWatch reports showed that landowners could become amateur scientists when faced with the challenges of proving the provenance of their illnesses to companies and uninterested regulators (Jalbert, Kinchy, and Perry 2014). The following report from a Pennsylvania cattle farmer describes how he organized and collated data during two and a half years of observing the declining health of his herd (for more on livestock complaints, see Bamberger and Oswald 2012):

> There was a leak/spill from a pit at the well-pad, and the spilled liquids went onto the field and also ran into the pond where my cows drank. The pit and well-pad are 75 feet from the pasture fence-line and 280 feet from the pond where my cows drink. My cows were bred at the time of the spill, and four months later when the 18 cows gave birth/aborted 10 of the calves were stillborn. This was highly unusual—in 18 years of breeding and raising cattle we had never had anything like this happen, we had close to a 100% live and successful birth rate.
>
> Out of the 10 calves born dead, the first five appeared to look normal although they were dead. The sixth calf had cloudy blue eyes that were very unusual. The 7th calf had the same cloudy blue eyes. The eighth calf had a cleft palate on the right side of its mouth, it never would have been able to survive/drink with that deformity even if it had been born alive. The ninth and tenth calves had pure white colored eyes. I contacted everyone I could think of, including the DEP [Department of Environmental Protection]. Although initially they seemed to be sympathetic, they ended by saying it was basically "farmer's luck."[28]

In this context, WellWatch became a tool for communities and individuals to share their testing methods and results. Grand Valley Citizens Alliance coordinated with GCM to take grab samples in and around the homes of those experiencing the worst health symptoms (GCM 2011). Odor log forms were shared through WellWatch so other communities could use them. Beyond air sampling, landowners also tested their blood for BTEXs (benzene, toluene, ethyl benzene, and xylenes): "Landowner had blood test done, and industry related chemicals such as benzene, toluene, xylene and triethlybenzene 125 and 157 were present in his blood."[29]

Landowners tracing the causes of farm animal deaths even tested their grain:

> Feb 20th 2011
>
> We also had a rash of animal deaths of many of our animals. Three and a half years ago, about in Nov 2008 we lost 4 horses in one week. The first horse we had had for only two weeks, then one day she was foaming at the mouth and then fell over dead. The vet could not diagnose it. A few days later, another horse was having severe hip problems. The horse was literally dragging its legs along and the vet diagnosed it with muscular issues. The horses then went down, its back legs were completely useless. Another horse went down; it was older, and the vet surmised it was old age related. The fourth horse was carrying a foal but wasn't going into labor—we had her induced, but the baby was so herniated it died. The vet could not explain this rash of horse deaths in such a short time period. We had the grain tested and it tested fine.[30]

WELLWATCH SOCIOLOGICAL FINDING 3: THE PRESENT STRUCTURE OF ENVIRONMENTAL SCIENCE HARMS IMPACTED LANDOWNERS AND RESEARCH

Focusing on Single Causes Rather Than on Aggregate Impacts. Testing grain, air, water, or other singular exposure routes for the causes of chemical diseases subdivides and artificially isolates experiences that are actually composite and interactive. As Linda Nash argues (2004, 2006), these traditional exposure study processes fail to capture the complexity of people's lived experiences. Complicated context is lost or obscured because exposure testing does not even attempt to effectively recombine these parts into the whole complex of lived experience. Rather, a lack of evidence of significant exposure via one route is often used to dismiss the possibility of harm, even when harms such as those described above are readily apparent (Murphy 2006).

The focus of regulatory bodies, industry, and tort law on pinpointing a single cause to establish responsibility is illusive, and arguably inappropriate, because it delays attention to remediation or a remedy due to lack of sufficient evidence, which works in industry's favor. Given the evidentiary requirements of agencies under weakened laws and judicial requirements for proving an uninterrupted chain of causation beginning with baseline evidence, tests were often insufficient to stimulate regulatory responses and prove adverse changes:

Feb 20th 2011

> We had no baseline water samples done before these wells were drilled and fracked. I called the DEP after our water went bad and I talked to DEP Official X, I feel she did absolutely nothing for us at all. . . . Both the DEP and the company A did water tests. Both entities said that there was obviously a problem with the water, but they didn't know what it was.[31]

Lags in Testing Results. WellWatch narratives illustrated how landowners can wait months for test results that shed little light on causation: "In April 2011 Company U came and tested my water well and several neighbor water wells—it seemed to be protocol. . . . After calling the company, I finally got the water report back in July from the April 2011 test."[32] Another report read as follows:

> Today May 18, 2011 about 3 to 4 years after the date I locked the gate 6 years after the frac liquid spill, at 10 AM COGCC Employee of the COGCC Rifle office came out to the property and she stayed for 3 hours, taking soil samples from the location and pictures of the spill location. She said she would get back in touch with me to inform me of the test results within 3 or 4 weeks—which is about how long the tests usually take.[33]

Exporting the research process to laboratories distances landowners from knowledge making about samples and places them in the position of receiving expert knowledge (J. Brody et al. 2007; Minkler and Wallerstein 2008). Landowners experiencing health problems report feeling dismissed by agencies:

> The unnecessary flaring of this well, has caused my son to have huge red and white welts, that itch and burn all over his body, as well as all of us having burning eyes, skin, and throats. Company M just tells everyone to call Employee M, and he in turn tells you to call the COGCC, and COGCC Employee, thinks it's a joke. He told my neighbor that when people call with complaints about skin issues he thinks, "Waahhh, Waahhh, Waaahhhh, Call 911." Yes, that is what he said. He also said he is to make as many complaints disappear as possible, to keep the budget in good standing. I have called COGCC Employee twice, and followed up with a detailed email, and no reply.[34]

A landowner also from Pennsylvania similarly reported a lack of responsiveness in describing how leaking waste pits contaminated one of his ponds:

> All the vegetation in the 2.5 acre pond died. Even the cattails turned yellow and died. While fishing, a friend and his daughter noticed the devastation. I observed that the color of the pond water wasn't right. The pond water turned dark and black. Then the bottom of the stream, that comes from the pond overflow, looked like black velvet. It covered the whole bottom of the stream. I showed this to DEP and Company G personnel. No one had any answers for me.[35]

This lack of regulatory support means that landowners are frequently put in the position of trying to make sense of their results by themselves:

> July 18, 2011: I called Company U for the water test results. He said the water was fine, and the contamination was due to natural occurrences, "from surface water and too much rain." . . . That day I also called the water well company back and they said they did not think that the contamination was naturally occurring. In their opinion the surface water would not have reached 300 ft underground to the depth of our water well. He also said that the water well people should not have pulled the pump from our water well and put it into the buffalo due to contamination issues. . . .
>
> Since 7/19/2011 I've been on the phone with different entities including Marcellus Outreach trying to get advice and assistance. I spoke to a chemist in late July and I showed him my test results.[36]

One clear academic finding from this work is that it would be beneficial for landowners if they were able to more easily consult with and find knowledgeable experts to help interpret and contextualize their test results (Allen 2003).

Models for engaged sociological and public health research were developed during the early 20th century by the Progressive Movement with organizations such as the Hull House (Sellers 1997; Murphy 2006). The Hull House's maxims have been summarized as residency, research, and reform (Wade 1967; Addams 1892). Rather than seeing these as separate activities, Hull House founders theorized that living in a place was the foundation of noticing and establishing research questions. Research results born from community experiences were intimately connected to advocating for reforms to improve public health. The acclaimed founder of industrial hygiene, Alice Hamilton, subscribed to these ideas and was the first to link lead exposure in the workplace to employees' neurological damage (Sellers 1997; Murphy 2006). The emphasis on laboratories as central for knowledge making has split communities from researchers, to the detriment of both (Murphy 2006). Organizations are emerging for network scientists, physicians, and engineers to study environmental health such as Physicians, Scientists, and Engineers for Healthy Energy and South West Pennsylvania Environmental Health Project.[37] Connecting tools like WellWatch to such organizations could enable a progressive scientific movement that focuses on studying and responding to the public and environmental health problems that emerge from our industrial practices.

WELLWATCH SOCIOLOGICAL FINDING 4: FLIGHT OF LANDOWNERS ENABLES A HISTORICAL ERASURE

For highly impacted residents faced with damage to their properties, health, and livelihood as well as regulatory failures, fleeing their homes was not only recommended by doctors, but was often the only viable solution:

> While working as an irrigator near a well, I stepped into a toxic cloud that knocked me off my feet and into semi-consciousness. This cloud was created by open topped condensate tanks. Since that moment I was unable to tolerate my property's air inside or outside my house. I had to wear a respirator to exist until I could move from the area. I lived there nine years prior to this incident without problems. My doctor diagnosed me as chemical intolerant due to this extreme exposure. He said I needed to move because I would never be able to regain tolerance.[38]

Another family in Colorado was also forced to leave their home:

> Yesterday we had those emissions in the air and we would get them at different times. I could not function. I felt like I was trying to come out of being put to sleep after a surgery . . . all day long. Today while I have some energy, I am packing our trailer (to leave for good). I wasn't able to do anything yesterday. It affected everyone in the house . . . in many ways. Dr. W is still trying to get the gas company to tell him what the chemicals are that we need to test for. To no surprise, they have not replied. I will let you know how things are as I know them.[39]

Four families from Colorado and two from Pennsylvania reported to WellWatch that they had to leave their properties. The *New York Times* reported one family's painful departure from their home and their neighbors' serious health issues (they eventually also left) in late February 2011.[40] Supplementing this powerful snapshot, WellWatch reports documented both families' declining health over time, giving important context to their final choices to abandon their homes:[41]

> Went through a tough week and the Dr spent 4 hrs at our house. He got a headache while he was here, too. 3 total chickens are dead. Found out that since we live in a low lying area, that we get all the junk in the air and it sits here. The whole family is having ailments. There was stuff coming out of the pipes at the tanks last night at around 10 PM and it smelled bad. We tried to get a pic, but it stopped. We are experiencing strange things (WellWatch report from one month prior to move).[42]
>
> [My son] has such severe rashes on last Wednesday, HIS 13TH BIRTHDAY, he couldn't go to the skating party! Thanks to Company X he was home, not having a great time. The other photos, are yesterday morning, Christmas morning, opening presents, while being so uncomfortable with his rash. We are going to see Dr. W on Tuesday, he worked with Landowner Y. Hopefully, he can get us some answers (WellWatch report from the second homeowner in December of 2010).[43]

These suffering families were brought together by working with WellWatch's local coordinator who helped them exchange their experiences and advised them both on- and offline. Before WellWatch was developed, four other families in Amos's neighborhood followed this fight-become-flight pattern and had to eventually leave their homes because they had become too ill. WellWatch created an important geographically located and publicly accessible archive of their experiences, records of their relationships to each other that could otherwise have disappeared when the families moved.

WellWatch's failure to remain available online and the loss of these testaments is a source of great sadness for me. In the conclusion, I revisit the importance of creating a permanent record of industrial activity to partially redress the multiple processes through which this industry's history is effectively erased from public record. The site's truncated online availability nonetheless confirmed ExtrAct's ethnographic findings. Its aggregated geo-located reports illustrated the potential of distributed multisited ethnographic inquiry to build knowledge of physical, environmental, and socioeconomic changes through networking and sharing personal experiences (Marcus 2000b). This knowledge has been severed from public consciousness in the petro-chemical present, because we have lost an embodied sense of how fossil fuels are produced. It has not always been this way.

Embodied relationships to key infrastructures and technical systems were vital to the creation of representative democracy, as illustrated in the following section.

Connecting Embodied Experience to Political Power

Coal miners viscerally experienced the Industrial Revolution, because they were subjected to high-risk labor deep beneath the earth. The disparity between their wages and the value of the coal they gathered gave birth to a new political will as well as a new form of power, the power to control the fixed linear structure that moved their product from mines along train lines to factories (Mitchell 2011). Collective action disrupting or threatening to disrupt *the* energy source for steam-powered industry earned coal miners the right to vote, a social safety net, and limitations on the working week and child labor (Mitchell 2011).

Working- and middle-class people today have comparatively little knowledge of, or influence over, resource flows. Early oil extraction, merely boring a hole in the earth to release oil using subsurface pressure, required fewer laborers for a shorter term than coal. The necessity to refine oil created a new node of power in the energy industry around which corporate industrial organizations coagulated, such as J. D. Rockefeller's Standard Oil. Rockefeller took advantage of geographically dispersed producers, who had little influence over production, by forcing them to use his transport network to have their product refined. "Standard" became the company brand because refining rendered geographically variable raw crude into a homogeneous product, further separating regions of production and consumption. Burying pipelines underground meant consumers had no need to know where their oil originated in this vertically integrated, opaque production process. As the oil industry developed internationally, emerging companies like Shell and BP adopted Standard's vertical integration and branding techniques, creating lasting economic niches that have persisted for more than a century (Yergin 1991: 56–95).

This stability was partly achieved by tying oil to "developed" countries' national security. Oil did not supplant coal across all industries until the 1950s; however, the industry's ties with security were forged much earlier. During World War I (WWI) oil became vital to reducing the strength of coal-based labor movements and mechanized warfare. Just prior to WWI, coal strikes by well-organized miners threatened the coal-dependent British Navy, prompting the UK government to convert its fleet to internal combustion engines dependent on oil and to sign a secret 20-year contract with Anglo-Persian (BP) in which the government then took a 51% controlling interest. Gasoline proved to be a deciding factor in WWI, which mechanized warfare at a previously unimaginable scale. Oil-supply shortages were crucial in Germany's defeat, after it failed to capture Russia's oil reserve in Baku.[44] However, all of the warring nations felt the shocks of limited oil supplies. The United States responded by virtually nationalizing its supply, under the authority of an oil czar, and by speeding the construction of oil pipelines.

Oil's role in national security had significant consequences for democratic development in the countries producing it. Democracy was systematically undermined in the Middle East and other fuel-rich regions (Vitalis 2007; Mitchell 2011). Unlike the chaotic free-for-all that characterized the early years of oil production, international oil mining by major companies used highly technical enclaves from which oil was extracted with as little as possible involvement from local surface inhabitants, via leases or other contractual agreements for mineral rights (Yergin 1991; Bowker 1994; Black 2000; Santiago 2006). These extractive enclaves divided citizens in western nations from their structural power as producers and divided inhabitants of areas of resource extraction from their legal power as citizens (Mitchell 2011).

The WellWatch report with which this chapter began exemplified the structural inequalities and environmental justice issues generated by this industry between resource extraction regions, the industry, and consumers (Bullard 1990; Hofrichter 1993; Brulle and Pellow 2006). The participant with health problems is told by her doctor "to get out of Dodge." The doctor leaves the area, as do the neighboring ranchers raising "high-end" horses, and her neighbor receiving royalties. Meanwhile she is too poor to move.[45] Low-income residents, whose lives are built around their homes and neighborhood, disproportionately bear the burdens of industry development and pollution throughout the United States and worldwide because they have no means to escape it (Allen 2003; Faber 2008). As one Colorado resident reported to WellWatch on his deteriorating health, he and his family can "try to stay away" from their ranch "to avoid breathing the air and the contaminants there from the well" "but, at a certain point I had no choice because this is where I live with my animals on the family ranch—I have nowhere else I/we can go, and I'm way too sick to work."[46]

Although the history of coal shows that physical awareness of and control over infrastructure creates social and political power, we frequently imagine that our technical systems could be designed no other way. This is in part due to relying on experts whose technical and engineering skills are aimed toward producing informatics systems that sideline and reduce embodied experience (see chapter 5). Subcontracting and leasing agreements and the industry's temporary relationship to landscapes and modular market forms enable risk management, by diffusing responsibility and awareness between different sections of the industry (Appel 2011). Wells can have many different owners in their operating lives. Scarce data are poorly organized for understanding the multifaceted, complex conditions this industry creates on the ground (Bishop 2013; Korfmacher et al. 2013; Maule et al. 2013). When responsibility and awareness of harms do spread between modules or across scales, the complex structures of ownership, oversight, and contractual arrangements create rich contests over culpability (OSC 2011). Attempts to study these corporate forms are thwarted by their structural formlessness. Constitutionally, this industry is made to be, and has been made, hard to study and track (Ferguson 2005, 2006; Watts 2005; Barry 2006; Appel 2011). The outcome of this structure is the lived experiences that landowners described in WellWatch. It is at this scale that connections must be wrought and reflected upon to reconnect this industry to place, and to understand the environments it creates, so as to give the industry form and to enable more recursive relationships for shaping its perpetuation. WellWatch offers a potential new medium for multi-sited ethnography where geographically distributed researchers and communities could collaboratively articulate the locally inflected but frequently shared social and environmental consequences of this industrial system (Altman 2008; Marcus 1995, 2000b).

WellWatch stories show that the conjoined environmental health and social injustice experienced by impacted landowners renders them powerless (Perry 2011, 2012, 2013; Bamburger and Oswald 2012, 2013). When test results or human experiences illustrate that harms are occurring, according to WellWatch reports, companies debate who is to blame:

> The fumes were really bad a month ago. I called Company S and complained, and one of their employees, the liaison, told me that allot [*sic*] of the fumes around where I lived were coming from Company X's wells.[47]
>
> We also have a Company V pipeline running through our driveway. Company N tried to blame Company V for the methane contamination in our water.[48]

Alternatively they pressure landowners to sign liability waivers:

> Company N recently said they would give me a water buffalo and they would bring water for three months then leave us with the equipment,

> but not bring more water to fill it after 3 to 6 months—but in order for this to happen we have to sign a document stating that they were not the cause of our problems. They said they would do this because they were trying to be "good neighbors."[49]

Another landowner in Pennsylvania reported a similar experience: "July 18, 2011 I called Company B for the water test results. He said the water was fine, and the contamination was due to natural occurrences, 'from surface water and too much rain.' . . . He then said, 'I'm being a good neighbor by providing you with water' re the buffalo."[50] This was also reported in Colorado after a pipeline leak into a local tributary of the Colorado River:

> Company X reportedly attempted to get the landowners to sign a document that implied that the landowners' domestic well had design issues/flaws that did not enable Company X to properly sample the well for contamination. The landowners reportedly did not wish to sign the document, and they contacted an attorney. In the interim, no water samples were taken of the landowners' domestic well—from the time the pipeline leak was first reported.[51]

Similar reports from people with no prior knowledge of each other who are from different regions undergoing gas extraction suggest a deep need for a system that more effectively monitors company and community interactions. WellWatch narratives suggest that one of the structural patterns preventing recursive grassroots engagement in shaping the development of this industry is that landowners, under pressure to address immediate threats to their health and safety, have little choice but to enter contractual agreements that defray corporate responsibility for those harms. Landowners are thus unable to shape the trajectory of gas development beyond their own individual cases.

Ethnographers and environmental historians have written about both the devastating effects that extractive industries can have on surface inhabitants and the violent environments they produce (Peluso and Watts 2001; Sawyer 2004; Tsing 2004; Zalik 2004, 2009; Santiago 2006; Watts and Kashi 2008). However, ethnographies take years to produce, and a study's view is frequently limited to that of a single researcher describing cultural conditions over a short time in a geographically limited area. Ethnographies rapidly become historical documents that circulate only within the academy, despite the political importance of their findings. Journalism is tasked with generating and amplifying stories such as those above, but it does not offer tools to empower citizens to share their own stories and analyze them to reveal structural relationships. Similarly, statistical analysis in social science can pinpoint trends and patterns, but it strips data of cultural and human context and relies on data that, given the tactics described above, can be difficult to obtain. Academic research results often fail to reach front-line communities (Brown and Mikkelson 1997; Allen 2003).

WellWatch suggests exciting new possibilities for combining multi-sited ethnographic study with grassroots experience through a medium that simultaneously supports academic analysis, individual empowerment, and social advocacy. Environmental health and justice work has stressed the need for scientific and social scientific research that incorporates, builds on, and amplifies the expertise of communities, rather than exporting data and modeling communities as "research subjects" (Brown and Mikkelson 1997; Allen 2003; Corburn 2005; Brody et al. 2007, 2009; J. Brown 2007; Minkler and Wallerstein 2008). One of our most promising findings was that the tool was useful not only for academics but also for individual landowners and community organizations. A second iteration of the tool could similarly forge ethnographically rich public environmental health research that increases individual agency and collective advocacy.

The Fossil-Fuel Connection

with coauthor Len Albright

We've got to make the fossil-fuel connection.
—THEO COLBORN (October 1, 2013)

The stories of landowners told on the WellWatch site (see chapter 9) and The Endocrine Disruption Exchange (TEDX) database of chemicals used in natural gas extraction (chapter 4) reveal often obfuscated connections between the two industrial revolutions; we rely on fossil fuels for energy and petrochemicals that form the infrastructural backbone of our high-energy, globally connected, resource-intensive society and economy. This chapter follows the connections between the energy and petrochemical industries to reflect on the downstream consequences of the boom in natural gas extraction. It argues that these industries must be thought of and tackled together as part of what Theo Colborn called the fossil-fuel connection if we are to fully comprehend and respond to the twin environmental human health consequences of late industrialism: climate change and endocrine disruption. At both the macroscale of global climate and the hyper-local scale of fetal development, building our economies and societies around fossil fuels and petrochemicals is transforming the future of life on earth. Social scientists, scientists, policymakers, and advocates need to connect the dots between these two threats to human health and address them by dramatically rethinking energy and chemical use and production.

In Memoriam

On December 14, 2014, at the age of 87, Theo Colborn passed away. Her passing was memorialized in scientific and popular publications alike. *Environmental Health Perspectives* published a reflection cowritten by Linda Birnbaum (the director of the National Institute of Environmental Health Sciences [NIEHS]), the journalist and author Elizabeth Grossman, and an endocrine disruption researcher, Laura Vandenberg. The piece describes the "watershed role" Theo played in environmental health research: "It is no exaggeration to say that endocrine disruption science brought about a sea change in how we consider chemical safety and that Theo was instrumental in bringing about this paradigm shift" (Grossman et al. 2015: A54).

Colborn was also in good part responsible for bringing about another sea change in public perception of the health and safety of fracking. Colborn's database, combined with the experiences of Amos and other citizens impacted by fracking, became the bedrock of a new way of viewing the practice as a potentially massive threat to public health. Fracking is now a household word (Lustgarten 2011a) and "fracktivist" is a new turn of phrase to describe the dynamic and growing national and international antifracking movement that is fighting for regulation and bans on the practice.[1] In the United States, bills to end fracking's exemption from the Safe Drinking Water Act have been introduced in Congress (*New York Times* 2009). New federally funded Environmental Protection Agency (EPA) studies found that hydraulic fracturing can impact drinking water under some circumstances.[2] Sundance Award–winning documentaries are shown on HBO and in theaters, and fracking is a regular topic covered by the most prestigious news outlets, including the *New York Times*.

TEDX's project influenced many of these outcomes, from proposed regulatory reform at the state and federal levels to popular media. Josh Fox, the director of award-winning documentaries on community experiences with fracking, *Gasland* and *Gasland II*, learned about fracking's potential health hazards from Colborn's research, and after meeting with the Damascus Citizens Alliance (Fox 2010, 2013). Based partly on research by TEDX, New York State placed a state-level moratorium on fracking in 2013 in order to further assess its health and environmental impacts (Klopott 2013). In 2014, New

York became the first natural gas–rich state to ban fracking based on public health concerns (Lustgarten 2014). Colborn's work helped stimulate the first EPA field study linking fracking to water contamination in Wyoming (EPA Region 8 2009). On the local level, Paonia and the North Fork Valley managed in 2012 to prevent public lands being leased to gas development (Heller 2012). Thus Colborn's research, which generated rich questions about the identity and usage of toxic chemicals by the oil and gas industry, has assisted in dramatically changing the information landscape around fracking. Subsequent public pressure and visibility around the issue forced leading companies to disclose the contents of fracking fluid.[3]

TEDX's example encouraged a boom in NGO and academic studies of the industry. The Natural Resources Defense Council and the Environmental Working Group (EWG), among others, are now studying oil and gas issues, building on the foundations laid by Colborn, TEDX, and the Oil and Gas Accountability Project (OGAP/Earthworks) (EWG 2008; Horwitt 2008, 2010). A research-based environmental NGO based in Washington, DC, EWG became involved with fracking issues and gas drilling's health effects after hearing a presentation by Theo, according to Dusty Horwitt, a lawyer who worked with the group. EWG released three research reports since 2008 on the hazards of natural gas extraction, focusing directly on chemical harms (Horwitt 2008, 2010; EWG 2009). The group's first report in 2008 added 150 dangerous chemicals to TEDX's list, building directly on Theo's work (EWG 2008).

Academic researchers across the country and internationally are now studying every aspect of the fracking boom, some building directly from Colborn's data. Epidemiological studies demonstrate increased rates of birth defects, preterm births, and other endocrine-related endpoints in communities near fracking operations (McKenzie et al. 2014; Casey et al. 2015). A Pennsylvania study assessing the health of newborn infants in gas production areas has recorded increases in low birthweights, fetuses that are small for their gestational age, and reduced APGAR scores (Hill 2012). People living near natural gas production areas also risk increased cancer and noncancer health hazards because of their exposure to carcinogenic benzene and trimethylbenzenes, aliphatic hydrocarbons, and xylenes, which have neurological and/or respiratory effects (McKenzie et al. 2012). Brain and neurologically damaging as well as endocrine disrupting chemicals were detected in TEDX air samples around natural gas operations, and a six-state air-quality study documented carcinogens and EPA criteria pollutants that exceeded health-based standards (Colborn et al. 2014; Macey et al. 2014). Though exempt from the Toxics Release Inventory (TRI) and Clean Air Act, oil and gas operations are the largest industrial producer of volatile organic compounds (VOCs) (Tollefson 2012; Weinhold 2012). Particulate matter linked to asthma and cardiovascular problems has been recorded at high levels near shale gas operations (ADEQ 2011; City of Fort Worth 2011; McKenzie et al. 2012; Moore et al. 2014). Health problems in humans, pets, livestock, and wildlife exposed to fracking chemicals reported within a six-state study included respiratory, gastrointestinal, dermatologic, neurologic, immunologic, endocrine, reproductive, and other adverse health results (Bamberger and Oswald 2012). Studies by the National Institute for Occupational Safety and Health revealed that workers in Colorado were exposed to unsafe levels of silica dust and carcinogens because basic workplace safety practices were not followed (Esswein et al. 2013, 2014).

Earthquakes in previously stable regions in Alabama, Ohio, and Texas have been linked to fracking and its injection of wastewater (Davies et al. 2013; Kim 2013; Hargrove 2014). Radioactivity has been documented in fracking wastewater entering watersheds in Pennsylvania (Stromberg 2013; Warner et al. 2013).

The promise that unconventional gas is a greener alternative to oil and coal is being strongly contested. Multiple comparative studies, from Cornell University, the National Oceanic and Atmospheric Administration (NOAA), and the University of Colorado, among others, show that methane lost at the point of extraction offsets the gains of lower carbon emissions at the point of burning the gas fuel (Howarth et al. 2011; Pétron et al. 2012; S. M. Miller et al. 2013; Howarth 2014). The MIT Energy Initiative goes to great lengths to address and dismiss Robert Howarth and colleagues' scholarship in its 2011 report (MITEI 2011: appendix 1A).[4] Colborn remarked on the relationship between climate change and natural gas extraction in the very last document she worked on, "The Fossil Fuel Connection":

> Scientists have proven that human induced climate change and global warming are from burning fossil fuels and are seeking alternative energy sources. Adding to the urgency is the evidence from other scientists who have proven that humanity is facing a more imminent threat to survival because of the overlooked alternative uses of the fossil gases that surface during drilling for oil and gas. Methane and a group of semi-gaseous liquids called the aromatics are both separated from the raw gas on the well pads and enter separate pipelines. The aromatics, including benzene and toluene, are comprised of six carbon atoms attached in a circle called the benzene ring. Chemists use this carbon ring as the

building block for the production of practically everything the infrastructure and economy of modern society is dependent upon. (Colborn 2014, 1)

Colborn and I first discussed the fossil-fuel connection in 2006 when I shared with her a literature review I had written on the history of synthetic chemistry and how synthetic chemicals were first distilled from coal tar—a by-product of England's first industrial revolution. Synthetic chemistry has been described as the Second Industrial Revolution as synthetic products derived from coal, oil, and later natural gas replaced plant-derived chemicals and produced entirely new materials such as plastics (A. Chandler 2005). We discussed the need to connect climate change and endocrine disruption research and advocacy. In 2010 we experimented with how the two issues might be brought together during reform of the Toxic Substances Control Act, and we drafted up an internal TEDX document called "The Fossil Fuel Connection." For the front cover of the document I excerpted two advertisements: one from the American Petroleum Institute (API) ("The People of American Oil and Gas Campaign") promoting natural gas and the other from the chemical company BASF ("Invisible Contributions. Visible Success") promoting use of its plastics. The API advertisement shows a young, white boy strapped into the backseat of a pristine SUV, clad in a soccer uniform awaiting a ride, presumably from his dutiful parent sitting in the front seat. The viewer takes the parent's perspective glancing back at the child, whose image is framed by two others that also serve to remind the viewer of fossil fuel's necessity for suburban existence: a mother and daughter changing the laundry and a snow-blanketed home, cozy thanks to central heating. Underlining how feminine, middle-class, white domestic life is enabled by "American Oil and Gas," the top of the ad shows three scenes from oil and gas extraction: a cross section of a massive arterylike pipeline, beside a bird's-eye view of an offshore drilling rig, and last (diagonally paired with the image of the mother and daughter changing the laundry) is a roughnecker, the man "animating" this machine—presumably for his family. He, like the young boy, meets the viewer's gaze: the boy all grown up, done with games and not waiting but working hard to make his island of home life possible. This ad engages the working middle and upper classes of white America by visually reminding them that the American dream of a heteronormal suburban family depends on fossil fuels. This view, while essentially accurate (the suburbs are a creation of fossil fuels), conveys nothing of the costs associated with that lifestyle, such as the violence associated with energy extraction, and the downstream environmental hazards that fossil fuel and petrochemical dependence create.

Downstream hazards, such as the endocrine disrupting potential of phthalate plastics, are portrayed ironically in the second ad produced by BASF. Like the oil and gas industry in 2008, the chemical industry was also under the lenses of scientific and social scrutiny about its human and environmental health hazards. BASF responded with a media campaign stressing the importance of its products to contemporary life. The ad makes this argument visually through a polyethylene umbilicus that is biomedically sustaining a newborn white baby. The baby, sleeping peacefully, is a picture of health, except for the bending tube connected to its belly. In the context of contemporary research on flexible plastics such as these IV tubes containing endocrine disrupting phthalate chemicals, there is a painful irony to the text interrupting the ad's plastic umbilicus, which reads "Invisible contributions. Visible success."[5] The aims of connecting these two images were first to stimulate thinking about how the boom in natural gas production is literally being embodied by fetal exposure to petrochemicals and second to link two industries whose fates are wound together historically, financially, and materially, but not frequently in the media.

Colborn's interest in the connection between these two industries focused scientifically on the use of endocrine disruptors in fracking and the fact that the fugitive emissions during fracking act as endocrine disruptors. TEDX's database predicted that chemicals used in natural gas extraction would be endocrine disruptors (see chapter 4). Dr. Susan Nagel from the University of Missouri, after a conversation with Colborn, decided to test fracking water and water around fracking sites for endocrine disrupting effects. She found estrogenic, anti-estrogenic, and anti-androgenic activity in water contaminated with fracking fluid (Kassotis et al. 2014; E. Webb et al. 2014). Similarly Colborn began her own grassroots air-monitoring studies, and she documented high levels of benzene, toluene, ethyl benzene, and xylene (BTEX) chemicals in the air around natural gas operations (Colborn et al. 2014). Her work on the fossil-fuel connection describes how

> direct exposure to benzene and toluene interferes with endocrine control of the brain, heart, nervous, and immune systems, and also causes immediate effects to the eye, skin, and nervous system similar to complaints from those living and working in gas fields. In addition, another class of ignored gas that escapes at the well head, the polycyclic aromatic hydrocarbons (PAHS) were found in a neighborhood immersed

among natural gas wells at concentrations 3 times higher than those detected in a study done in New York City where pregnant woman wore back pack air monitors. (Colborn 2014: 2)

Colborn's work on the fossil-fuel connection mentions in passing that natural gas is the feedstock for many petrochemicals. This chapter picks up this thread to explore how the boom in natural gas extraction is leading to a boom in chemical production, and I examine the consequences of the present regime of imperceptibility about the connection between these two industries. The fossil-fuel connection is not made by a single thread but is instead a fibrous network of connections that embody a material, political, social, and economic infrastructure that is actively fighting to grow and sustain itself. This chapter dissects these connections: (1) the dependency of chemical production on fossil fuels to (2) the Möbius strip of petrochemical use in fossil fuel extraction; (3) the political and scientific machinations preventing the study of chemically induced harms and climate change; and (4) the environmental, economic, and political problems associated with investing in natural gas as the bridge fuel to low-carbon economies. To fully understand and dismantle these fossil-fuel connections, I argue for (1) new forms of grassroots research, described as civic science; (2) new forms of community-centered governance; and (3) new modes of academic community and regulatory collaborations. Networked, public interest, and grassroots research is required for us to collectively comprehend and reshape the products of our industry.

The Boom in Ethylene Production

Although it is not often a focus of fracking debates, one industry that benefits significantly from the boom in fracking is the petrochemical industry, because unconventional gas reduces the cost of a vital constituent of petrochemicals: ethylene (IHS 2011: 34). Ethylene is the most produced petrochemical in the United States and the world. It is the major constituent of a vast range of chemicals such as polyvinyl chloride, trichloroethylene, and polyethylene glycol (IHS 2011: 34). Ethylene derivatives are used in basically every familiar chemical product, from personal-care products such as shampoos, detergents, and soaps to paints, plastics, coolants, antifreeze, tire rubber, industrial solvents, adhesives, corrosion inhibitors, and textiles. The chemical is used in every industrial sector from apparel and textiles to beverage, food products, computers, electronics, paper, machinery, metal manufacture, and pharmaceuticals.[6] Ethylene is also used as a plant hormone, because its endocrine disrupting properties can artificially force fruits to ripen (Chaves et al. 2006). Synthetic ethylene production enables unripe, hardier fruits to be picked and shipped vast distances to markets, through fossil fuel–intensive supply chains, rather than being picked when ripe and consumed locally (Yeoshua 2005: 135).

Seventy percent of the cost of ethylene production derives from natural gas, both as a source of energy and as a feedstock (IHS 2011: 34). Production is extremely energy-intensive, requiring large amounts of natural gas to be burned to produce steam. The steam is used to crack natural gas by heating it to 750–950°C. Cracked gas is rapidly cooled and repeatedly compressed and distilled. The reduced cost of natural gas is attracting ethylene production back to the United States from the Middle East and Asia (IHS 2011: 29). Before the gas boom, the last cracking facility to be made operational in the United States was in 2001, but now numerous companies, including Dow Chemical, Royal Dutch Shell, Chevron Phillips, Exxon-Mobil, and BASF, are building cracking capacity in the United States (IHS 2011: 29, 30):

> Royal Dutch Shell is planning to build a world-scale ethylene cracker with an integrated derivative unit in the Appalachian region of the United States. . . . "With this investment, we would use feedstock from Marcellus to locally produce chemicals for the region and create more American jobs. As an integrated oil and gas company, we are best-placed in the area to do this." (Marvin Odum, President of Shell)
>
> Marcellus ethane is the most competitive feedstock for petrochemicals in the US, so it makes sense to use it there, rather than add to its cost by transporting it across one-third of the country, then sending derivative products back up to the Northeast. (Iain Lo, Vice President of New Business Development and Ventures for Shell Chemicals). (IHS 2011: 31)

From the Gulf of Texas to the Louisiana Chemical Corridor, the fracking boom is creating a cracking boom.

Cracking plants bring jobs and prosperity, according to industry PR, but they also create local air pollution and exacerbate environmental justice issues (Allen 2003; Bera and Hrybyk 2013). Ethylene plants emit an array of hazardous toxic pollutants, including carbon monoxide and nitrogen oxides (VOCs that can form harmful tropospheric ozone) and carcinogens such as benzene (S. Fitzgerald 2012; Frazier 2013). Communities on the fence-line of cracking and petrochemical production plants report numerous illnesses, from increased asthma rates to heart disease (Allen 2003; Subra 2014) and can experience systematic environmental injustice (Auyero and Swistun 2009). For those bearing the costs of industrial development—fence-line communities, gas-patch residents, and consumers of gas and petrochemicals—the fracking and cracking booms further entrench the petrochemical, fossil fuel–driven Anthropocene both in terms of the corporations involved and the subsequent environmental and health problems of climate change and endocrine disruption.

The world's largest and oldest oil companies are among the world's top 10 ethylene producers: Exxon Mobil, Chevron Phillips, Royal Dutch Shell, and Total (Breisford 2014). Exxon and Chevron evolved out of the first integrated oil company Standard Oil, and Royal Dutch Shell was one of Standard's first international competitors (Yergin 1991). Other major chemical manufacturers that have built market permanence, Dow Chemical and BASF, are also major ethylene manufacturers reinvesting in the American chemical industry. These century-old companies are as wealthy and long-lived as many nations. Only 25 countries in the world had a higher GDP than Shell's earnings in 2012.[7] Of the 150 largest economic entities worldwide, 58% of them are companies, and of the top four, three are international oil companies, Shell, Exxon, and BP.[8] What sorts of academic, bureaucratic, and social transformations are needed to adequately monitor, research, and respond to corporations and industries that span nations and exceed them in economic scale? Should we develop academic departments for Dow and BP studies similar to the academic fields devoted to studying nations? How can we grapple with environmental health problems such as climate change, endocrine disruption, and chemical contamination due to fracking that transcend biological, legal, and social boundaries? While some have proclaimed fracking as a revolution, empirically it is no such thing. It entrenches the old guard, the industries, consumer economies, environmental and human health hazards, and the deep and expanding inequalities between regions of extraction, production, and consumption that pattern modernity.

The Möbius Strip of Chemical Production and Consumption

The connections between the fossil fuel and chemical industries are hard to bring to the surface in part because of the regimes of imperceptibility that both industries are able to create. One way in which these connections surface is in the experience of everyday individuals whose lives veer off course due to chemical exposures.

2-BE: MAKING OIL AND GAS EXTRACTION FEASIBLE AND UNMAKING HISTORY

2-Butoxyethanol (2-BE) originates from ethylene, being produced through a chemical reaction between ethylene oxide and butanol. The story of TEDX's database began with 2-BE and Laura Amos. It is fitting that the end should also begin here. The chemical 2-BE is experimentally correlated with the adrenal tumors that Theo described in her 2002 memo to the Bureau of Land Management. Laura read that memo and connected the contamination of her water well during fracturing to her own tumor. Although the company fracking the well initially denied that the virtually odorless, colorless, and tasteless chemical had been used in the area, documents provided later to the Colorado Oil and Gas Conservation Commission (COGCC) investigators by Encana confirmed that 2-BE was in fact used in the fracturing procedures "coincident with" the contamination of her well water.[9] Because it was impossible to prove her exposure to 2-BE and its role in causing her tumor years later, Laura settled a suit with Encana, enabling her family to move from their contaminated home. Encana's price for this settlement was a nondisclosure agreement. For the rest of her life, Laura must be silent about the well contamination, her illness, and the loss of her family home. These agreements prevent information from circulating beyond individuals' cases so illness experiences cannot be aggregated across communities. Such aggregated information could provide evidence of systematic links between companies and hazards structurally produced by their operations.[10] Nondisclosure agreements strategically thwart social communities' abilities to coalesce around shared disease or other experiences due to industrial hazards (P. Brown et al. 2003; Dumit 2006; Efstathiou and Drajem 2013). This isolates and disempowers individuals.

Nondisclosure agreements prevent information about the potential hazards of 2-BE from circulating and aggregating. In contrast, 2-BE itself is cycling in a material feedback loop. It is both used in fracking natural gas and is itself made from natural gas. This exemplifies the paradoxical, little-examined, and unsustainable looping between these two industries, where chemicals derived from natural gas are required to gather more natural gas to make more chemicals, forming a sort of Möbius strip. And just as described in chapter 1, chemicals are required to manage the adverse impacts of other chemicals in the fracking process, for instance because acids are needed to clean pipes, oxygen scavengers are required to stop the activity of acids before they corrode the pipes, so 2-BE as a surfactant has a second major industrial use: it is used to clean up oil spills.

2-BE is also a chemical additive in Corexit, an oil dispersant used in the 2010 Deepwater Horizon oil spill (Lustgarten 2010; Wang 2010).[11] It is one of the two active ingredients in Corexit 9527, the first dispersant used in the Gulf to break up the oil gushing from BP's well.[12] Corexit's surfactant properties allow it to lower the surface tension of oil, speeding dispersion. Unfortunately, the hemolytic effects of 2-BE (in which red blood cell membranes are broken down) make it particularly toxic to sensitive tissues such as fish gills (Dugan 2010). Furthermore, Corexit's human toxicity was suspected after its use in the 1989 Exxon *Valdez* oil spill, when cleanup crews experienced health problems (Lustgarten 2010; Schor 2010). Public outcry, largely due to the newfound awareness of the possible health hazards of glycol ethers, motivated the EPA's requirement that BP choose a less toxic dispersant. It is unknown exactly how much Corexit 9527 was used in the Gulf before this ban. The replacement product, Corexit 9500, did not contain 2-BE but is banned for use in the United Kingdom because it failed toxicity evaluations. Corexit 9500 also rated among the most potentially toxic of eight dispersants rapidly evaluated by the EPA in 2010 (EPA 2010; Judson et al. 2010).[13] Despite these concerns, 1.9 million U.S. gallons of Corexit products were used in the Gulf.

Why was Amos not on TV telling her story and warning people who lived near the Gulf? She had been silenced by a nondisclosure agreement. Because cases like Laura's have never gone to court and her experience was not correlated with the experiences of those exposed to Corexit during the *Valdez* cleanup, 2-BE was presumed to be safe. BP did not perform extensive health studies on Corexit or 2-BE before using it in the Gulf because the company was not required to do so. With so little directed study of the toxicity of these products, the material safety data sheet (MSDS) for Corexit 9500 says, "No toxicity studies have been conducted on this product."[14] Therefore it is impossible to evaluate with any certainty the long-term effects of dispersants used in the Gulf. Corexit's producer, Nalco Holding Company, initially refused to make the composition of both products public, arguing that the information was a trade secret. Chemical compositions were eventually shared with the EPA after Corexit was used in such bulk that its maker could be forced to release its ingredients based on the Toxic Substances Control Act (Schor 2010).

These problems exemplify the issues that plague attempts to study the gas and oil industry and allied petrochemical industry. As described in chapter 1, petro-state dynamics protect the security of the companies over the security of communities and environments in which they operate (Watts 2005). Although the Deepwater Horizon incident concerned crude oil, it illustrates, in a condensed form, the tumult of issues occurring across the country related to natural gas extraction. This industry shapes the future by editing the past, by controlling what information is archived and aggregated. Divers who took samples for NOAA in the Gulf during and following the spill were assured of Corexit's safety, told it had a half-life of 90 minutes in water, and were not supplied with recommended protective equipment during dives (Kirby 2013). That summer those divers started experiencing symptoms consistent with dispersant exposures: "nausea, headaches, fatigue, memory loss, and blood in the eyes, nose, and stool" (Kirby 2013). Three years after the spill, despite evidence of Corexit and oil exposure in their blood and tissues and persistent health problems, they were, like Laura, unable to definitely prove connections between their illnesses and exposures in order to receive compensation. The Government Accountability Project, a nonprofit watchdog group, gathered similar stories from fishermen, cleanup crews, and community members exposed to Corexit (Devine and Devine 2013).[15] By distributing the oil and making it harder to see, the dispersants made the oil spill visually less appalling while simultaneously creating a host of harder-to-pinpoint, distributed health risks to humans and the larger Gulf ecosystem that we presently lack scientific and regulatory systems to aggregate and address.

To respond to these gaps, we need to start studying the material and the social processes of the gas, oil, and petrochemical industries together. The entire production and consumption cycle for a chemical like 2-BE should be examined, in order to question the long-term sense of unsustainably manufacturing a chemical that is used to further unsustainable energy production practices, and to manage the ecological and human health disasters they create. This material inquiry needs to be contextualized in the life experience of people like Laura who are exposed to such chemicals and whose experience can tell us about their potential hazards. A third level of inquiry should look at the political, economic, and legal forces that contain information about the adverse effects of such chemicals and prevent research on their environmental and health effects.

Grassroots Research

Grassroots research is required to study both the impacts of these chemicals and the industries that use them. My own research trajectory has been transformed by the need for generating grassroots data and for improving citizens' abilities to gather much-needed information about their environments and this industry. A fly-over ban during the Gulf spill prevented direct imaging of the oil threatening the coastlines of five states. My colleague at MIT's Center for Civic Media, Jeff Warren, had recently invented a way to generate satellite-like images without a satellite or plane, which he called grassroots mapping. This involved flying a cheap digital camera from a helium balloon or kite using a long piece of string. During the Deepwater Horizon spill, Warren worked with Louisiana Bucket Brigade's Shannon Dosemagen to organize grassroots maps of contaminated gulf waters and shorelines. These grassroots maps formed the largest publicly owned and community-made archive of images of the extent of the damage (Dosemagen, Wylie, and Warren 2011). Field science machinery evolved for Big Science is ill-suited to distributed public interest research (see chapter 5). Not only is it costly and dependent on expert operators, it also achieves state and industry goals of extracting and centralizing data, thereby removing communities from the research process (Wylie et al. 2014). Grassroots mapping suggests an alternative approach: low-cost, do-it-yourself (DIY) tools that empower the public to do their own data gathering and analysis and to leverage digital media for sharing and archiving the results. These DIY tools support the spread of environmental research at a community level.

I, Warren, Dosemagen, and four others founded a nonprofit around this idea in the fall of 2010. Public Laboratory for Open Technology and Science, or Public Lab, extended open-source software principles to hardware, creating a publicly available repository of tools for community-based environmental research.[16] Building on ExtrAct, we are using grassroots data gathering to produce research tools that are affordable, that are easy to use, and that provide high-quality data. Public Lab's grassroots maps have better resolution than available satellite images and have been integrated in Google Maps and Google Earth.

Our tools are enabling communities to take the lead in generating images of their environment rather than relying on government agencies or companies. Projects like Public Lab, TEDX, and ExtrAct are creating important new possibilities for retooling field sciences to improve our ability to recognize and act on emerging environmental health problems (Wylie et al. 2014). For example, Citizens for a Healthy Community and TEDX collaborated on a study of the North Fork Valley's air quality in which hikers wore VOC- and PAH-sensing backpack units (Colborn et al. 2014). It was this study that found elevated levels of neurotoxic chemicals around gas extraction sites (Colborn et al. 2014). Communities are stepping up to fill the gaps left by regulators and academics (Pennigroth et al. 2013; Jalbert, Kinchy, and Perry 2014).

The first citizen science peer-reviewed study of unconventional oil and gas was published in 2014. Based on air samples collected with buckets by communities in six different states, it showed that the levels of eight VOCs exceeded federal guidelines. The most common compounds to exceed acute and long-term exposure thresholds were the carcinogen benzene, the neurotoxic hydrogen sulfide, and a broad-spectrum toxicant, formaldehyde (Macey et al. 2014). We are now working with community organizations in Wyoming, New Mexico, and Texas to develop low-cost ways of visualizing hydrogen sulfide contamination.[17] Combined with platforms like WellWatch, these tools can provide wholly new high-quality data generated by and for communities so they can visualize and make tangible their changing environmental conditions, then take action on them.

Grassroots research around the oil and gas industry may be meaningless, however, if it cannot lead to regulatory change. Currently grassroots research is unable to scale in the same way the industry can to influence federal and state research agencies, as the following story about 2-BE and petrochemical politics illustrates. Promising headway was made in 2016 when EPA's science advisory committee (coauthored by Shannon Dosemagen of Public Lab) released a report, "Environmental Protection Belongs to the Public," that called for integrating citizen science into every aspect of work at the EPA (NACEPT 2016). Unfortunately, the 2017 inauguration of Donald Trump, who promised to dismantle the EPA, makes it highly unlikely this vision of a citizen science led EPA will be realized.

2-BE, CONTINUED: PAVILLION, WYOMING, AND THE CAPTURE OF SCIENCE AT THE STATE LEVEL

TEDX's findings crystallized resistance to drilling in Pavillion, Wyoming, a rural farming and ranching community where the water wells (including drinking water wells) were contaminated during natural gas extraction. There, one woman suspected that her respiratory and cognitive issues, which started as flu-like symptoms and degenerated into a seizure disorder, were related to contamination of her well water by fracking. Theo's presentation reportedly enabled the woman to connect her and her family members' illnesses—her son's liver disease, her daughter's miscarriages, and her grandchildren's cases of high fevers and subsequent kidney failures—to exposure to the same contaminated water. Her exposed neighbors similarly experienced respiratory ailments and loss of the sensation of taste and smell, neuropathy, and cognitive difficulties.

For 10 years, Pavillion's community repeatedly asked the state to investigate severe changes in their water. The state consistently responded it had neither the budget nor the staff to conduct an investigation and did no sampling. Industry conducted limited case-by-case water-well testing but did not use a uniform test across all the wells or systematically check whether the shared aquifer was contaminated, so residents were unable to see if they shared contamination. After industry testing found nothing, a Pavillion rancher paid an independent company to test the water and sent the samples to an out-of-state lab. These tests showed that his well was contaminated. Frustrated at the state's lack of response, Pavillion community members approached the Powder River Basin Resource Council, who sent a regional organizer, Deb Thomas, to meet with them. Thomas was experienced with contamination from drilling in her local community of Clark, Wyoming. She convened a community meeting that brought together families and their industry test results and reviewed the independent findings by the out-of-state lab. After critiquing the industry's disparate test methodologies and its failure to perform a Basin-wide hydrological study as well as hearing residents' frustrations with state unresponsiveness, Thomas suggested residents approach EPA Region 8. When EPA researchers invited Thomas to present her and the community's findings at Region 8's Denver headquarters, Thomas reported they were visibly upset, urging her to wait for the federal election outcome with the expectation that they would then have more latitude to start an investigation.

The subsequent EPA Region 8 draft report of its investigation (begun in 2008) identified breakdown products of 2-BE in 11 water wells and determined that the hydrogeology[18] of the region had been contaminated during gas development (EPA Region 8 2009). That the EPA had the regulatory authority to perform this study under Superfund laws was partly due to TEDX's work. In response to TEDX's database, the EPA began in 2008 to compile a list of chemicals used in natural gas extraction. The list states "several different types of drilling fluids, containing several hazardous compounds, are used to install gas wells" (EPA Region 8 2009: 7). Furthermore, this EPA study cites and uses TEDX's data: "Additionally . . . TEDX . . . has compiled a list of chemicals used in natural gas development in Wyoming. While the TEDX list is comparable to the EPA Study List, it adds several metals that may be found in compounds used in gas well installation and are as follows: aluminum oxide, arsenic, cadmium, copper, iron, lead, mercury, nickel, vanadium and zinc"

(EPA Region 8 2009: 8).[19] The study could be performed under Superfund regulations in part because of the potential presence of these chemicals: "The Superfund Chemical Data Matrix (SCDM) is a list of benchmark values used in the evaluation of National Priorities List (NPL) sites under the Hazard Ranking System (HRS). The following chemicals are found in both the EPA Study List and the TEDX list [and have an] SCDM associated with them as well (EPA Region 8 2009: 8)." The heavy metals included on this list are the same as those found in my TEDX research on Soltex in a Greenpeace analysis of heavy metal additives being used in gas drilling mud (see table 10.1).

The EPA was unaware of which specific chemicals from this list to concentrate on in their aquifer tests because the industry maintained inadequate records during gas development in the region. Therefore EPA researchers used a scattershot testing approach, first identifying spikes that occurred in various test procedures and then comparing the results against chemical libraries in order to identify the chemical signatures in the water. The resulting list of contaminants found in EPA-tested water in Pavillion contained arsenic, methane, adamantanes, 2-butoxyethanol phosphate, and caprolactum. Adamantanes are "naturally occurring hydrocarbons found in crude and gas condensate" (EPA Region 8 2009: 17). Their presence points to aquifer contamination by gas drilling. The EPA study additionally found that "it is possible for 2-butoxyethanol to react with natural occurring phosphates to produce 2-butoxyethanol phosphate" (EPA Region 8 2009: 17).

Although the company involved had installed water filtration systems for families in Pavillion, it is unlikely these would effectively remove glycol ethers like 2-BE (Fox 2010). Wyoming families had been potentially exposing themselves for years to 2-BE and other chemicals because they believed the filtered water was safe. The Pavillion follow-up study is significant because EPA evidence that 2-BE and adamantane contamination derived from gas drilling (there is presently no competing cause for the contamination) would provide the first proven, government-documented case of fracking contaminating a watershed and placing human health at risk. It would give the lie to industry's oft-repeated claim that "there are no documented cases of aquifer contaminations due to hydraulic fracturing" and call into question the veracity of a document published by the Interstate Oil and Gas Compact Commission that compiled statements made by oil and gas regulators in 10 states, including Wyoming,[20] reassuring the public that "no documented cases of groundwater contamination from fracture stimulations" have occurred in their states.[21] As the EPA has previously been unable to study fracking, and its review of hazardous chemicals used in natural gas extraction relied on inde-

TABLE 10.1. Hazardous chemicals in hydraulic drilling fluids with a Superfund Chemical Data Matrix (an SCDM) value

Chemical Name	*SCDM Drinking Water (January 28, 2004) Concentration in μg/L (Maximum Contaminant Level)*
Benzene	5.0
Toluene	1,000
Ethyl benzene	700
Xylene	10,000
Naphthalene	20
1-Methylnaphthalene	20
2-Methylnaphthalene	150
Fluorenes	1,500
Ethylene glycol	73,000
Formic acid	73,000
Methanol	18,250
Ethylene glycol monobutyl ether	18,000
Aluminum oxide	36,000
Arsenic	0.057
Cadmium	5
Copper	1,300
Hydrogen sulfide	10
Iron	11,000
Lead	15
Mercury	0.63
Nickel	730
Vanadium	36
Zinc	11,000

Note: The SCDM is a list of benchmark values used in the evaluation of National Priorities List sites under the Hazard Ranking System. The listed chemicals are found in both the EPA study list and the TEDX list and have an SCDM associated with them as well. The possibility of their presence in drilling fluids prompted an EPA investigation into the Pavillion watershed.

Source: EPA Region 8 (2009: 8, table 1).

pendent organizations like TEDX, it is not surprising that there have been no prior official confirmations of water contamination due to fracturing.

Recognizing the threat posed by the EPA's precedent-setting study, Region 8's first study was criticized vociferously by the oil and gas industry and the state of Wyoming. EPA staff were harassed by members of Congress supporting the oil and gas corporations (Lustgarten 2013). This pressure, along with budget cuts, reportedly led to the EPA's 2013 decision to hand over the follow-up study to Wyoming's Department of Environmental Quality. That study was to be funded directly by Encana, the company accused of contaminating the aquifer in the first place (Lustgarten 2013).

Because the state was put in control of the investigation, there has been no material progress in terms of identifying or remediating the contamination issues in the groundwater. The state has instead engaged in a cycle of review and further research, moves that stall remediation and drain community organizations. The state took a year to hire new experts to review existing data (already collected and analyzed by the EPA experts). Rather than remediating residents' water, the state has developed a cistern program in which they receive water from underground cisterns filled with water trucked in from Pavillion's municipal water-well system. However,

> there has never been an identification of what the aquifer contamination is, how large the plume or plumes are, or how they're moving through the Pavillion area. No-one can say if or when the contamination will impact other drinking and stock water wells in the Pavillion area; including the Pavillion municipal water wells. (Personal communication with Deborah Thomas, February 10, 2015)

A second draft report released in 2011 confirmed the first study's findings, but it did not lead to any significant regulatory change (EPA Region 8 2011) and, instead, recommended another round of research work. As these reports have been "drafts," nothing has been formally published in citable format, keeping the case "inside the state" and muting the study's potential to set an important public precedent that fracking contaminated a community's groundwater.

This decade-long struggle in Pavillion illustrates how industry's superior financial capacity and manpower outmaneuver its opponents, exhausting the community with endless research and applying relentless pressure through legislators, lobbyists, and skillful PR. Forcing yet further studies if it is dissatisfied with findings contrary to commercial interest is the same tactic that industry has used to contest other industrially related hazards: tobacco, endocrine disruption, and climate-change research (Oreskes and Conway 2010). Conventional "objective" science is unequal to its responsibility for community and environmental health because it is required to determine hazards but is insufficiently certain for closing debates. Further research can always be pursued, and every scientific study has its limitations (Beck 1992). These cases underscore the need for proactive social science to study corporate activity and influence on regulatory science.

We have to transition away from the period of Enlightenment science that divided the social and natural world into different disciplines, where the natural scientists studied the physical world and the social scientists the human-built one. If we ever did live in a universe that was so simplistically parsed, we certainly do not now. The products of human-made industry have transformed the biochemistry of the planet (McNeill 2000). I have reached toward the term "civic science" to describe a mode of research that seeks to connect the practice of science (defined as a collaborative enterprise to study how we and the world around us interoperate) and the practice of democracy (defined as the collective effort to build collaborative governance that ensures human rights).

To break these fossil-fuel connections, citizen science data and community-based research need to be more directly tied to governance. Tools for studying these industries should be developed collaboratively by communities, academics, and regulatory agencies rather than as proprietary projects for individual nonprofit organizations. Both the EPA and the NIEHS are encouraging community-based participatory research with targeted environmental justice grants.[22] These funding mechanisms are an important step forward. However, the academy could be doing far more to design and develop open-source infrastructure that could be used by state agencies. A tool like WellWatch could be developed by academics and communities and used not only as a check on state agencies but also *for* state agencies to aid legislators making policy and encourage public participation in democratic processes. Open-source licensing of nonproprietary software is an entirely new way for agencies to develop their database and mapping tools that could dramatically lower agencies' infrastructural development costs and make such tools available to the public in exciting ways. Through tools like Syncscraper and WellWatch, civic organizations could be sustained by voluntary contributors, much like Wikipedia, with academic institutions helping to design and sustain such projects. To forestall the industry use of nondisclosure agreements and strategic lawsuits against public participation that aim to quash research, web tools like ExtrAct should be protected by research exemptions similar to those developed by the National Institutes of Health that protect research subjects' identities.[23]

Civic science ties together research and action to improve shared economic, environmental, and social conditions. Current systems of governance are based on mass-media technologies that broadcast information from research and regulatory agencies. Citizens interact with regulatory agencies in a one-to-one relationship, and the direction for research agencies is set by professional experts or at the macro-congressional level in a top-down fashion. Open-source systems alter these dynamics, enabling the public to take a much more active role in creating software and hardware for governance, contributing data and analysis to research projects and directly funding research relevant to their communities. At both state and federal levels, open-source systems can be leveraged for governance in clear ways, for reduced costs and increased public support of governance infrastructure, and for better public reporting and data analysis. Some governments are already embracing this. For example, Taiwan's "gov" effort uses open-source tools and creates online public forums for participatory decision making (Roubini and Tashea 2014). Open-source tools can enable governance to be more mobile and targeted, for example, transforming disaster response by enabling the public to augment federal agencies in providing and directing assistance. Using a Public Lab tool called MapMill during Hurricane Sandy, OpenStreetMap's Humanitarian Team collaborated with the Federal Emergency Management Agency (FEMA) to organize 6,000 online volunteers to sort images of storm damage. The project helped to rapidly identify damaged areas and target FEMA's relief effort at no cost to FEMA for either the software or the volunteers' efforts (Crowley 2013; Munro et al. 2013).

This example can be expanded upon to build publicly supported and developed software, creating participatory and recursive communities to support web-based archival, decision making, mapping, voting, and research and analysis tools. In parallel, through sites like WellWatch, the public can become active participants in monitoring and researching physical industrial infrastructures. Public funding can similarly be rethought to create crowdfunding opportunities and processes.

As chapters 2, 5, and 6 showed, our current scientific process and regulatory structures are in a large part responsible for obscuring the negative impacts of the boom in natural gas. Recent studies by Public Accountability Initiative (PAI) shed light on the corporate influence over and conflict of interest surrounding academic research around oil and gas.[24] PAI found undisclosed financial conflicts of interest in supposedly independent research from the University of Texas (Connor, Galbraith, and Nelson 2012a) as well as plagiarized research that regurgitates industry reports in a study from the University of Buffalo (Connor, Galbraith, and Nelson 2012b). This kind of industrial engagement in producing favorable science is vital to sustaining corporate interests.

Chemicals and companies are tied together and those relationships must be studied together. Making these industrially related environmental harms manifest requires a more active social science directed at studying industrial systems. In particular, science and technology studies (STS) occupies an important disciplinary space that examines the fusion of nature and culture that occurs through scientific research and technological development, and the confusions and possibilities their interactions create (Haraway 1991, 1997). STS can increase its relevance by embracing the "in practice" methods explored in this book to actively construct alternative forms of technoscience that address contemporary sciences' systematic power asymmetries (Wylie et al. 2014). History of science and sociology of science have been vital in revealing how the impossible ideals of "objective science" are skillfully manipulated by industrial interests with sophisticated PR and research infrastructures (A. Brandt 2007; Oreskes and Conway 2010). Similar to the ways in which climate scientists have found themselves under professional and personal attack for sharing their research, researchers studying fracking have come under intense PR and media attacks (Oreskes and Conway 2010).

The American Petroleum Institute and the Independent Petroleum Association of America developed aggressive PR campaigns, such as the website Energy in Depth (EID), to discredit people perceived to be antifracking activists and scientists (Matz and Renfrew 2014). EID contemptuously dismissed the peer-reviewed research of Theo Colborn and Anthony Ingraffea as "junk science" (Horn 2011; Lomax 2012; Massaro 2012; Matz and Renfrew 2014). WellWatch received similar treatment: in April 2014, an industry spokesperson and executive director of the West Slope Colorado Oil and Gas Association indirectly characterized WellWatch as an "echo chamber" as listed here:

> Social media and online platforms are wonderful tools for society but not when they simply attract and create echo chambers for people who gravitate toward, or benefit from, taking extreme positions. . . . Extremism and activism will continue taking on more technological complexity masking itself with a veneer of academic credibility. . . . All we can do in response is maintain a ready willingness to engage and address in a balanced way the real questions average people have about western Colorado's energy businesses. (Webb 2014)

The indirect structure of this response creates deniability about its slanderous intent. It cleverly separates the reader from WellWatch contributors, by suggesting that users are not "average people" with "real questions," but rather they are "activists" and "extremists," with presumably fake questions. There is no more pejorative and frankly dangerous way to characterize people in contemporary culture than "extremist." The term connotes terrorist groups like the Islamic State—violent, ideologically motivated people irrationally opposed to the American way of life. Framing like this attempts to alienate readers from tools like WellWatch, to make them think, "I don't want to be associated with people like that." Contrary to this frame, WellWatch reports show that contributors were working-class families, were farmers and ranchers experiencing terrifying, confusing, and life-altering loss of property, clean air, water, health, and peace of mind. They pose reasonable, incisive questions, such as why are there no afterburners on rural wells?

The oil and gas industry is deeply threatened by the possibility of platforms that systematically document their harms to human and environmental health, hence the final part of this attack, dismissing WellWatch as "a veneer of science." It is as tragic as it is predictable that public interest research should be demeaned as "junk science," because the industry has dismantled public and environmental health protections, thereby forestalling systematic monitoring and study of their activities and impacts and necessitating sites like WellWatch (Kosnik 2007; Faber 2008; EWG 2009). Rather than admit that the industry created a vacuum of knowledge around its practices (for instance, by refusing to provide basic information about the constituents of its chemicals to health-care providers in emergency situations), this quote ends with the false promise that "all we can do in response is maintain a ready willingness to engage and address in a balanced way the real questions average people have about western Colorado's energy businesses." The suggestion that the industry is somehow powerless, that "all" it can do in response "is maintain a ready willingness" not to *answer* but "engage" people's questions disingenuously allows it to play the victim's position in order to gain readers' sympathies. In fact, the industry is not, in practice, ready and willing to answer landowners' questions, and it has a well-honed arsenal of media, political, scientific, and legal responses to those who threaten its viability.

Unfortunately, unlike industry, gas-patch communities are not networked into the university system. Scholars who work with community organizations often face difficulty in their academic jobs: "Scientists who partner with SMS [social movements] also experience ambivalence if not outright suppression; there are significant career costs associated with doing research that may be at odds with industrial interests, the dominant research programmes, or simply the rhetoric of objectivity" (Hess 2004: 421).

Conrad Volz, an assistant professor at the University of Pittsburgh who started FracTracker (a nonprofit that maps fracking data), produced a study of gas industry wastewater contamination. Volz resigned after the university sought to restrain his "advocacy work":

> He said that his beliefs concerning environmental advocacy and public health clashed with those of the University. Conrad "Dan" Volz Jr. said that he is resigning over a "difference of opinion between the University and me of the proper advocacy in the world of public health." Volz, the director of the Center for Healthy Environments & Communities at the Graduate School of Public Health, is an open critic of Marcellus Shale drilling. He said the University has not allowed him to openly voice his dissent, although he did not cite specific instances. (Friedenberger 2011)

While Volz resigned because his citizen science work was restrained, MIT academics working with industry had no such problems. Researchers at the industry and MIT university collaborative, MITEI, were permitted to develop technologies that enabled gas extraction in collaboration with, and funded by, oil and gas companies and oilfield service companies, and they authored policy papers advocating fracked gas as "the bridge fuel." Why are the former MITEI head Ernest Moniz's supportive statements framed as "policy recommendations" and not "advocacy," while Volz's dissenting statements are counted against him as advocacy?

Nonprofit–Academic–Community Coalitions

Truly transformative work in universities can be fostered through encouraging researchers to actively work with communities and to design for them and nonprofit organizations. The novel Social Science Environmental Health Research Institute (SSEHRI) at Northeastern University, founded by Phil Brown, the researcher who theorized popular epidemiology (see chapter 3), has collaborated closely with three nonprofits, the Silent Spring Institute, Toxics Action Center, and Public Lab. Such collaborations allow for real-world research projects to develop at the undergraduate and graduate levels. Toxics Action Center, a nonprofit that has built the capacity of New England communities to resist and remediate toxic hazards for the past 25 years, holds its annual conference at Northeastern University. This conference brings together people from throughout New England who are managing different but overlapping issues such as pollution from incinerators, power plants, and landfills. The presentation of Public Lab tools at this conference led to an undergraduate student research project that worked with Providence, Rhode Island, residents to balloon-map an illegally operating construction and demolition facility. The project evolved a new approach for using balloon maps to estimate the volume of waste piles.[25] The student, Holden Sparacino, and I then applied this method to help Chicago communities mapping waste piles of Petcoke from tar sands refining (Erbentraut 2015).[26] Holden went on to build a career in citizen science as the outreach manager of the Dickinson College Alliance for Aquatic Resource Monitoring (ALLARM), which supports community watershed monitoring in Pennsylvania and throughout the Chesapeake Bay region. As this example demonstrates, applied work in STS with community organizations and nonprofits can create new career paths for students and new, stronger networks between communities and institutes of research and education.

Humanities and social sciences and funding agencies can do much more to shine a light on nonacademic career routes and to support career development beyond the academy. NIEHS has begun actively supporting such collaborations by providing a training grant to SSEHRI and Silent Spring Institute to develop transdisciplinary training in environmental science, environmental sociology, and STS within a university and a nonprofit organization. Community collaborations are often incorrectly viewed as biased activism or advocacy, whereas working with corporate funding or creating pipelines to for-profit careers is not. This misconception is doubly unfortunate, given that much academic research is publicly funded. Additionally, universities' intellectual property provisions mean that publicly funded research can be effectively privatized by universities. Researchers, communities, and nonprofits should attend to developing memoranda of understanding aimed at providing guidelines for shared projects where data and results are co-owned and shared in the public domain (Wylie et al. 2014). For these reasons, focusing on developing low-cost tools that are open-source and creating provisions that enable public interest research to remain in the public domain are vital underpinnings of these coalitions. Without the forging of connections among communities, government regulators, and universities, the oil, gas, and chemical industries will continue to develop research programs and policies that support their own rather than public interest.

MITEI released its finalized report on the future of unconventional gas in 2011, with the overall message that unconventional gas is the all-important bridge fuel that will guide the United States toward a low-carbon economy (MITEI 2011). Moniz was appointed as Barack Obama's Secretary of Energy in 2013 (D. Chandler 2013; Mufson 2013). Obama's 2014 State of the Union speech echoed MITEI's opinion, unequivocally supporting gas as the bridge fuel:

> One of the reasons why is natural gas—if extracted safely, it's the bridge fuel that can power our economy with less of the carbon pollution that causes climate change. Businesses plan to invest almost $100 billion in new factories that use natural gas. I'll cut red tape to help states get those factories built, and this Congress can help by putting people to work building fueling stations that shift more cars and trucks from foreign oil to American natural gas. My administration will keep working with the industry to sustain production and job growth while strengthening protection of our air, our water, and our communities. And while we're at it, I'll use my authority to protect more of our pristine federal lands for future generations. (Obama, in Plumer 2014)

Moniz and Obama's aim was to close and better regulate coal-fired power plants, making way for the country's conversion to gas-burning plants and further entrenching the gas industry's infrastructure (Farney 2013). Moniz's first few acts in office accelerated reviews of liquid natural gas facilities in Louisiana with the aim of developing North American exports of natural gas (Efstathiou and Snyder 2013):

> The increase in domestic natural gas production is expected to continue. This May, the Energy Department announced that it has conditionally authorized the second proposed facility—the Freeport LNG Terminal on Quintana Island, Texas—to export domestically produced liquefied natural gas (LNG) to countries that do not have a Free Trade Agreement with the United States. And we will expeditiously work through the remaining applications, reviewing each one on a case-by-case basis to ensure that all approvals are in the public interest. (Moniz 2013: 3)

It was hoped this export market would spur further gas development in the United States and raise natural gas prices from their historic low levels,

levels caused by the sudden U.S. gas glut, itself the result of the intense and rapid expansion of fracking (Silverstein 2013).

Many problems are emerging with the argument that natural gas is the bridge fuel to a low-carbon economy, despite wide-scale media campaigns to promote the idea: "'Here's a question for you,' proclaims Exxon's 2014 TV advertisement with a rhetorical flourish since the company then supplies the answer. 'When electricity is generated with natural gas instead of today's most-used source [read coal], how much are CO_2 emissions reduced? Up to 30%? 45%? 60%? The answer is up to 60% less.'"[27] First, in the 20 years after it is released, methane (in natural gas) lost throughout natural gas production and transport is over 84 times more potent as a cause of global climate change than the CO_2 released when that natural gas is burned (Myhre et al. 2013). Life-cycle analysis of natural gas production shows that the high levels of methane lost in production make natural gas no more "green" than coal (Ingraffea 2011; S. Miller et al. 2013). Studies vary on the degree of methane released during production and transport of natural gas, varying from 1% to 9% of total life-cycle emission (Cathles et al. 2012; Howarth et al. 2012; Pétron et al. 2012; Skone 2012; Tollefson 2013; Brandt et al. 2014). One study argues that to be greener than new coal-fired power plants, fugitive methane emissions during natural gas production need to be lower than 3.2% (Alvarez et al. 2012). Debate now rages over precise emissions levels and how to control them. What is clear is that actually monitoring the degree of emissions across natural gas production requires a systematic real-time monitoring system of all wells and pipelines, a need that is far from realized, as recent studies of methane lost in city pipeline networks show (Caulton et al. 2014; Phillips et al. 2013).

The second problem with the "natural gas is green" argument is that low prices for natural gas, which also led to cuts in prices for oil, encourage higher energy consumption and reduced investment in renewal resources such as wind and solar (Davis and Shearer 2014; McJeon et al. 2014). Using five different energy-economic models, researchers examined the impact of increasing natural gas supplies on greenhouse gas emissions. They found that, without energy policies specifically requiring 50% cuts in greenhouse gas emissions, the increased availability of natural gas would do little to nothing to reduce overall carbon emissions by 2050: "The authors' study is the most robust evidence yet that expanding supplies of natural gas will not help us to avoid climate change and manage the transition to renewable energy sources in the absence of an effective climate policy" (Davis and Shearer 2014: 437).

In the U.S. context, such effective climate policy is as yet unrealized, particularly as politics promoting natural gas and protecting fossil-fuel industries have retarded development of renewable alternatives. Economic models tend to treat the market as a naturally occurring force. In reality, markets are shaped by political and social forces. This is made very clear in the contrast between the developmental timelines of offshore wind energy and unconventional gas extraction over the past 15 years. While natural gas production has boomed since 2000, particularly with benefits of supportive policies in the 2005 Energy Policy Act, development of offshore wind energy has crawled. It is not just that low gas prices discourage investment in renewables; policies supportive of natural gas and political activities connected to fossil-fuel interests have created numerous obstructions to developing renewables. The case of Massachusetts's Cape Wind Project is illustrative.

Aiming to be the first large-scale offshore wind farm in the United States, the Cape Wind Project was proposed, just as fracking was coming into common use, in 2001. After 14 years of legal wrangling, agency reviews, and heated controversy, the project was effectively dead in 2015. The wind farm would have generated 75% of the energy needs renewably for Cape Cod, Massachusetts, an area particularly vulnerable to climate change. Two primary and interrelated obstacles forestalled the development of this important experiment in large-scale renewable energy: (1) unclear and changing regulatory jurisdiction at the state and federal levels over approval and regulation; and (2) concerted campaigning by fossil fuel–invested interests to tie the project up in legal battles over issues such as jurisdiction (Seelye 2013; T. Casey 2014). While the oil and gas lobby with its connections to regulators and regulatory agencies was able to manage legal obstacles to fracking (see chapter 1), offshore wind developers had no similar well-funded and connected lobby (Motta 2015). The project was subjected to drawn-out debates over who had jurisdiction to regulate the project and how the project's environmental impact should be assessed. Initially the Army Corp of Engineers (the Corps) was by default the lead agency. Unfortunately, the Corps lacked the jurisdiction, resources, or experience to conduct the Environment Impact Statement (EIS) required by the EPA. In 2005, the EPA rejected the Corps' draft EIS as "inadequate," in what two former directors of the Massachusetts EPA Office of Environmental Review describe as an "unusual" occurrence (*EV World* 2005; Motta 2015). A review of similar EISs found it was more common for the EPA to ask for further information rather than deem EISs inadequate, a status that required the agency to begin again. "In effect, the EPA said that the Corps lacked legal authority to do what it was legally obligated to do"

(Motta 2015: 127). Overall, four years was lost to an agency that had been set an impossible task.

The 2005 Energy Policy Act seemed on its surface to solve this "regulatory gap" as it moved management of offshore wind development to the Mineral Management Service (MMS) under the Department of the Interior (Motta 2015). MMS seemed a logical choice, given its experience with regulating offshore oil and gas (the division came under severe criticism during the Gulf oil spill for its lack of effective management of offshore oil). Despite having jurisdictional authority, MMS dragged its feet on the Cape Wind Project, deciding the entire EIS needed to be redone. It did not produce a second version of the EIS until 2009. The Union of Concerned Scientists pointed out that

> more than three and a half years have elapsed since the Energy Policy Act of 2005 (EPAct 2005) called for pending offshore wind energy projects, including Cape Wind, to move forward expeditiously—independent from, but concurrent with, the development of an overall framework for renewable energy development on the OCS. Yet progress on these fronts under the Bush Administration was incremental, at best. (Reid, Greene, and Nogee 2009: 2)

Meanwhile the 2005 Energy Policy Act created new exemptions from the EIS process for unconventional gas extraction. The act revised the EIS process of oil and gas extraction to place the burden on the public to prove that a full environmental review is required if operations have a surface disturbance of less than five acres, drilling is occurring on an already existing pad (within five years), or if drilling is occurring in an already developed field (approved within five years). Each of these provisions allows for a dramatic expansion of gas or oil extraction without a review of the environmental health impacts (Kosnik 2007; EWG 2009).

In contrast, the Cape Wind Project did not receive approval until April 19, 2010 (the explosion precipitating the Gulf spill occurred the next day), after Senator Ken Salazar took over as the Secretary of the Department of the Interior with a mandate to

> change the direction of the Department and to restore the confidence of the American people in the ability of their government to carry out the functions under [his] charge. That confidence had been seriously eroded by well-publicized examples of misconduct and ethical lapses. This kind of fundamental change does not come easily, and many of the changes we have made have raised the ire of industry. In the past

16 months our efforts at reform have been characterized as impediments and roadblocks to the development of our domestic oil and gas resources. But we have not, and we will not, back down on our reform agenda. We have been making major changes at MMS, and we will continue to do so. (Salazar 2010)

Unfortunately, the Cape Wind Project still faced another major hurdle: consistent resistance of vested fossil-fuel interests to the project, resulting in legal battles, until in 2015, the two major electricity suppliers, who had agreed to purchase 50% of the project's energy, broke their contracts due to unending delays (Crimaldi 2015). The Cape Wind Project received strong push-back based on aesthetic concerns of wealthy homeowners, particularly from Nantucket Island. A leading figure in this was William Koch, a billionaire whose fortune was made in fossil fuels and petrochemical production. A vocal opponent to the project, he donated $5 million to the Alliance to Protect Nantucket Sound and, as the alliance's chairman, he waged a 12-year battle to prevent the project from moving forward (Seelye 2013). Koch's resistance to Cape Wind is one small part of this industrialist family dynasty's resistance to climate change. Koch-affiliated foundations donated about $26.3 million to climate-change countermovement organizations between 2003 and 2010 (Brulle 2013; D. Fischer 2013). Cape Wind's drawn-out and expensive death by attrition is now a cautionary tale for other investors in U.S. offshore wind development.

Viewed in the context of Cape Wind and the recent science on carbon emissions, the boom in natural gas extraction is delaying and undermining a transition to renewable energy, and instead reasserting the status quo: an energy-intensive, petrochemical-based consumer economy marked by increasing social, economic, and environmental disparities.

INDUSTRY SELF-REGULATION AND PATINAS OF TRANSPARENCY

With no community–regulatory–academic networks to act as a buffer, industry is able to extend its own self-regulation. Industry has opted for self-disclosure of chemicals used in fracking rather than mandatory disclosure. This is an extremely effective tactic in that it appears to provide transparency but serves to sever two linked issues for which TEDX and OGAP advocated since 2006: disclosure of chemicals used in natural gas extraction *and* an effective monitoring system.

Many mapping and database projects have emerged in the wake of WellWatch. Tellingly, only FracFocus, an industry-funded web database, has thus far been enshrined by state agencies as the place for companies to meet new regulatory requirements for disclosing the content of frack fluids (Konschnik, Holden, and Shasteen 2013; Soraghan 2013a, 2013b). FracFocus's information is voluntarily provided and not checked for accuracy and completeness. David Hess studied the efficacy of voluntary industry disclosures and found that self-disclosure frequently becomes little more than a PR activity when there are no requirements for accuracy, completeness, and responsiveness. It does not substantively change industry behavior (Hess 1999, 2001, 2007b, 2008).[28] A 2013 Harvard Law School study evaluates FracFocus as useless for comparative research or even for ensuring that regulatory requirements are met. Database items cannot be compared laterally, thus preventing searches across records. There are uneven requirements for disclosure of chemicals and uneven reporting by companies that sometimes report that chemical compositions are trade secrets and at other times disclose the composition of the very same product (thereby invalidating protections via trade secret laws). Also there is no time horizon dictating when disclosures have to be filed. Importantly, there is also no process for contextualizing this data with other information about gas extraction, such as complaints filed or grassroots experiences (Konschnik et al. 2013). In 2015 FracFocus did make some steps to increase the site's transparency, such as offering data in a machine-readable rather than PDF format; however, problems of uneven reporting, trade secret protections, and limited search options remain (Hurdle 2015). A collaboration between academics and communities would have produced a very different disclosure process: one that centered on linking complete and timely disclosure to the reduction of environmental and human health threats by enabling preemptive discussions about whether hazardous chemicals were necessary, on a case-by-case basis, depending on the locus of the oil and gas operation.

Self-regulation has already proved to be a failure in this industry. In 2003, the three largest fracking companies promised the EPA that they would stop using diesel in specific situations when fracturing is carried out directly within underground sources of drinking water.[29] In addition, the Energy Policy Act in 2005 required a permit for using diesel in fracking.[30] Congressional investigations revealed that companies both continued to frack with diesel and failed to seek required permits to use diesel[31] between 2005 and 2009:

> The congressional investigation finds that oil and gas service companies have injected over 32 million gallons of diesel fuel or hydraulic fracturing fluids containing diesel fuel in wells in 19 states between 2005 and 2009. In addition, the investigation finds that no oil and gas service companies have sought—and no state and federal regulators have issued—permits for diesel fuel use in hydraulic fracturing, which appears to be a violation of the Safe Drinking Water Act. (Energy and Commerce Committee Democrats 2011)

Nonetheless, more information-rich monitoring of large-scale industries has effectively changed industry behavior in some cases. The TRI is considered the EPA's most effective regulatory action, particularly because some polluters voluntarily reduced emissions when they saw aggregated emissions data (Fung and O'Rourke 2000; Karkkainen 2001; EPA 2003). Natural gas wells are currently exempt from the TRI.

Inequity, Climate Change, and Endocrine Disruption

As societies and as a species, humans are collectively experiencing the social, economic, and environmental outcomes of the past two centuries of the fossil fuel–driven industrial revolutions: first, the use of fossil fuels for motive energy; second, the distillation of synthetic chemicals from fossil fuels. Modernity's mass production and consumption generated three interrelated problems: climate change, chemical toxicants, and widening social/economic inequality. Each of these is connected in part to our fossil-fuel petrochemical infrastructure and the ineffectiveness of our present regulatory and research systems to identify and address structural inequalities and hazards generated and sustained by this system.

In the pursuit of the control of nature, modern humans developed new infrastructures and feats of engineering and the sciences, both physical and social, and we are currently struggling to understand the consequences of our industrialism. The vast machines of meteorology are producing statistics such as the following: nine of the 10 warmest years on record have occurred since 2000; in 2012 Arctic ice sank to its lowest level on record; and carbon dioxide levels in the atmosphere reached their highest level in 650,000 years (Edwards 2010).[32] With global industry we have produced global chemical and biological changes. Pollution from plastic, a star of the second industrial revolution, while described and used as disposable, is ubiquitous the world over: "five trillion pieces of plastic, collectively weighing nearly 269,000 tonnes, are floating in the world's oceans" (Eriksen et al. 2014; Milman 2014). Micro-plastics are now systemic parts of the marine food chain (M. Cole et al. 2011). Many plastic components are hazardous because they can act like hormones and disrupt fetal development (Liboiron 2012). Plastics are just a drop in the synthetic chemical soup now circulating between our bodies and our environments: more than 200 different chemicals are now found in the blood, fats, and tissues of the postindustrial human (EWG 2005; NHANES 2009). The consequences of these exposures are just beginning to be understood but they include altering the expression of DNA, inducing cancer, and changing reproductive, neurological, and immunological development.[33]

Technological progress since the 1950s promised to produce a thriving middle class, a four-day work week, to close the gap between developing and developed nations and support democratic development. However, rather than a flourishing modernist dream, research has documented worsening social and economic inequalities (Fry and Kochhar 2014; Piketty 2014). Wealth is increasingly concentrated in the upper echelons of U.S. society. As of 2012, the top 1% of income earners in the United States accounted for 42% of total wealth, while the bottom 90% of the American population collectively owned only 22.8% (Saez and Zucman 2014: 22). The bottom 45% have zero or negative wealth due to debt (Saez and Zucman 2014: 24). Moreover, developed democratic societies seem indelibly tied to the persistence of authoritarian regimes rich in fossil-fuel resources (Peluso and Watts 2001; Ferguson 2005, 2006; Watts 2005; Watts and Kashi 2008; Mitchell 2011).

Rather than addressing these concerns separately, we must develop new forms of research, technology development, and governance that are capable of connecting them and systematically reimagining the fossil-fuel petrochemical infrastructure. The conclusion of this book examines how the connections between these two industries have been obfuscated and how we may transform our research strategies to remake our environmental and social relationships.

CONCLUSION

Corporate Bodies and Chemical Bonds

A Call for Industrial Embodiment

Those things which we see with our eyes and understand by means of our senses are more clearly to be demonstrated than if learned by means of reason.
—GEORGIUS AGRICOLA (1546)

The science of metallurgy and mining did not emerge from theoretical structural chemistry, but rather from the physical experience of mining and forging metals. Through studying mines, miners, and metalsmiths across early Renaissance Europe, Georgius Agricola was the first to systematize the embodied knowledge developed by miners and metallurgists across centuries. Similarly, the practice of fracking emerged from drillers' physical experiences of attempting to increase well productivity in the real world, not in laboratories (Montgomery and Smith 2010).

In these cases, embodied experiences gained credence and became the grounds upon which to build new practices. Conversely, the embodied experiences of landowners and workers whose lives and landscapes are transformed by the shale gas industry are readily dismissed, silenced, and forgotten. Their stories of loss—of health, property, livelihood, social stability, even control over their own bodies—are deemed anecdotal, not scientific. Their exposures and illnesses are impossible to prove through traditional scientific methods.

Mike Edelstein (2003) described the physical and mental trauma experienced by communities and individuals impacted by toxic contamination as "inherently uncertain."

However, to residents experiencing the hazards of this expanding industry, harm is not "inherently uncertain." They *know* it through their *senses* and from their *experiences*, and they *learn* it through sharing their *histories*. As readily as fracking developed through tireless field testing to optimize the process for the broadest profit margin (Begos 2012; Shellenberger et al. 2012), Rick Roles, as one who suffers the costs and reaps little reward from fracking, was well placed to recognize its risks. This book began with Rick's fantasy of giving a taste of his embodied experience to an industry PR person by "breaking his jaw and letting him drink it." The "it" in this case was condensate, an industry waste product that the spokesperson purports was nothing but "guar gum and water." Rick implicitly asserted that the body is the proving ground where PR is tested against material risk and consequence.

This conclusion calls for an interdisciplinary effort toward industrial embodiment to record, aggregate, and remediate the material risks posed by this industrial activity so that the lives of those it harms can be improved. To build a networked understanding of our shared *corporate bodies*—the industrial systems that enable, sustain, and also endanger contemporary societies—environmental health research must begin with the embodied experiences of individuals. Building on Haraway's "Cyborg Manifesto" (1991), which urges feminists to embrace that our bodies are now both organic and technical systems, this conclusion asks us to build methods of accounting for the organic, bodily worlds built through industrial systems.

Bodies are not solely biologically determined. They are socially shaped. Definitions of what counts as a body depend on historical and cultural factors. Indeed research on embodiment finds that postmodern bodies are "spacing out" to become something people actively and self-consciously construct.[1] Based on this literature and inspired by the call for a "new materialism"[2] in academic work focused on connecting social critique to material change, this conclusion pushes further to ask, does expanding our notions of bodies to include industrial systems better account for industry's environmental and social impacts? Forming open-source informatics networks and supporting grassroots place-based research can help us collectively recognize ourselves as part of corporate bodies whose future is collectively determined by the shared physical and social relationships formed through networks of production and consumption.

The chemistry of endocrine disruptors connects us materially to the twentieth century's petrochemical fossil fuel–driven industries, forming "chemical bonds" that reveal our entangled biological connections to our environments. We cannot extract ourselves from these relationships by walling off our bodies into some separate space. Instead we are constituted by our interconnections (Haraway 1997: 37). To responsibly build our industries we need methods to trace, connect, socially reflect on, and shape these bonds that draw us each into new relationships of risks and possibility. How are we to account for chemicals that change the male-female sex ratio in a community or impact the rate of preterm birth (D. Scott 2010)? Presently science and technology assist in obfuscating these infrastructural connections, disabling effective collective understanding and oversight of these vital industries (Murphy 2006; Hess 2007a, 2015; Frickel et al. 2010; Frickel and Edwards 2014).

Merging critically applied and politically engaged environmental health and social science research can help draw this industry back into a space of civic contemplation (Faber 2008, 2011). "Civic" describes the collective body of common interest. A civic science rather than "simply serv[ing] the state" enables communities to "question the state of things" and investigate issues of collective urgency to shape them to preserve the commons (Fortun and Fortun 2005: 50). In pursuit of civic science, I argue for an informatics of industrial embodiment that enables us to have more participatory and recursive relationships with infrastructures that shape our everyday lives and embodied experiences (Kelty 2008). Our lack of awareness of our energy and chemical infrastructure is a product of strategies enacted by energy extractors and producers that *disembody* key infrastructures, making them into a heterogeneous entity that cannot be collected into a socially, historically, and geographically coherent whole that could be accountable for its actions (Mitchell 2011).

This book traced these processes of disembodiment. The conclusion summarizes and systematizes them so that future social and environmental health scientists can apply this model to other extractive industries and infrastructures. Methods of embodied industrial research are developed that counter corporate techniques that disconnect the oil and gas industry from places, people, and its own history. How can we better recognize, support, and teach these methods to form a corporeal civic science that enables us to democratically account for the bodily consequences of industry?

Why Embodiment? What Is a Body and Why Should We Start There?

Rick's fantasy of forcing a PR person to physically share the risks he experiences relates three important properties of bodies.[3] First, they are materially connected to the world around them. Second, they shape and are shaped by the environments through which they travel, socially and physically. Bodies have histories that can be traced. They occupy space in measurable ways using a variety of metrics. However, bodily boundaries are porous, contested, extensive, and socially constructed. For example, a mother's body extends to her child during its development. An amputee's prosthesis becomes part of her body, both neurologically and socially. Third, bodies are therefore loci for exerting and impressing social and physical power (Foucault 1979).

Modern democratic states inscribed in law and socially normalized the idea of the bounded, rights-bearing individual body (Foucault 1979; J. Scott 1998). But scientific and technological development, history, and our own experiences show that bodies' boundaries are more flexible and porous than this (Landecker 2010; D. Scott 2010). As constellations of cells, organs, microbes, hormones, water, nitrogen, and oxygen that circulate blood, respire air, taste, and smell, bodies are sustained by and help sustain their "environments." The distinction between where a body ends and where an environment begins is a matter of scale and contested boundaries (Coole and Frost 2010). Research on microbiomes shows our bodies are constituted and co-inhabited by millions of other unseen organisms. These organisms travel within, on, between, and around us, extending beyond the corporeal to inhabit our environments (Betts 2011; Lax et al. 2014). Families carry similar microbial populations, comprising and sustaining a shared ecosystem (Dominguez-Bello et al. 2010). Freezing eggs for in vitro fertilization and cultivating stem cell lines from tissue samples change the boundaries of where our bodies begin and end (Landecker 2010). What (historically privileged) vision tells us is an exterior world, other senses and processes—touch, taste, smell, respiration, digestion, fever and pain, and technological intervention—remind us is a continuum with our bodies (Merleau-Ponty 2008). Bodies are complex selection systems adapted to make use of available materials useful for sustaining or reproducing our systems. They are vulnerable to materials to which we are unaccustomed and ideally expel those experienced and known as harmful.

Bodies are "legacy systems" organized in part by processes that historically enabled self-perpetuation and/or reproduction. Theoretically and on a molecular level, every living organism can trace back its ancestry. Bodies literally have history, though we require technoscience or other cultural mechanisms to divine it. For example, immune systems respond to germs, leaving antibodies behind that are both a hallmark of prior exposure and a system for recognizing and organizing future responses to a pathogen. Babies inherit from their mothers the bacterial cultures that inhabit digestive systems and help process food (Dominguez-Bello et al. 2010). Viruses leave their genetic imprints on human DNA, sending their encrypted genetic materials across bodily and species boundaries.

Bodies are also "feedback systems"; they process, transform, and expel materials that shape and are shaped by the worlds through which they move. As birds deposit plant seeds with their excrement, our bodies consciously and unconsciously select and support the life cycles and evolution of other lifeforms (Pollan 2001). Bodies' feedback mechanisms are vital aspects of both constituting and preserving the "self." For instance, hormonal mechanisms remind us we are sated with food. By technically manipulating food biochemistry with sucrose substitutes, like high-fructose corn syrup, that do not trigger satiation and artificially flavoring food in the 20th century, we began playing a whole range of sensorial tricks on our bodies that dislocate feedback loops evolved to help human bodies eat nutritious foods in limited quantities (Stanhope et al. 2009; Yang 2010; Page et al. 2013). These ongoing "self-experiments" baffle our feedback systems. They shift our biological legacies in open-ended, unpredictable ways and have so far produced not only low-cost foods but also rising obesity and type 2 diabetes, along with swathes of mono-cropped, pesticide-intensive, government-subsidized fields of patented corn and concentrated economic power in the hands of industrial agriculture and processed food companies (Johnson and Gower 2008; Lynch and Bjerga 2013; Ng et al. 2014).[4]

Thus bodies are complex environmental records that document how the world moves through them, and they leave traces, whether it be chemicals lodged in fats or virus DNA. Our bodies hold clues that can help us understand changes in our environmental human health (Betts 2011). There are no better sensors or indicators for our lived world than our bodies. It is time for environmental and social science to take this seriously and to apply what we can learn from it to counter manufactured hazards.

In addition to environmental science focused on the physical characteristics of body environments like the microbiome and exposome (the totality of human environmental exposures after conception) (Wild 2005; Rappaport and Smith 2010), social science needs to *listen carefully* to people's self-reported experiences and connect them to larger social, political, and economic changes.

Focusing on the body does not imply ignoring or eliding a person's reported experiences. People's unique biological and social histories "attune" them to different experiences. Chemically sensitive individuals are more apt to register and recognize their bodies' responses to chemical exposure (Shapiro 2015). Moreover, people can become attuned to recognizing bodily experiences, as an asthmatic may recognize triggers in an environment (Csordas 1994; Choy 2012a). Such sentinels' reports are frequently dismissed as the ramblings of hysterics, the irrational convictions of "nervous nellies," or baseless preoccupations of people who are peculiarly sensitive or otherwise pathological (Edelstein 2003; Allen 2004; Murphy 2006). We need to develop collective literacy around experiential phenomena so we have the facility to recognize, articulate, archive, and act on these embodied experiences (Shapiro 2015). It is particularly important to generate scientific and regulatory responses when embodied changes are collectively experienced and reported (P. Brown et al. 2004; P. Brown 2007). Such information should produce new collective questions for shared research.

Rick dreams of fighting this disregard in an embodied fashion, through physical violence to assert the reality of his experience. His anger was directed at the PR person, the corporately designated and paid social and political broker between the industry, the state, and the community. The spokesperson's assertion that frack effluents were "just guar gum and water" exemplifies the official industry dismissal of Rick's embodied experience. Rick knew that the PR person was wrong because (1) he lived surrounded by oil and gas extraction; (2) he was sick; (3) he was talking with his sick friends and neighbors; (4) his livestock was sick; and (5) he had occupational exposures while "roughnecking" (working) on rigs. Rick is an experiential witness to the adverse physio-social impacts of unconventional gas extraction. Despite over 15 years of communities expressing environmental health impacts associated with fracking, there is still *no formal mechanism for aggregating and collectively studying their experiences*. To expand on the body metaphor, there is no effective feedback mechanism to limit the activities of this corporate body. Rick could not occupy the privileged and privately funded position of spokesperson. His body and experience were excluded and dismissed.

Some may feel disquieted by Rick's fantasy of assaulting the PR representative with contaminated water. However, the interpersonal violence he imagined was physically direct and preventable. The violence done to Rick's body, on the other hand, resulted from different, persistent, long-term processes (Nixon 2011). No one forced him to drink condensate. Had that happened, he could have sued the culprit, established a point of exposure, and been treated and received restitution. Because we have organized tort law around the idea of rights-bearing individuals, we have relatively well-developed, though frequently ineffective, systems for accounting for person-to-person violence. We lack similar systems to manage collective, systemic violence perpetuated directly or indirectly in the pursuit of profit (D. Scott 2010; Nixon 2011).

The damage Rick and others endure came from undocumented changes in the air they respired through their skin and lungs and the undisclosed pollution of the water they drank. Who bears responsibility? The drilling company? The fracking company? The leaseholder? The negligent workers who spilled frack fluid on his property? The chemist who designed frack fluid for its effectiveness in releasing gas from shale beds and who neglected to consider its impacts on surface life? Congress for exempting fracking from the Safe Drinking Water Act?

The agents of Rick's damage are impossible to pin down (Beck 1992). And this brings us back to the third relevant property of modern bodies: in modernity we developed a host of tools for tracking and tracing, disciplining, and socializing bodies (Foucault 1979). Unfortunately, Rick's opponent has no body. Unlike Rick's named, addressed, historically continuous and geographically transcribed body, the "industry" (comprising corporate "persons") is amorphous, multiple, heterogeneous, and ahistorical. We lack feedback technologies to track the legacy systems that perpetuate this industry and make the physical hazards it creates tangible, open to regulation, and hazardous to its continuity in the same way that its hazards are tangible and hazardous to Rick's survival (Fortun 2012).

However, the oil and gas industry does have a physical infrastructure. It is a circulatory system through which the world passes and is reconstituted. Its operations occur in particular places and are performed by particular people. It creates links between them and involves processes like extraction, circulation, refining, and emission, and it functions, in some senses, like a physical body. Under U.S. law, corporations have legal standing as people with constitutional rights to freedom of speech among other things (Hartmann 2002; Nace 2003; Wiist 2011). However, beyond the provision of rudimentary taxation systems, the "industry" lacks any mechanisms of recognizing either itself as a body or its physical, daily impacts on the organisms who inhabit it and social and physical environments it interacts with. Without this embodied information, it is extremely hard to call the industry to account for its assaults on others' bodily livelihoods. How and why is this very material, very physical, eminently traceable industry that is infrastructural to contemporary society so hard to connect to both physical place and history?[5] How has it been socially disembodied?

Informatic Techniques of Industrial Disembodiment

This book empirically evidences two overarching approaches to disembodiment used in shale gas production: informatic and material. Together, they make the oil and gas industry hard to aggregate and to trace historically and geographically. Informatic techniques of disembodiment manage how the industry can be known and how it knows the world. These practices determine its relationship to history and enable it to shape the future. This book describes three techniques of informatic disembodiment seen in the oil and gas industry's development of the shale gas boom: agnotology, synoptic data-gathering systems, and PR practices.

AGNOTOLOGY

Agnotology is defined as the process of structured forgetting. It shapes what can be known about something by effecting what is or is not archived (Fig 2005; Proctor and Schiebinger 2008). Fracking's exemption from the Safe Drinking Water Act (EPA 2005) is the primary example of this practice, discussed in chapter 1. The Halliburton loophole made it impossible for the Environmental Protection Agency (EPA) to track and monitor fracking. In seeking to preserve themselves by protecting intellectual property, oil-field services companies created a regime of imperceptibility in which the tools and infrastructures of environmental protection, in this case the EPA, actually function to make hazards less perceptible (Murphy 2006; Frickel and Edwards 2014).

Nondisclosure agreements such as Laura Amos's are another tool of industrial agnotology. Laura is unable to definitively prove a link between her cancer and 2-Butoxyethanol (2-BE) exposure. She is forced to sign away her constitutional right to free speech in order to afford to leave her contaminated space. This systematically excludes her and others' stories from the archive. It wipes from history not the physical threat of harm (the contamination is still there) but the ability to socially mark her and her families' bodies and property, as well as the ability to connect her story with others in the future. It erases history and shapes the industry's future knowability and thereby accountability.

As also discussed in Laura's case, a third agnotological technique that helps to erase the industry's history is the use of subcontractual arrangements and corporate acquisitions (Appel 2011). Subcontracts mean that the histories of multiple operators at one well site are separated, making it very challenging to tell a combined coherent narrative about the place where they converge. Corporate acquisitions also create holes in history where the new well owner can deny knowledge of and/or refuse responsibility for what happened at the site before the change in well ownership (Fortun 2001). These agnotological informatic practices make it impossible to trace what happened at one well, or to one family, and to connect that information with others' experiences.

SYNOPTIC AND EXTRACTIVE INFORMATIC SYSTEMS

Synoptic approaches to information gathering that extract, centralize, and internalize knowledge and decision making, as discussed in chapters 5 and 8, are another important informatic technique of disembodiment. Seismic and satellite imaging exemplify this process in three ways. They (1) export data from regions of extraction to centers of calculation (Latour 1990); (2) reconfigure workers' bodies as mechanical and replaceable; and (3) are developed by engineers who have little contact with regions of extraction or meaningful understanding of the political and social consequences of the tools they develop.

Modern states and corporations come together in their shared interest in synoptic forms of data collection and representation (Chandler 1977; J. Scott 1998). To represent a space synoptically is to organize data so that one can view, from a single point (e.g., a map or a schematic diagram of a building) a landscape or system as a whole. This is the point of view of the architect or the manager who operates with a simplified and abstracted representation of the world that highlights only the features germane to their interests. A seismic map is controlled for surface noise and focused on the subsurface. It does not represent the surface because to do so would muddle the picture. Modern states develop such synoptic tools to manage their populations and territories as described in chapter 8 (Campbell-Kelly and Aspray 1996; J. Scott 1998; Bowker 2005; Manovich 2002; Foucault 2007). For both states and corporations, the mapping practices entailed within satellite imagery and seismic data collection enable calculation and decision making at a distance. The satellite is synoptic when working seamlessly because it offers a bird's-eye perspective over vast territories without having to transgress physical boundaries. Satellites collect abundant information while those who are imaged are kept unaware. They are excellent tools for spying on neighboring states or people without visibly violating territorial boundaries. The development of satellite monitoring during the Cold War is not incidental. Satellites enable those who receive and therefore control the processed information to plan the future of faraway places without involving their residents (Farman 2010).

Seismic data similarly exported by oilfield services companies become treasure maps through which oil majors make plans for how they will configure social, political, and physical landscapes in order to access their subterranean resources. They help oil and gas prospectors decide where they will seek leases, from whom, what configuration of wells will maximize extraction, what legal and political barriers exist, how they can organize supplies to this region to support development, and which service companies will act as subcontractors.

These synoptic systems effectively disembody the industry from locations of extraction because inhabitants there have little awareness and even less control over the development of these plans that will literally shape their futures. Chapter 7 exemplifies this process with Ohio's eradication of home rule, or local control of zoning for mineral resource development. Political wrangling happened years in advance of well development without the awareness of suburban inhabitants whose lives and livelihoods would be impacted by the activities of extractive industries. Synoptic data-gathering tools are crucial to the industry's ability to shape experiences of time in regions of extraction, particularly the experience of acceleration as plans made at a distance rapidly unfold.

Chapter 5 describes how the technical design of synoptic information systems modularizes workers. These capital-intensive cyborg systems are designed to fit the bodies of workers, and workers are entrained to use these systems via simulations. A worker's individuality is irrelevant. The mind and history of John or Jane Citizen is not vital to operations. Instead, a choreographed role is developed into which any similarly trained worker can be slotted. Workers are organized to perform their tasks without having to understand the other parts of the whole system in order to keep it operating. This is agnotological too, because the workers become replaceable, like machine parts. Each worker-component can be calculated, trained, managed, and replaced. The historical particularities of individual worker bodies are removed.

Those designing these technologies and analyzing their resulting data experience similar detachment, as investigated in chapters 5 and 10. Engineers are rarely trained to understand the social-economic implications of their work, and they do not see how technical systems enable different structures of power (Noble 1977).

PUBLIC RELATIONS PRACTICES

PR practices are not agnotological (enabling forgetting and shaping what's forgotten). Rather, PR shapes what is known, who knows what, and how they know about it (A. Brandt 2007; Oreskes and Conway 2010). This book describes a number of interoperating PR tactics: (1) the third man technique, which uses a trusted expert to expound a company message because that person's seeming independence and social capital promote consumer trust (A. Brandt 2007); (2) industry-related think tanks and research institutes that strategically produce research to create doubt about research on the industry's health and social hazards (Oreskes and Conway 2010); and (3) industry lobbying strategies and groups such as the American Petroleum Institute and American Chemistry Council that employ economic capital to promote favorable policies (Rampton and Stauber 2000).

Unlike the neoliberal academy, large-scale corporations take social science very seriously. While espousing ideals of individual choice and free markets filled with rational consumers, corporations fund advertising, research, and high-profile public figures in order to influence public perception of their products and to create and maintain markets (A. Brandt 2007). These practices are structured to tie the industries' interests to those of consumers and to create self-identification (A. Brandt 2007; Dumit 2012). For example, *Drill Here, Drill Now* ties shale gas development to national security. The "Mature Region with Youthful Potential" campaign tied shale gas development to economic progress and job growth in the Marcellus Shale region (IOGCC 2005). Corporations also create histories that tie them to states' economic and social histories (Fortun 2001; Rajak 2014).

Such campaigns are supported by industry spokespeople who are different in form and content than the replicable bodies of the mechanical workers inside their technical systems. Spokespeople are valued for who they are, for their individual social capital, life histories, and their ability to stand as experts. Experts, such as Secretary of Energy Ernest Moniz, frequently have revolving-door histories with academia and industry, allowing them to act simultaneously as disinterested experts and industry consultants, board members, and shareholders (Bender 2013; Connor and Galbraith 2013). Moniz helps build the industry's market by accelerating approvals for liquid natural gas terminals that expand U.S. export markets and by advocating for switching to gas-fired power plants (Silverstein 2013). Such figures help create industrial niches that perpetuate the industry.

Experts connected to industry, particularly those who serve on panels to evaluate the health and safety of industry-related practices, are of immense value. As occurred with the 2004 EPA review, scientific and technical experts lend credibility to research and enable agnotological outcomes favorable to the industry's continued development.

PR disembodies industry, making its interest and influence very hard to trace. It manufactures doubt and builds myths and promises without having to produce deliverables. Its campaigns focus citizen-consumers emotively on their self-interest. They promote domestic oil and gas development on the grounds that it will help protect the homeland from terrorist violence, an agnotological move that ignores the historical collusion between western governments, the oil industry, and autocratic regime formation in regions of energy extraction (Bowker 1987; Peluso and Watts 2001; Hodges 2004; Ferguson 2005, 2006; Santiago 2006; Mitchell 2011). The U.S. role in creating Saudi Arabia's repressive monarchical autocracy in order to secure access to oil is well documented. It was described as America's "oil colony" in 1947 by the U.S. Secretary of State and was the homeland to Osama Bin Laden (Vitalis 2007: 31; Mitchell 2011; Nixon 2011). For all the rhetoric of oil producing national security, reliance on this system has driven humanity into a self-perpetuating state of permanent war in the name of security (Bennis et al. 2007; Mitchell 2011; Houen 2014). While natural gas promises to end our dependence on foreign (Middle Eastern, Venezuelan, Nigerian, etc.) energy supplies, we ignore that the same companies, the same tools, the same practices shape the extraction of this resource in the United States. In other words, these same companies (particularly oilfield services companies) are working on and profiting from domestic and international frontiers. Indeed the violence and instability of foreign energy resources are used to justify the sacrifice of gas-patch communities in the United States. The structural and transnational forms of violence inherent in this lucrative industry are made harder to see because the public expects laws and monitoring systems to protect them, along with elected and appointed regulators ready to defend the public.[6] How can we ensure that these legacy industrial systems do not perpetuate the economic, social, and political asymmetries that have historically appeared wherever this industry operates (Ferguson 2005, 2006; Santiago 2006; Mitchell 2011)?

Material Practices of Disembodiment

Informatic tools of disembodiment are supported by material practices that physically separate areas of extraction from centers of calculation and consumption as well as sites of extraction from surrounding social and physical landscapes. These material practices manage the oil and gas industry's relationship to place (Bowker 1987, 1994; Ferguson 2005, 2006; Appel 2011).

ENCLAVING

As described in chapters 4, 6, and 7, the practice of enclaving in oil and gas extraction was developed to extract oil efficiently for export while only minimally interacting with troublesome surface inhabitants (people). Enclaves build their own roads, run their own electricity, and operate behind "no trespassing" signs on a 24-hour basis. They disembody by reshaping landscapes through physical, legal, and social barriers. Enclaves try to operate in separate space-times. Life on the inside is organized around rapidly developing wells and drilling. The twin pressures of efficiently using high-cost machinery and the technical processes of well drilling, which must continually recycle drilling muds to keep the well walls stable, accelerate these procedures. Once established, a well operates 24/7 (Bowker 1994; Santiago 2006; Appel 2011). When not in the rush of technical development or production, or when gas prices are too low to make development cost-effective, the well pad is dormant, land-locked in idleness. Alternating acceleration and idleness creates temporal as well as physical disjunctures with other land-use patterns (Tsing 2004).

MOBILE, MODULARIZED LABOR

While the well pad itself is designed to remain for the term of the lease, the rest of drilling's physical infrastructure is mobile and modular (Appel 2011). Man camps house workers who sleep in bunks; they have no physical space for themselves and move in a cycle between spaces. The camps themselves are designed as mobile homes that can move from place to place so workers can be in the same space anywhere in the world. Using mobile migratory labor obscures the role of place and shared embodied experience in shaping workers' histories and bodies. Laborers generally work two weeks on, two weeks off, commuting to distant parts of the country, making it hard for them to connect with the extraction region's inhabitants or with each other. This makes it very difficult for workers to form a shared understanding of their conditions. Local workers are brought in to do low-level work, including driving trucks and stocking and delivering materials. These positions do not enable them to envisage the whole system. Furthermore, migratory workers tend to be missed by epidemiological studies because of their temporary residency. Social science research has repeatedly illustrated the role of workers' wives, partners, and families in documenting environmental health problems (Brown and Mikkelson 1997; Allen 2003). Housewives, partners, and mothers in communities are well positioned to recognize emerging health problems by sharing stories. They realize when problems like birth defects or rashes are common. By building an industry around migratory workers, the formation of place-based community connections is thwarted. This makes popular epidemiology, one of the primary ways in which toxic contamination is effectively recognized, much more challenging (Brown and Mikkelson 1997). Place-based environmental health problems may well occur, but they are more difficult to recognize because place-based shared experiences and community are severed.

LEASING

This modular mobility characterizes the whole organization of fracking down to the very land-use contract structure of leasing. The term "leasing minerals" connotes a fundamentally inaccurate picture. In most circumstances when an object is leased, it is used and given back. Minerals are not leased. They are extracted and removed. The land is not returned in the same or even close to the same condition as it was before the lease. Furthermore, the long-term ramifications of that extraction that continue to exist in land and water are not borne by the leasing company, because it does not own the physical space. The practice of "remediating" surface acreage is the final piece of this historical erasure, creating a surface mask suggesting that land can be returned close to its former condition. As landowner stories repeatedly show (chapter 6), even surface remediation is rarely done correctly. Many states have literally forgotten where wells are because of poor requirements for well documentation and monitoring (Pennsylvania Department of Environmental Protection 2000). The EPA's research showed that these abandoned orphan wells create routes in which fracking chemicals and buried hazards can resurface (Horwitt 2011).

The instrumental design, development, and use of chemical tools also obscure industry's connection to place. Biocides, acidizers, and 2-BE are all used in fracking to form and stabilize connections between the surface and subsurface. Biocides clear surface water of bacteria that might block the well, acidizers clean inside pipes, and 2-BE breaks up thick fracking fluid so gas can come to the surface. However, each of these chemicals also has material consequences for surface life. They can be poisons. Focusing on them as instruments in a particular task ignores their consequences for life. These risks are manifold and go far beyond the initial point of use. First, they include the risks of manufacturing such chemicals. In 2011 two companies manufacturing fracking chemicals exploded in Texas and Louisiana, respectively (Stengle 2011; Burdeau and McConnaughey 2011). Second, housing large volumes of such chemicals in urban areas have resulted in explosions, acid clouds, and evacuations (Frolik 2006; Hrin 2012). Third, bulk-transporting concentrated chemicals occurs along poorly maintained roads and railways,[7] and, fourth, there are continuing impacts, such as bio-accumulation, magnification, and recombination, once chemicals are used and in circulation (EPA Region 5 2014). Storage of wastes in open-air pits releases breakdown products into the air. Inadequate or faulty infrastructures cause fracking and other waste fluids to be released into waterways and soils (Horwitt 2011; Beans 2013; Jacobson 2015).

MATERIAL MIMICRY

Mimicry is the final material method of disembodiment examined in this book. Many chemicals came into common circulation as mimics (Wylie 2011b). Synthetic dyes replaced plant-based dyes and smuggled in totally different production methods (J. Beer 1959; Travis 1993). Plastics also mimicked available products. Consumers misrecognize such mimics as "just like" glass or plant-based dyes, but entirely new production processes are instituted (Fenichell 1996). Any product claiming to mimic an existing tool must be approached with great caution and its production processes carefully compared and evaluated (Taussig 1993). Many people leasing early in the Marcellus play believed that the wells they discussed with landmen would be just like existing shallow oil and gas wells (Grow, Schneyer, and Driver 2012). This misrecognition led them to sign leases quickly without considering the industrial infrastructure associated with the production of shale gas versus conventional gas.

Taken together, these material and informatic techniques of disembodiment disrupt our collective ability to correctly perceive this industry's history and its relationship to physical places and people—our shared corporate bodies. They actively prevent the industry's embodiment, where embodiment is establishing traceable connections to physical places and people to create historical and geographic coherence so that the industry's future development can be shaped through disciplinary measures.

Informatics of Industrial Embodiment

These disembodiment strategies are legacy systems developed by the oil and gas industry to perpetuate its immense profitability by externalizing costs, limiting profit distribution, and rendering itself infrastructural (Mitchell 2011). The shale gas boom in the United States provides new opportunities to embody this industry, to make it democratically accountable, by creating feedback systems. I argue that this can be achieved through developing informatics of embodiment that trace and tie the industry back to place, consciously connecting the humans who inhabit this system and methodically gathering narrative data for decentralized analysis from multiple points of view. These techniques would enhance our understanding of this system and track its impacts on the physical and social well-being of people and places.

While this may sound like far-fetched idealism, it is important to recall that historically humans have not had such dislocated, disembodied relationships to their technical systems; embodied relationships to key infrastructures were vital to creating representative democracy as illustrated in chapter 9's brief history of coal and oil development. Gas and oil's structural invisibility actively impedes the grassroots formation of political power derived from influence over vital infrastructures. Communities along pipelines are unaware of the hazards beneath their feet. City pipeline infrastructures are so old that they leak massive quantities of methane (Phillips et al. 2013). Workers who become ill are not able to recognize that their illnesses might be work-related. Even if they do have Rick's insights, workers often lack the political and social infrastructure to investigate and settle their claims. This problem scales up to whole communities where cities are outbid for rights to their watersheds, leaving them powerless to protect their natural resources and intervene in shale gas extraction (Buchanan 2006; Lofholm 2006; Miller 2006; Spaulding

2006a). An embodied awareness of this infrastructure is urgently needed. Creating media to connect these communities and their experiences could bring energy production back into civic contemplation by enabling community involvement in gas development. The section that follows describes how tracing these relationships using informatics of industrial embodiment effectively develops feedback systems for corporate bodies.

TECHNIQUES OF INDUSTRIAL EMBODIMENT

Becoming a Beagle—Tracking and Collecting Surprises. Metaphors shape how we organize knowledge and perceive the world (Lakoff and Johnson 2003). As discussed in chapter 2, many scientific discoveries are framed around vision; geniuses have great insights or they reveal hidden truths. The scientific fixation on vision is bound up with ideas of objectivity that separate the observer from the observed (Haraway 1991; Daston and Galison 2007). This dichotomy is logically problematic. It perpetuates the illusion of being able to establish boundaries between two positions that are impossible to systematically maintain. Nor does this separation accurately account for the material processes through which new forms of knowledge or new material arrangements are made (Deleuze and Guattari 1987).

The social theorists Gilles Deleuze and Félix Guattari describe an alternative approach to both science and engineering that begins with physical experience (1987). Physical knowing does not generate a separation between the observer and observed. Rather, the two inhabit the same space, and one must identify the connections or "relationships of becoming" between things that at a glance appear separate. Deleuze and Guattari illustrate this co-location by contrasting two different ways of defining territory. The first way of knowing is built around synoptic tools like maps. A land is discovered, explored, and mapped. The map travels to the metropole where a state marks its newfound property by a boundary line drawn on the map. This line is then inscribed socially in the physical landscape. They describe this as "royal science," the process of separating form from content. Conversely, a second way of marking a territory is typified by dropping seeds into a river, waiting a while, then following the river to see wherever the seeds have planted and calling that one's territory. This second way engages with the environment by following how changes (seeds) travel through a system. It acknowledges that one can come to understand how a terrain operates only through its interconnections. While the first method represents a space and attempts to impose boundaries on it, the second interacts with an environment by inducing change in the system and following it to understand how the system is composed. The second method is required to understand how a new element, seeds, or chemicals, in the case of oil and gas extraction, interact with and transform the systems through which they travel.

As argued in chapter 2, Theo Colborn implemented the latter method. I playfully called this beagling. She stopped trying to organize matter by the philosophy that toxic chemicals=cancer and instead started following and gathering up problems that appeared to be scattered, like accidents. Taken together these accidents told a story about a shared exposure to a flow of chemicals through a food chain and a shared vulnerability through similar biological systems and ecological positions. Colborn's beagling, or following the scent, took place from within the landscape rather than from above it. Her interdisciplinary aggregation of cases made the transition to a mode of research that followed relationships of becoming. Following and gathering these previously unimagined relationships formed by endocrine disrupting chemicals across the terrain of the Great Lakes mapped a complex interconnected system that had been systematically occluded by academic disciplines that separated information.

This is the first method of embodied research described in this book. It is a way of knowing that is less about standing back to see a terrain and fit it into a known pattern than following surprises and putting them together in a new configuration. Following Deleuze and Guattari, I imagine this as physically similar to the experience of building the first arch from pieces of stone. The physical skill needed to construct unlike pieces of rock together to stabilize a new flow of force across them into an arch requires integrated senses. As Agricola describes them, connections are felt and then seen. The finished arch can then be analyzed, studied, and diagrammed, and stones can be perfectly cut, so that arches can be readily and easily made. However, making this new form possible required a different process of knowing and making that follows connections to bring new possible arrangements of matter into being. Exploratory and embodied science made endocrine disruption visible. This is the kind of science best suited to perceiving the system-transforming dynamics of shale gas production.

BUILDING ON RELATIONSHIPS OF BECOMING

While mimicry is a method of disembodiment for those duped by what is mimicked, it can also be a powerful method of embodiment by revealing previously obscured relationships. The anthropologist Michael Taussig describes mimicry as the "nature culture uses to make second nature" (1993: xiii). This encapsulates the idea that mimicry, which at first glance seems to be a method for copying (reproducing the same), is actually a powerful force for social and material change. The phenomenon of endocrine disruption rests on mimicry, on the signal produced by a synthetic chemical being similar enough to the natural one to be misrecognized as an endogenous chemical signal, potentially changing the biological fate of the developing fetus. I refer to these errant chemical signals as chemical bonds. Like the seeds traveling through terrain in the earlier example, these chemicals create new relationships and connections across places. They bond people and organisms that are exposed into new relationships across and in a shared system based on shared vulnerability. This has been described as "chemical or toxic trespass" based on the logic that these chemicals transgress boundaries (Doyle 2004), but they actually follow and describe connections across perceived boundaries. For instance, estrogen and estrogen receptors evolutionarily make up the most ancient hormonal system (Thornton 2001). Hence, although mammals, fish, and birds may have diverged evolutionarily, they share biochemically similar estrogen-response pathways and hence a shared susceptibility to estrogen mimics.

This discovery of biochemical relationships can be both positive and negative. In our present system, where these bonds are created and systematically rendered imperceptible by techniques of disembodiment, they are indeed a form of bondage that torque a person's life off biological and social course (Bowker and Star 2000). However, these shared risks and susceptibilities can also describe new potential forms of social organization (Beck 1992). Such relationships of becoming might be leveraged to undermine this industry's insistence on modularization, individuation, and structured forgetting.

COLLECTIVE COMMUNICATION OF SITUATED KNOWLEDGE

To create a shared recognition of the phenomenon of endocrine disruption, Colborn brought together researchers connected to each individual piece of the puzzle to share data at the Wingspread conference. Organizing these researchers and separating them from their normal disciplinary habits to listen to each other's work created a social space in which they could, for the first time, recognize interrelationships between their data. With their findings put together, they could articulate a new, coherent story that no one individual had ever articulated. The formation of the Wingspread Group and their collective statement literally embodied endocrine disruption. It created a heterogeneous group of professionals who now saw themselves as interrelated and researching part of the same problem. Their collective knowledge came not from being distanced observers but from their specific connection to a particular field of knowledge, and a newfound recognition of its connection to another specific or situated knowledge (Haraway 1988; Harding 2004). Networking these situated knowledges created its own disruptions: a pack of researchers calling for change in how disciplines are organized, how data are structured, and how toxicology is studied and applied.

HEIRSHIP—A FOCUS ON THE FUTURE THROUGH A CONNECTION TO THE PAST

Unlike disembodied and purportedly value-neutral objective science, embodied research is oriented to explore how research questions and methods are culturally based and politically shaped. Starting from the premise that we coconstruct our environments and communities through our scientific, social, and technical interventions, HEIRship regards humanity as the inheritors of our technical choices and evaluates our activities based on how they influence the next generation. HEIRship prioritizes the rights of the next generation, from an ecological perspective, one that sees humans not as masters of their environments or separate from them, but rather as recursively, organically bound to them through a family tree. It values rapid responses to emerging hazards, particularly those that may affect fetal development. HEIRship differs from laboratory-centric "basic" research because it investigates the consequences of human engineering on humans—in the field biologically, socially, and ecologically.

While informatics is normally counterposed to embodiment, with data being extracted from physical objects to be manipulated in computation (Haraway 1991), The Endocrine Disruption Exchange (TEDX) use of HEIRship (as described in chapter 3) attempts to capture how databases can enable embodiment by creating historical consistency and comparison across fields. HEIRship is fundamentally an archival practice that operates by following (beagling) and collecting surprises, then arranging them by their connections (the database structure does not predate the data) in order to recognize relationships of becoming. Lorraine Daston has described how databases are sites for research because they enable new lateral patterns to emerge across archived materials (2012). HEIRship aims not to necessarily answer existing research questions but to generate new questions.

As a proactive form of research, HEIRship connects multiple kinds of actors to shared questions simultaneously so that they can begin investigating them in their different arenas (Fortun 2004). Rick Roles could take the TEDX database to his neighbors, Josh Fox could use it to begin his film narration, Susan Nagel could begin studying endocrine disruptors in water sources, and the EPA could begin examining fracking under Superfund designation.

HEIRship describes both the scientific and socially produced data gaps caused by regulatory exemptions, such as the problems with material safety data sheets and the use of trade-secret laws that made studying shale gas's chemical hazards impossible. Fusing these two forms of data analysis, TEDX enabled both scientific and political partners of the organization to begin working on the different attributes of this problem. The database of potential health effects of chemicals used in gas extraction became therefore a platform for "sound advocacy" (see chapter 3), a map through which to begin contesting, questioning, and researching the safety of unconventional extraction from many perspectives simultaneously.

Royal science and technology development are tuned to state and corporate interests in tracking and instrumentally mapping citizen-consumers and resources. As discussed in chapter 8 the development of computer databases and maps is bound up with the state in developing tools to count, tax, and map citizens (Campbell-Kelly and Aspray 1996; Manovich 2002; Bowker 2005). With the emergence of the Internet, decentralized or participatory maps and databases became possible and can be used to contest formal maps (Farman 2010). Presently, however, the informatics of the state are being channeled into protecting the state, for example, by aiming to sort terrorists from nonterrorists. The National Security Agency's PRISM program, as an example, illustrates the huge scale of state-based informatics (Gellman and Poitras 2013; Greenwald and MacAskill 2013). Similarly, the use of the web's metadata properties, including the traceability of online presence, enables corporate counting, classification, and individuation of consumers (Levy 2010; Vaidhyanathan 2011; Madrigal 2012).

By focusing purely on economic measures of corporate activity and conceptualizing them as chiefly economic rather than social, scientific, and environmental actors, we relinquish the ability to evaluate their activities from the perspective of civil society, or to understand the vast industrial systems that employ people and shape environments worldwide. From a social science perspective, HEIRship would create informatics systems that embody corporate actors rather than citizens, recording coherent histories for corporations, connecting them to places and people, and improving our ability to hold them socially and politically accountable. ExtrAct's shift to informatics was inspired by Colborn's use of databases, described in chapters 3 and 4. ExtrAct began from embodiment by focusing on collecting and networking communities' experiences and knowledge.

WITNESSING AND DEVELOPING EXPERIMENTAL ENVIRONMENTAL SCIENCE

Importantly, both Colborn's database and ExtrAct (see chapter 6) began their research with the lived experiences of people who are encountering industry in their everyday lives. In environmental science as it is currently configured, such "findings" are frequently dismissed as anecdotal. This word is an absurdity when examined in the context of both experimental research and legal paradigms. Representative democracies built legal systems around the idea that humans can credibly witness and report on events. Testimony is not disregarded because it is anecdotal. Indeed, experimental science is also built around the concept of witnessing, as research in science and technology studies shows. The experimental program is built around the idea that humans can gather to watch an experiment, observe the phenomenon, report upon it, and come to an agreement about what they have seen (Shapin and Schaffer 1985). Why do we accept such witnessing in the lab and disregard it in the field? In part because we fail to recognize that the field is a space of experiment.

One central problem in building an environmental science focused on connecting industrial hazards to environmental and human health issues is that we currently develop knowledge about industrial harms in laboratories where chemical behaviors are modeled. Laboratory results in model organisms or testing regimes are used to establish regulatory limits, without a thorough feedback system documenting the actual behavior of those chemicals in the environment to determine whether or not the initial model is accurate or protective of human health (Murphy 2006; Frickel and Edwards 2014). An environmental health science built around gathering and connecting embodied experiences would be a first step toward systematically monitoring the large-scale, industry-based environmental experiments to which we are currently subject. Epidemiology attempts this kind of process through cohort studies, but because it is not organized by exposure route, that is, connecting industry workers and communities living near fracking, or networking communities as active developers of research programs, it misses large bodies of potential information. Instead, epidemiologists are left to work with abstract public health statistics and insufficient data about exposure. And communities are left to experience their problems in isolation.

Embodied experiences should be gathered and networked to understand whether they have a systematic basis. Specifically, informatics networks should be generated to interconnect workers, thereby counteracting frayed community knowledge produced by migratory labor. Similarly with work in experimental sciences, any surprises should become the departure point of further research (Rheinberger 1997).

PARTICIPATORY AND RECURSIVE DATABASES AND MAPS

Gas and oil activities need to be reconnected to both physical places and histories via the development of maps and databases that enable communities to begin taking an active role in describing and shaping the infrastructures that produce their social and physical realities. Properly designed platforms like WellWatch could connect individuals with similar problems to help them form communities. These could then connect those individuals to experts who would offer advice and make visible systemwide problems. These databases should be public, open-source, and collectively maintained through academic-community collaborations. Taken together such systems could articulate corporate bodies by connecting diverse people whose lives and livelihoods are shaped by these structures in order for them to (1) understand their structural interdependence and (2) enable them to project possible alternative arrangements (DiSalvo 2009). If industry contamination damages property, then that damage should be socially marked and people should be properly recompensed, enabled to move and connected with a community of similarly displaced persons who can together watch-dog the industry and advise others who may be forced into similar circumstances in the future. Our system presently displaces these costs onto impacted landowners, encouraging them to mimic industry practices, that is, to hide damage done to their property so they might be able to resell it and escape (Wylie and Albright 2014). This divisive thinking perpetuates, obfuscates, and devalues injury rather than analyzing it, accounting for it, and remembering it.

It is this tangible record, a valuation of embodied witnessed impacts, that WellWatch tried to gather. It is the networked and diverse impacts of all the different fracking chemicals that TEDX's database attempted to capture and make legible. We have informatics tools available to predict and recognize these chemical bonds and to gather and collectively visualize the grassroots hands-on experiences of workers and residents. Organized through online databases, we can build participatory and recursive relationships to infrastructures like the energy and chemical industries (Kelty 2008). As occurred with the earlier history of coal, we can develop a shared infrastructural awareness and work toward reshaping that infrastructure to reduce environmental health hazards and increase social and political accountability.

GRASSROOTS SCIENCE

To more fully engage citizen-consumers' agency in creating industrial systems, people should be empowered to make knowledge about this system through grassroots research (Wylie et al. 2014). Citizens can do much more than simply report on their experiences. They can gather data systematically, through structured observation or by using scientific tools as basic as phone cameras (Corburn 2005; Ramirez 2014; Wylie et al. 2014). We need to empower ourselves to be the researchers and engineers of our own conditions by designing tools for grassroots research (Brown and Mikkelson 1997; Allen 2003; Ottinger 2010). We already see these trends in health and environmental controversies, where communities organize their own research, fundraising, and even laboratories (Allen 2003; Corburn 2005; P. Brown et al. 2006; Ottinger and Cohen 2011). Organizations like Public Lab and TEDX, along with communities, are becoming proactive in filling in the gaps in our understanding of industrial systems through local air- and water-quality monitoring efforts (Colborn et al. 2014; Macey et al. 2014). Making the most of this ability requires creating systems for data validation that make use of digital metadata to time-stamp and GPS findings, store raw data, create chains of custody, and prove tool calibration, similar to methods used in meteorological or ornithological citizen science (Fritz et al. 2009; Edwards 2010; Marris 2010; Timmer 2010; Khatib et al. 2011). Humans can expand their social-technical sensorium to track and collectively study industrial systems just as we have "natural" ones. Rather than exporting data, the products of this research would be owned by communities and licensed for use in public interest research (Wylie et al. 2014). Aggregated results from such systems should be available to companies, but these systems should not be owned or operated by vested industry interests for whom obscuring, minimizing, or erasing damages is beneficial.

To enable inhabitants and laborers in regions of oil and gas production and consumers of oil, gas, and related synthetic chemicals to scientifically study and democratically account for the embodied impacts of these industrial systems, I argue that we need to develop an informatics of industrial embodiment: (1) to develop and sustain open-source, decentralized informatics systems that network communities in extraction zones; (2) to begin research with individuals who have everyday experiences of living and working amid extraction industries and empower them to study and share their experiences; and (3) to actively connect these communities with academics, regulatory agencies, journalists, and lawyers in order to form responsive research and monitoring networks. Such networks should be protected as sites of research rather than legally constrained as websites, and they should be funded publicly as part of the public interest in protecting environments and health and civil liberties.

Clearly, there are barriers to these sweeping changes within contemporary governance structures, including secrecy, efforts to prevent coordination between agencies, and the influence of large-scale industries within government agencies, universities, and among regulators. However, it is important to note that citizens do not have to ask permission to express subjective experiences or to do research on their private property or on public property. Nor are there barriers preventing open-source research tool and software development in the public interest. Similarly, academics are free to work with communities and many agencies have mandates for public engagement. Public libraries, nonprofit organizations, and science museums could make excellent partners for further developing grassroots environmental health research.

Movements are aggregations of small-scale acts that evoke change from the ground up. The goal of the projects and methods described here is to create spaces that facilitate such embodied recognition and collaboration to transform shared physical and social energy infrastructures. If civil society is to monitor and thoroughly account for an industry as vast, distributed, and politically, technologically, and economically influential as the oil and gas industry, then the academy, bureaucracies, and advocacy organizations must transform themselves to rise to this challenge. Participatory and recursive databasing and mapping offer novel opportunities for such transformations. They enable deeper civic engagement in embodied knowledge development and industrial monitoring. The question remains: Can we make the academic, social, and political investments to realize these possibilities?

Agricola, the 16th-century "father of metallurgy" stressed the importance of demonstration and experience in coming to know a substance. He argued that though "many persons hold the opinion that the metal industries are fortuitous and that the occupation is one of sordid toil, and altogether a kind of business requiring not so much skill as labour," the metal industries are actually grounded in both physical skill and natural philosophy stemming from a physical awareness and understanding of surface and subterranean landscapes (1556: 1). He listed seven arts and sciences of which the miner must be knowledgeable: natural philosophy, medicine, astronomy, surveying, mathematics, architecture, and law. Remarkably, medicine is listed as the second most important skill that the miner should possess so that he might be able "to look after his diggers and other workmen that they do not meet with those diseases to which they are more liable than workmen in other occupations, or if they do meet with them, that he himself may be able to heal them or may see that the doctors do so" (1556: 3–4). This prescient insight and prescription is sadly unfilled today (Esswein et al. 2013, 2014).

NOTES

As WellWatch and Landman Report Card are no longer online, materials from these websites can be found on Sara Wylie's website: https://sarawylie.com /publications/fractivism-corporate-bodies-and-chemical-bonds/.

Preface

1. Chapter 1 further discusses the chemistry of these fracking fluids. More thorough reviews of the history of fracking can be found in Montgomery and Smith (2010) and Shellenberger and colleagues (2012).

Introduction

1. *Condensate* refers to liquid hydrocarbons that can be produced along with natural gas. Here Rick was using the term more generally to refer to natural gas waste fluids, including frack fluid or condensate that might be mixed with chemicals used in fracking.

2. Throughout this book I mix pseudonyms for those who wished to be de-identified with the real names of those who have consented to be identified in order to protect the identities of those who prefer to remain anonymous. Most of the people and places whose stories are included in the book are widely known and published in other materials.

3. Colborn died on December 14, 2014. Chapter 10 reflects on her passing and the lasting consequences of her work on the health impacts of fracking. Additionally in 2016 I coauthored an article, "Inspiring Collaborations," with Deborah Thomas, Kim Schultz, Susan Nagel, and Chris Kassotis, reflecting on Colborn's cross-disciplinary influence.

4. The phrase *little, yellow, different* is an advertisement tagline for a painkiller that was satirized in the cult movie *Wayne's World.*

5. OGAP was founded in 1999, and it merged with Earthworks in 2005. http://www.earthworksaction.org/about.

6. The material safety data sheet for Soli-Bond was received from Rick Roles via https://web.archive.org/web/20040610174459/http://www.toxic-totnes.org.uk/thechemicals.html.

7. This exemption has been colloquially termed the "Halliburton loophole" as chapter 1 investigates.

8. Colborn (2007), appendix A, page 4.

9. For more on this story from the perspective of her friend and colleague, see Meixsell (2010).

10. It is hard to know exact royalty rates. It depends on contract terms and minerals rights ownership. There is an avid royalty rumor mill about earnings, but it's hard to get to the truth of such claims.

11. Resource extraction is historically tied to underdevelopment. Academic research on underdevelopment begins with Walter Rodney's seminal study *How Europe Underdeveloped Africa* (1972). Ferguson (1994) and Comaroff and Comaroff (2000) further explore the relationship of neoliberal capitalism to African underdevelopment. This anthropology and critical theory tradition counters economics discourse describing Africa and the Middle East as "resource cursed" and unable to fully appropriate the rents of their mineral resources due to in-country political corruption (Auty 1993; Ross 2006; Humphreys, Sachs, and Stiglitz 2007) and advising free-trade and democratic development to improve such nations' economic positions. These arguments do not account for why gas and oil development flourishes in the most politically unstable, brutal, and undemocratic "developing" nations (Ferguson 2005, 2006). An examination of how oil and gas extraction sites are separated from the surrounding social, economic, and physical landscapes through corporate and nationally controlled enclaves reveals how lucrative extractive industries' violent creation and protection of stable environments for extraction succeed in and depend on destabilizing other local economies, social relationships, and land traditions (Bowker 1987; Peluso and Watts 2001; Hodges 2004; Ferguson 2005, 2006; Santiago 2006). Watts (2003, 2005), Ferguson (2005), Watts and Kashi (2008), and Appel (2011) further develop oil and gas extraction's relation to the destabilization of democratic development and the continuing legacies of underdevelopment, particularly in regions of resource extraction.

12. There is insufficient space to thoroughly discuss the history of natural gas or oil development. Indeed it is frequently hard to separate the two industries as they codeveloped. However, there is a small historiography on the gas industry that centers on the relative invisibility of this fuel as opposed to oil in terms of political, popular media, and academic attention (Castañeda 1993, 1999; Castañeda and Smith 1996). It is interesting to note that the term *natural gas* was first developed to refer to gas extracted from the ground rather than gas "produced" through the biological decay of waste and piped into cities in the early industrial era for illumination.

13. Environmental history is a branch of history that investigates how physical environments of water, air, and earth are actively shaped by human habitation (and vice versa). For a review of the literature in environmental history, see Worster (1990, 1992), Cronon (1995), Steinberg (2002), and Stine and Tarr (1998).

14. The suffix *-ship* when appended to a noun forms a new noun that indicates the possession of a state or condition, i.e., fellowship or the possession of a craft or a skill, i.e., scholarship (see the *Oxford English Dictionary*).

15. For instance, the nonprofit organization FracTracker maintains a map of oil and gas wells throughout the United States: https://www.fractracker.org/map/. Harvard

University maintains a similar map of fracked wells: http://worldmap.harvard.edu/maps/FrackMap. The Natural Resources Defense Council maintains a map of pipeline spills: https://www.nrdc.org/onearth/spill-tracker. Earth Justice maintains a map of fracking related accidents and spills across the United States: http://earthjustice.org/features/campaigns/fracking-across-the-united-states?gclid=CjoKEQjw9r7JBRCj37PlltTskaMBEiQAKTzTfDpIJDEKcTo4VqsBT81KQY7BbMaIg55qgtGo_izH6lIaAoKS8P8HAQ.

16. The center has since changed directorship and been renamed the Center for Civic Media. See https://civic.mit.edu/.

1. Securing the Natural Gas Boom

An earlier version of this chapter appeared in Appel, Mason, and Watts (2015).

1. http://www.slb.com/about.aspx.

2. https://web.archive.org/web/20140608141053/http://www.slb.com:80/about/who.aspx.

3. Oilfield services companies early in their development established noncompete principles that they would never directly compete with oil majors by leasing and developing minerals (Bowker 1994).

4. For more on the history of fracking, see Cooke et al. (2010).

5. Micro-seismic measurement sensors are placed in wells adjacent to the well being fracked in order to "listen" to and interpret the direction and spread of fractures as they form (Shellenberger 2011).

6. Popular accounts of the fracking boom tend to stress the role of independent wildcatters over big companies and federal funding (Zuckerman 2013). However, more detailed historiographic work shows that federal funding was vital for the development of fracking and horizontal drilling (Shellenberger et al. 2012). Additionally, though fracking enabled small companies to participate in the U.S. boom, the oilfield services companies involved such as Halliburton and Schlumberger are the world's largest and oldest.

7. Earthworks, "Hydraulic Fracturing—What It Is," https://web.archive.org/web/20171013102856/https://www.earthworksaction.org/issues/detail/hydraulic_fracturing_101#.Wf-w9RNSzEZ; Earthworks, "Acidizing," http://www.earthworksaction.org/issues/detail/acidizing#.U4I_3C8inBM.

8. Research on chemicals used in fracking was part of Northeastern University's Hydraulic Fracking Research Cluster. An undergraduate chemistry student, Bakkar Hassan, researched the use of acids, surfactants, corrosion inhibitors, and oxidizers as part of this group.

9. Here intellectual property was protected, as opposed to the physical property of pipelines and wells investigated in Watts's work in Nigeria (Watts 2005). However, in the United States, developing scientific research favorable to the industry achieved a similar result to the use of physical force in Nigeria and the continued development of fossil-fuel resources.

10. MIT Executive Education Program, "Schlumberger Building a Dialogue for Innovation," https://web.archive.org/web/20100609151900/http://mitsloan.mit.edu/execed/pdf/schlumbergercasestudy.pdf.

11. MIT News 1999.

12. MIT News 2007.

13. See the MIT Energy Initiative "About" page from 2010: https://web.archive.org/web/20100316171648/http://web.mit.edu/mitei/about/index.html.

14. https://web.archive.org/web/20100412113555/http://web.mit.edu:80/mitei/about/members.html.

15. https://web.archive.org/web/20100412113555/http://web.mit.edu:80/mitei/about/members.html.

16. See https://web.archive.org/web/20120511125855/http://web.mit.edu/mitei/research/spotlights/mit-math.html.

17. This first chapter discussed the 2010 draft report released by MIT. Chapter 10 reflects its updated report, released in 2011. I focus on the 2010 version because of its impact in framing the emerging debate about fracking (MITEI 2010; MIT News 2010; MIT Press Release 2010).

18. See MIT News 2010. "MIT releases major report: The Future of Natural Gas Study finds significant potential to displace coal, reducing greenhouse gas emissions": https://web.archive.org/web/20130922134515/http://web.mit.edu/press/2010/natural-gas.html.

19. MIT News 2010.

20. Oreskes and Conway (2010) discuss how circulating white papers discounting climate change influenced policymakers but were dismissed by scientists.

21. *Regime of imperceptibility* is a term drawn from Michele Murphy's (2006) discussion of the Sick Building Syndrome, but similar terms and concepts were also developed by Ulrich Beck, whose study of "risk society" argues for seizing the "means of perception" in order to examine "the blank spots of modernization" (1986: 61). The philosopher Giorgio Agamben's *Homer Sacer: Sovereign Power and Bare Life* ([1995] 1998) also develops a similar concept of "zones of indistinction" to describe spaces outside of the law, spaces that are treated as an exception to the rule of law but are created by that very law. The term is usually applied to spaces like Guantanamo Bay. Gas-patch "regimes of imperceptibility" might also be described as "zones of indistinction" where the inhabitants lose their agency to act on or describe their physical and environmental conditions because of laws that hamper information gathering. I employ Murphy's term because it combines Beck's and Agamben's formulations and it applies to both legal and scientific "blank spots" on the map.

22. Personal narratives are frequently found in toxics literature because the life course of the narrator is frequently torqued by toxic exposures. See Steingraber (1998), Antonetta (2002), Davis (2003), Edelstein (2003), and Walley (2009, 2013).

23. The impact of the split estate in this boom is detailed in *Split Estate*, an Emmy Award–wining documentary film by Debra Anderson, released in 2009.

24. Colborn 2007, October 31, appendix A, p. 4.

25. Colborn 2007, October 31, appendix A, p. 4.

26. Rather than being "found," frontiers are actively made. Research in environmental history and anthropology argues that as the identification of a resource highly

valued by distant markets or societies supports the forceful removal or legal dispossession of local inhabitants, it restructures the economic logic of an area around the valued resource and thereby destabilizes and undermines other local ways of life (e.g., the Dutch Disease), and it accelerates the rate of local social and economic change as the competition to rapidly capture a highly valued resource builds (White 1991; Coronil 1997; Black 2000; Watts 2003; Sawyer 2004; Tsing 2004; Santiago 2006; Zalik 2008, 2009).

27. For an excellent economic analysis of how the gas boom affected Laura's region more broadly, leading to increases in crime, traffic accidents, road damage, spiraling rents, housing shortages, and arguably the "Dutch Disease" (in which the overdevelopment of one resource marginalizes other regional economies), see BBC Research and Consulting's regional economic analysis (2008). Colorado, particularly the western slope, is no stranger to boom-and-bust economies. The environmental historian Richard White describes how the U.S. West developed through an internal colonialism process around resource extraction for the eastern and European markets. This undercut western states' sovereignty by leading to the development of federal bureaucracies to manage western lands, resources, and people (White 1991).

28. For more on the tactical use of gag orders by corporate lawyers, see Nader and Smith (1998).

2. Methods for Following Chemicals

1. For an excellent review of the policy and scientific controversies in regulating endocrine disrupters, see the special issue of *Environmental History*, edited by Roberts and Langston, entitled "Toxic Bodies/Toxic Environments: An Interdisciplinary Forum" (2008). Articles by Linda Nash, Barbara Allan, Sarah Vogel, Frederick Rowe Davis, and Arthur Daemmrich discuss the regulatory and scientific issues related to endocrine disruption. For history of science accounts of the field of endocrine disruption, see Krimsky (2000) and Langston (2010). Colborn et al. (1997) provide a general introduction to the study of endocrine disrupters. Vogel's history of science perspective (2008a, 2008b) gives a general review on the development and regulation of BPA.

2. Literature in both environmental history (White 1995; J. Scott 1998) and history of science (Pauly 1987; Oudshoorn 1994) detail how modernity's separation of nature and culture led to scientific management of both natural and social environments in which engineers and scientists sought control over nature by managing and rationally ordering biological and social systems.

3. See http://sciencewatch.com/ana/st/bis/09sepBisSoto/.

4. On the laboratory as a sacred space for knowledge production, see Haraway's analysis of how Oncomouse™, the first patented organism, is a modern Christ figure sacrificed to save humanity (1997). For more on the laboratory as privileged space for knowledge production, see Latour and Woolgar (1979), Shapin and Schaffer (1985), Latour (1987), and Rheinberger (1997).

5. For a thorough review of low-dose effects and nonmonotic dose response in EDCs, see Vandenberg et al. (2012). Particularly, hormonal feedback loops are not the only mechanisms that account for nonmonoticity of EDCs.

6. The experience of personal exposure, described by Brown as the "exposure experience," is a strong theme in sociological and cultural studies of toxics as a moment of personalizing and politicizing toxic exposure as a trespassing of the body's boundary through which one's physical and social sense of place and life trajectory can be reconfigured (Adams et al. 2011). Further reading in this area from a first-person perspective includes Sandra Steingraber's *Living Downstream* (1998) and Susanne Antonetta's *Body Toxic: An Environmental Memoir* (2002). Also see Chris Walley's *Exit Zero* (2013) for an anthropological and personal perspective on this issue in the context of Chicago's deindustrialization.

7. In 2014 Janet Heasman and Chris Wylie jointly received the Society for Developmental Biology Life-Time Achievement Award: https://web.archive.org/web/20150315223839/http://www.sdbonline.org/sites/SDBe-news/Fall2014/WylieHeasman_Lifetime.html.

8. This analysis is inspired by Deleuze and Guattari's concepts of royal and nomad sciences in *A Thousand Plateaus: Capitalism and Schizophrenia* (1987).

9. STS scholars have extensively analyzed the importance of visualization in scientific research (Latour 1990; Lynch and Woolgar 1990; Tufte 1997; Dumit 2004; Daston and Galison 2007). Tied to the epistemological distinction between the observer and the observed, visualizations are key to the production of scientific knowledge because they help create this perceived separation (Burri and Dumit 2007).

10. Science studies scholars have argued that debates about endocrine disruption threaten to reinforce normative sex–gender distinctions, particularly as popular media tend to fixate on feminized males as oddities (Roberts 2007; Scott 2009; Di Chiro 2010). However, rather than repeating popular media's mistake of anthropomorphizing other species' reproductive strategies and experiences, endocrine disruption and scientific research on it complicate normative perceptions of sexual development in humans and other animals (Anway, Leathers, and Skinner 2006; Anway et al. 2006). As I analyze EDCs it seems to further support arguments that sex and gender distinctions are culturally and environmentally constructed by revealing and producing remarkable developmental fluidity between sexes.

11. For more on the cultural and historical construction of hormones as a system of biological regulation and particularly how controlling hormones became a vital strategy for medicine and pest management, see Oudshoorn (1994) and Wylie (2011b).

12. In *A Thousand Plateaus* (1987) Deleuze and Guattari contrast two archetypes of scientific work: royal and nomad sciences. Royal sciences are characterized by applying a template to shape the world according to preestablished laws and norms, while nomad science follows and organizes traces to create new theories or material forms in the world. Nomads operate without an established territory or recognition of formal structures because their work does not fit within established fields.

13. See the Wingspread Conference Center website: http://www.johnsonfdn.org/aboutus/about-johnson-foundation-at-wingspread.

14. A full list of the attendees can be found in Colborn and Clement (1992: 7–8).

15. Social movement research shows that changes in a person's worldview often begin not from learning new facts but rather by changing the questions they pose to begin making sense of a situation (Harding 2000; Dumit 2012).

16. The historian of science Sheldon Krimsky (2005) reviews the growing use of weight of evidence (or strength of evidence) studies in both regulatory and legal evaluations. Colborn's first use of the weight of evidence for endocrine disruption is an example of what Krimsky calls "aggregating diverse evidentiary modalities" in which different kinds of evidence from different research methods converge or triangulate on a common cause, recommendation, or problem. The combination of perspectives enables researchers to articulate a common problem that could not be perceived from one lens alone. Importantly Colborn builds the field of endocrine disruption by helping researchers, through a weight of evidence approach, to realize that they each share part of a common problem.

17. Gordon Research Conference (2010), https://web.archive.org/web/20100418232735/http://www.grc.org/programs.aspx?year=2010&program=envendo; 2011 international conference on endocrine disruption research hosted by the Endocrine Society (U.S.). There have been 9 Copenhagen Workshops on Endocrine Disruption: http://www.reproduction.dk/cow2017/general-information.html.

18. The 1997 summit meeting for the environment leaders from the G8 countries adopted a declaration encouraging international coordination in research efforts around endocrine disruptors: https://web.archive.org/web/20080719100239/http://ec.europa.eu/environment/docum/99706sm.htm.

19. The Chemical Heritage Foundation held a meeting in 2008 with scientists and social scientists called "New Chemical Bodies": https://web.archive.org/web/20160712211125/http://www.chemheritage.org/community/store/white-papers/studies-in-sustainability/new-chemical-bodies.aspx.

20. See the iSmithers conference (2011) for polymer industries: https://web.archive.org/web/20110109091027/http://www.ismithers.net/conferences/XED11/endocrine-disruptors-2011.

21. This quote is from the Endocrine Society webpage on the history of EDCs and their scientific statement: https://www.endocrine.org/topics/edc/what-edcs-are/history-of-edcs. The Endocrine Society statement is republished in Diamanti-Kandarakis et al. (2009). It is also available online: https://www.endocrine.org/-/media/endosociety/files/advocacy-and-outreach/position-statements/all/endocrine disruptingchemicalspositionstatement.pdf?la=en.

3. HEIRship

1. In 2012 TEDX's office doubled in size and rented a neighboring office as the nonprofit expanded. Though Colborn passed in 2014 the organization continues to have an office in Paonia, Colorado. TEDX's website is: https://endocrinedistruption.org/.

2. TEDX (2005).

3. Broadening fields of corporate and scientific histories, historians have recently contributed many detailed and well-documented studies of how growing and protecting markets for products over the latter half of the twentieth century involved actively constructing scientific research institutes, funding science in the corporate interest, and skillfully manipulating public perceptions through science. See particularly Rampton and Stauber (2000), Rosner and Markowitz (2002), A. Brandt (2007), D. Davis (2007), Oreskes and Conway (2010), and Boudia and Jas (2014).

4. Responding to the lack of institutional support for communities, particularly communities living with environmental injustice and the growth of corporate-funded research, environmental health and justice movements have developed alternative methods of science. Barbara Allen's excellent study of community activism in the Louisiana Chemical Corridor in *Uneasy Alchemy* (2003) describes feminist approaches to scientific research. Breast cancer activists have called for and started "Labs of [Their] Own" such as the Silent Spring Institute (Brown et al. 2006). Alternative lower-cost research tools and methods have also been developed for community use, such as using buckets to take air samples (Ottinger 2010; Ottinger and Cohen 2011).

5. National Institute of Environmental Health Sciences (2010).

6. With the growth of endocrine disruption research and regulation, progress is being made to integrate cradle-to-grave toxicity studies into early stages of product design, particularly through programs like that of the University of Massachusetts Lowell Center for Sustainable Production (https://web.archive.org/web/20090105220408/http://sustainableproduction.org/) and the Toxics Use Reduction Institute (http://www.turi.org/).

7. Merchant Research and Consulting, "World BPA Production Grew by over 372,000 Tonnes in 2012," http://mcgroup.co.uk/news/20131108/bpa-production-grew-372000-tonnes.html.

8. For a full history and analysis of dispute about BPA as an EDC, see *Is It Safe?*, by Vogel (2012).

9. HEIRship shares commonalities with the precautionary principle (SEHN 2000), which argues for evaluating hazards based on their potential to cause harm rather than waiting until harm is irrefutable. It builds on the precautionary principle by actively attending to the role of companies in shaping scientific controversies to create doubt either by shifting the weight of evidence in their favor (as exemplified in the BPA case) or by creating enough doubt that the weight of evidence cannot be settled (Krimsky 2005). TEDX's HEIRship enables researchers to tease out subtle differences that shape research outcomes, such as the strain of rat or kind of feed, between studies in order to actively deconstruct industry science.

10. CWD has since been updated. For the most up-to-date version, see the TEDX website: (https://web.archive.org/web/20090105220408/http://sustainableproduction.org/).

11. See EWG (2005) and the Human Toxome Project, http://www.ewg.org/sites/humantoxome/. This project tested blood, urine, breast milk, and other human tissues for 304 chemicals in collaboration with the nonprofit organization Commonweal. This work builds on the Centers for Disease Control's National Health and Nutrition

Examination Survey (NHANES), which began human biomonitoring in the 1970s and has since scaled up to monitoring 246 chemicals, far fewer than the over 80,000 chemicals presently used commercially in the United States (Altman 2008; NHANES 2009, 2013). The EWG study participants eschewed privacy and shared their results online in order to facilitate public attention to the issue and to enable community engagement in the research effort (Altman 2008).

12. Pesticide Action Network, http://www.panna.org/; Collaborative on Health and the Environment, http://www.healthandenvironment.org/.

13. Investor Environmental Health Network, http://www.iehn.org/home.php.

14. For the Senate bill S.2828-Endocrine Disruption Prevention Act of 2009, see https://www.congress.gov/bill/111th-congress/senate-bill/2828. For the House bill (H.R.4190), see https://www.congress.gov/bill/111th-congress/house-bill/4190.

15. S.1361—112th Congress: Endocrine-Disrupting Chemicals Exposure Elimination Act of 2011. www.GovTrack.us. 2011. November 5, 2017. https://www.govtrack.us/congress/bills/112/s1361.

4. Stimulating Debate

1. ProPublica, "Fracking," http://www.propublica.org/series/buried-secrets-gas-drillings-environmental-threat.

2. Chesapeake Energy (2011); FracFocus Chemical Disclosure Registry, http://fracfocus.org/.

3. https://web.archive.org/web/20110101134727/http://www.halliburton.com/public/projects/pubsdata/Hydraulic_Fracturing/fluids_disclosure.html.

4. Environmental Defense Fund, "Our Mission and Values," http://www.edf.org/page.cfm?tagID=362; Scorecard, http://scorecard.goodguide.com/.

5. Scorecard added a digital informatic component to the already existing research and advocacy networks within the toxics and environmental justice communities. Along with the emergence of citizen science efforts within environmental health and justice research (discussed in chapter 3), this movement built itself through networking impacted communities via hubs such as Lois Gibbs's Citizens' Clearinghouse for Hazardous Waste (later renamed the Center for Health and Environmental Justice), and the American Northeast Toxic Action Center. Such hubs, which serve to support grass-roots community organizations, are discussed in chapter 6.

6. Despite the initial success of Scorecard, the site encountered troubles in creating continued support for the project in the nonprofit sector. The site's original developers, the Environmental Defense Fund, passed control of the site to Green Media Tool Shed in November 2005. The site languished until it was recently acquired by Good Guide, under the leadership of its original developer, Dr. Bill Pease (http://scorecard.goodguide.com/about/txt/history.html). However, as of 2017, much of the information on the website has not been updated since 2002–2003 (http://scorecard.goodguide.com/about/txt/FAQS.html). Chapter 8 examines why the site and others of its kind may have problems achieving sustainability.

7. See chapter 1 for more information on this exemption.

8. Scorecard, "How This Website Identifies Health Hazards of Toxic Chemicals," http://scorecard.goodguide.com/health-effects/gen/hazid.html.

9. http://www.cas.org/about-cas/cas-fact-sheets. Internet Archive: https://web.archive.org/web/20170222022856/http://www.cas.org/about-cas/cas-fact-sheets.

10. http://www.cas.org/about-cas/cas-fact-sheets. Internet Archive: https://web.archive.org/web/20170222022856/http://www.cas.org/about-cas/cas-fact-sheets.

11. U.S. National Library of Medicine Toxicology Data Network, http://chem.sis.nlm.nih.gov/chemidplus/; Centers for Disease Control, "Organic Solvents," http://www.cdc.gov/niosh/topics/organsolv/; PAN Pesticides Database, http://www.pesticideinfo.org/Search_Chemicals.jsp; MSDS Online, http://www.msdsonline.com/; Scorecard, "Chemical Profiles," http://scorecard.goodguide.com/chemical-profiles/; https://endocrinedisruption.org/audio-and-video/fracking-related-health-research-database/frackhealth-database.

12. Environmental justice demands that environmental risks and benefits should be equally distributed and not disproportionately distributed based on racial, ethnic, or class differences. See Bullard, *Dumping in Dixie* (1990), for theorization of the term. Also see Cole and Foster (2001), Adamson, Evans, and Stein (2002), Lerner (2005), and Pellow and Brulle (2005).

13. For excellent research on the limitations of epidemiology in the study of toxics, see Allen (2003, 2004), L. Nash (2004, 2006), and Steven Wing's "The Limits of Epidemiology" in Kroll-Smith (2000).

14. Earthworks, "Community Health Risk Assessment: A Case Study," http://www.earthworksaction.org/publications.cfm?pubID=111; https://web.archive.org/web/20130728133021/http://cogcc.state.co.us:80/Library/PiceanceBasin/WestDivide4_14_04summary.htm.

15. Heiman (2004).

16. This tactic of using indeterminate science to inappropriately dismiss a community's environmental health experiences from industrial exposures is documented in Allen's study *Uneasy Alchemy* (2003).

17. Internet Archive: https://web.archive.org/web/20140826055702/http://endocrinedisruption.org/chemicals-in-natural-gas-operations/pit-chemicals. TEDX's drilling and fracking chemical database is also online: https://endocrinedisruption.org/audio-and-video/chemical-health-effects-spreadsheets.

18. Here Sumi is citing Dennis Webb, "Tempers Flare over Barrett Pit Fires," (Glenwood Springs, CO) *Post Independent*, December 9, 2006.

19. The sociologist and psychologist Mike Edelstein began investigating the psycho-social impacts of living with toxic exposures in *Contaminated Communities* (1989/2003). He describes experiences similar to those of Rick Roles: the stress, loss of quality of life, and daily challenges when one loses trust in the safety of his or her environment.

20. Changes in the questions people ask about the world around them, such as Rick wondering whether his friends' illnesses were related, can indicate the beginning of significant changes in a person's worldview (Harding 2000).

21. For more on biological sentinels, see Keck and Lakoff (2013).

5. Industrial Relations and an Introduction to STS in Practice

1. See M. Fischer (1999a) for a review of the field.

2. Other research that could be considered "in practice" or "constructive" approaches to STS include Ron Eglash's platform for culturally informed mathematical practices (1999; Eglash et al. 2006); Lucy Suchman's ethnographic collaborations on the construction and design of technology (1987; Suchman et al. 1999; Suchman and Bishop 2000); Helen Verran's collaborative work with aboriginal Australian communities to develop learning and knowledge sharing systems (http://www.cdu.edu.au/centres/ik/ikhome.html; http://www.cdu.edu.au/centres/yaci/projects.html; Verran and Christie 2007); Katherine Hayles's (2002) and Chris Kelty's (2008) experiments in stretching the boundaries of written STS work; Mike and Kim Fortun's Asthma Files Project for integrating communities of researchers and patients managing asthma (Fortun et al. 2013); and Bruno Latour's Mapping Controversies platform for STS research to map and share emerging scientific controversies (http://www.demoscience.org/). Carl DiSalvo practices STS in his community-based robotics design projects (DiSalvo 2009; DiSalvo and Lukens 2009). Chris Csikszentmihályi calls this kind of work "applied STS." Others have described the practice as "critical making" (Ratto, Wylie, and Jalbert 2014). Given the focus on my work on STS I opt for the term *STS in practice* to describe a method of critical making specific to the field of science and technology studies rather than design, art, architecture, or engineering.

3. See http://www.media.mit.edu/about/academics. The theorist Nicholas Negroponte (1995) envisioned the Media Lab as a space for realizing digital futures in which information rendered in digital bits would create intelligent, responsive environments and tools. See Stewart Brand's history of the Media Lab (1988).

4. http://www.media.mit.edu/sponsorship/sponsor-list.

5. http://www.cutter.com/meet-our-experts/baudoinc.html.

6. Historical research by David Kaiser shows that as MIT's "history has shown, no one model holds a monopoly on virtue. All patronage involves a delicate balance between opening up new opportunities and mortgaging intellectual autonomy" (2010: 31).

7. There is a long tradition in anthropology of calls to study the central institutions of contemporary society, and this functions as a key charter for anthropological STS. The Late Editions series was a call to do this work of puzzling out contemporary society with interlocutors placed within central institutions; see Marcus (1995b, 1998, 2000a).

8. For more on Bowker's analysis in *Science on the Run* (1994), see chapter 1 of that book, as well as the introduction to my dissertation (Wylie 2011a).

9. STS researchers have analyzed the emergence of cybernetic systems and cyborgs, human-machine assemblages allowing humans to expand their sensorium to learn about environments in which no human body could survive unaided (Haraway 1991; Mindell 2002, 2008; M. Fischer 2009; Helmreich 2009).

10. http://www.britannica.com/EBchecked/topic/336408/Leonardo-da-Vinci/59785/Anatomical-studies-and-drawings.

11. Hannah Appel (2011), an anthropologist of the oil industry, argues that corporations' now ritual focus on "safety culture," protecting the individual workers' bodies, obscures how the political-economic structure of the work (particularly migratory, low-pay positions) makes workers' home lives insecure.

12. Regulations on notifications about seismic imaging activity vary by state and revolve around mineral ownership. For example, surface owners in West Virginia and New York who do not own minerals need not be notified of seismic imaging on their properties (https://web.archive.org/web/20101226233451/http://wvsoro.org/resources/advice/advice9.html; Welsh 2008).

13. On the social construction of technologies, see also Bijker et al. (1987), Bijker (1995), Oudshoorn and Pinch (2003), and Noble (1977).

14. In 1920 the federal government passed the Land Leasing Act, enabling the government to lease land for oil and gas extraction in return for a 5% royalty rate for 20 years (White 1991: 399). This law dramatically changed the role of government in the West by preserving the government as the largest western landowner and ensuring that it would be a continual and active participant in shaping western land use. Approximately $10 billion annually goes to the federal government from oil royalties (Savage 2008). The Department of Interior's Minerals Management Service has faced multiple charges of corruption in the collection and distribution of money from royalty payments, such as the undercharging of industry to the tune of $4.4 million (Savage 2008). A 2008 Inspector General report detailed the Royalty in Kind Department's "culture of ethical failure" characterized by "substance abuse and promiscuity" (Savage 2008). The Bush-Cheney administration further solidified this leasing arrangement in 2003 by instructing the Bureau of Land Management and the Forest Service to remove barriers to leasing, particularly in the Western Rockies (AP 2003), leading to a massive growth in natural gas development on federal lands: "The federal Bureau of Land Management approved nearly 42,000 drilling permits from fiscal year 2001 to fiscal year 2008—almost two-and-a-half times as many as in the previous eight years. Oil and gas drilling became the priority use for these lands" (Kenworthy 2010).

15. There is an extensive academic literature detailing the links between extractive industries and violence from dispossession of land, social upheaval, and militarization to environmental contamination (Appadurai 1990; J. Nash 1993; Coronil 1997; Peluso and Watts 2001; Watts 2003; Sawyer 2004; Tsing 2004; Ferguson 2005, 2006; Ong and Collier 2005; Santiago 2006; Watts and Kashi 2008; Zalik 2008, 2009; Strauss, Rupp, and Love 2013; Appel, Arthur, and Watts 2015).

16. Another MIT professor, Joseph Weizenbaum, followed Agre by arguing against the triumph of what he called "instrumental reason" in *Computer Power and Human Reason* (New York: Penguin, 1984).

17. Many AI researchers had similar realizations about the problems with their own work. See Terry Winograd and Fernando Flores, *Understanding Computers and Cognition: A New Foundation for Design* (Norwood, NJ: Addison-Wesley, 1987).

18. Csikszentmihályi thought of this practice as "applied STS." I avoid the term *applied* because I see both written and material practice as an intervention. The question

for me is whether the intervention seeks to deconstruct and reveal existing power relationships or to offer and develop new possible power relationships. Also, applied work has been historically constructed as "second class" when compared to theoretical or basic research. I don't think there should be a hierarchical distinction between "in practice" and "deconstructive" modes of STS research. They ought to both inform the other recursively.

19. The definition of tactical media is from the Next 5 Minutes (N5M) conference (1999): http://www.tacticalmediafiles.net/n5m3/pages/FAQ/faq001.htm#1.2. These conferences were crucial for organizing and theorizing tactical media. Many artists, particularly collectives, are now associated with tactical media, such as the Yes Men (http://theyesmen.org) and the Critical Art Ensemble (CAE 2001). Other groups such as Pre-emptive Media have critiqued the creep of surveillance into individuals' daily lives. Their projects include one that misuses bar-code scanners and radio-frequency identification (RFID) readers to illustrate how consumers' data are being passively gathered. For more on this field, see CAE (1996, 2001); Garcia and Lovink (1997); Lovink (2002, 2008); G. Meikle (2008); Ratto et al. (2014); and Thompson and Sholette (2004).

20. http://www.nyu.edu/projects/xdesign/feralrobots/.

21. http://www.anninaruest.com/thighmaster/index.html.

22. Jascha Hoffman, "Carbon Penance," http://www.nytimes.com/2008/12/14/magazine/14Ideas-Section2-A-t-002.html.

23. With the support of Professor Susan Silbey and the MIT Chemistry Department, we unsuccessfully applied for two grants for this project. As described in chapter 10, I returned to this concept of developing low-cost, open-source DIY tools for environmental health research through the Public Lab, a nonprofit I subsequently cofounded.

24. http://www.knightfoundation.org/about/.

25. The Knight Foundation is a nonprofit started by the Knight brothers, John S. and James L., to distribute some of the profit from the Knight-Ridder Media company (sold in 2006 to McClatchy, another newspaper company).

26. The historian of science Hanna Rose Shell has written on photography, camouflage, and reconnaissance (2012).

27. The entire ExtrAct team developed our tools and ideas. ExtrAct was composed of the creator of the Awesome Foundation (http://awesomefoundation.org/), Christina Xu; the Oil and Gas Accountability Project (OGAP) research director, Lisa Sumi; OGAP's Public Health and Toxics Campaign Director, Jen Goldman; ExtrAct's programmers, Dan Ring and Matthew Gordon; its web designer, Matt Hockenberry (who also cofounded Sourcemap); as well as an excellent group of undergraduate researchers and interns.

28. We also recognized that just as a text can be misread, any tool can be detoured, hacked, and transformed from its intended purpose. Chapter 7 discusses ExtrAct's experience planning for potential "misuse" of our tools.

6. ExtrAct

1. See http://www.sourcemap.com/.

2. Holmes and Marcus's (2008) and Marcus's (2000a) concept of a "para-site" builds on Michel Serres's (1982) theoretical work.

3. In a good example of para-ethnography Susann Wilkinson and colleagues taught employees in hospitals and the Veteran's Affairs to do their own ethnographies in order to foster different workflows and learning organizations (see Cefkin 2010).

4. de Certeau (1984: 37).

5. Such futures would particularly concern the development of energy systems. Sheila Jasanoff and Sang-Hyun Kim have referred to collective projections for future social and technical systems as "social technical imaginaries" (2009, 2013).

6. Janet Murray's introduction to *The New Media Reader* describes how the rhizome metaphor gave the new field of digital-media design and its "humanist project of shredding culture" "a radical new pattern of meaning, a root system that offered a metaphor of growth and connection rather than rot and disassembly" (Wardrip-Fruin and Monfort 2003: 9). ExtrAct took up the metaphor in this spirit as a provocation to imagine alternative ways to grow connections between gas-patch communities, lateral connections that would form a shared root system. Like rhizomes in poor soils, these root systems would anchor communities to influence and control the flow of mineral resources from beneath their feet and perhaps help redistribute mineral extraction's wealth and risks in a more equitable way.

7. In 2012 Gwen left OGAP to run successfully for local county commissioner: https://web.archive.org/web/20120219012649/https://gwenlachelt.com/about-bio.php.

8. A popular media version of this story is told in the Hollywood film *Promised Land* (2012), directed by Gus Van Sant.

9. Information about TXTMob is available from the Institute of Applied Autonomy website: https://web.archive.org/web/20160416101955/http://www.appliedautonomy.com/txtmob.html.

10. For more on social movement networking through digital media, see Jeff Juris's *Networking Futures* (2008). Juris discusses how anticorporate, antiglobalization movements evolved networked information and media systems, or "informational utopias" such as Indymedia, in order to express their political values such as direct democracy, egalitarianism, and horizontal collaborations. ExtrAct, particularly through WellWatch (see chapters 8 and 9), also experimented with building networks to foster such values. For more on the relationship between digital media and democracy, see Bholer (2008).

11. Hirsch was later subpoenaed by New York City for records of TXTMob messages (Moynihan 2008). The site has also been described by developers of the popular messaging service Twitter as inspiring the development of Twitter. See "How TXTMob Influenced Twitter," on the Next Generation Labor website: http://nextgenerationlabor.wordpress.com/2011/04/20/how-txtmob-influenced-twitter/.

12. *National sacrifice zone* is a phrase with its own history. See Steve Lerner's book *Sacrifice Zones* (2010).

13. OSHA, "Hydrogen Sulfide," https://www.osha.gov/SLTC/hydrogensulfide/standards.html.

14. http://www.gcmonitor.org/. For more on communities' use of bucket brigades, environmental justice, and health research, see Ottinger (2010) and Ottinger and Cohen (2011).

15. Odors from the well have been documented in the local newspaper. See Mayeux (2011).

16. For more on Dee's story, see Lofholm (2007).

7. Landman Report Card

1. See chapter 6.

2. *The New Corporate Activism* by Edward Grefe and Marty Linsky (1995) is an example of industry capturing Saul Alinsky's work through corporate approaches to countering and benefiting from grassroots advocacy.

3. Mock-up or paper versions of design tools are a common practice in participatory design. Through their use, you collectively figure out design issues in a flexible form that lacks the rigidity and finish of a digital representation (see Ehn and Kyng 1991).

4. A "play" is defined as "a set of known or postulated natural gas or oil accumulations sharing similar geologic, geographic and temporal properties, such as source rock, migration pathways, timing, trapping mechanism and hydrocarbon type" (IOGCC 2005: 4).

5. Tom Wilber (2012) provides a compelling and detailed journalistic account of the fracking boom in the Marcellus play, focusing on the contamination of groundwater in Dimock, Pennsylvania. The experiences of Dimock residents parallel those of the landowners whom this book documents in Garfield County, Colorado, and in Pavillion, Wyoming. For contemporary reporting on the Marcellus Shale boom see Applebome (2008).

6. A subterranean theme of this chapter (discussed further in the conclusion) is time, particularly what one's ability to shape another's experience of time, through acceleration and deceleration, reveals about power relationships. Anna Tsing (2004) has argued that the unfolding of a frontier changes the experience of time for people tied to that location: they are subjected to an acceleration in the pace of social and physical change, creating both surprise and a feeling of loss of control. That the ExtrAct team became subject to these dynamics illustrates that we were now not simply observers; rather we were participants in this frontier, accelerating with it and bound to respond to its dynamics.

7. The name is inspired, of course, by OGAP, which advised this group on how to organize.

8. See LRC Appendix: LRC report card 21.

9. Skrtic (2006: 7, figure 1).

10. Ohio General Assembly Archives, 1997–2014, http://archives.legislature.state.oh.us/bills.cfm?ID=125_HB_278.

11. In 2003 and 2004 Niehaus represented Ohio's District 88 in southwestern Ohio, as shown on this map: https://web.archive.org/web/20080522000744/https://www.sos.state.oh.us/SOS/upload/elections/maps/OEhouseDist.pdf. In 2005 he became the senator for the fourteenth district of the Ohio Senate. Both areas are well outside Ohio's gas-producing regions, as shown on the following U.S. Geological Survey map: https://web.archive.org/web/20100203183349/https://www.dnr.state.oh.us/Portals/10/pdf/pg01.pdf.

12. Archive available: http://archive.is/http://www.followthemoney.org/database/StateGlance/candidate.phtml?si=200435&c=399708.

13. Report authors included Bradley J. Field, the Appalachian and Illinois Basin Director from the New York Department of Environmental Conservation, who acted as chair; Rick Bender from the Kentucky Department for Natural Resources; C. Edmon Larrimore, Maryland Department of the Environment; Michael L. Sponsler from the Ohio Department of Natural Resources; David C. Hogeman from the Pennsylvania Department of Environmental Protection; Bob R. Wilson of the Virginia Department of Mines, Minerals and Energy; James Martin from the West Virginia Department of Environmental Protection; Al Clayborne from the Illinois Department of Natural Resources; and Bruce Stevens from the Indiana Department of Natural Resources (IOGCC 2005: 2).

14. See https://web.archive.org/web/20111204064133/http://www.damascuscitizens.org/about.html. They later renamed themselves Damascus Citizens for Sustainability: http://www.damascuscitizensforsustainability.org/.

15. Their optimism proved well grounded because in 2014 New York State, by order of the governor, became the first shale gas–rich state to ban the practice of fracking due to environmental and human health concerns (Lustgarten 2014).

16. http://www.wvsoro.org/.

17. *Russell v. Island Creek Coal Co.*, 389 S.E.2d 194 (1989), Supreme Court of Appeals West Virginia Decision No. 19104, http://law.justia.com/cases/west-virginia/supreme-court/1989/19104-5.html.

18. See https://web.archive.org/web/20130210234517/http://ewg.org/reports/pfcworld; or http://www.ewg.org/research/pfcs-global-contaminants/pfoa-pervasive-pollutant-human-blood-are-other-pfcs.

19. Digital Media Law Project, http://www.citmedialaw.org/.

20. Digital Media Law Project, "What Is a Defamatory Statement," http://www.citmedialaw.org/legal-guide/what-defamatory-statement.

21. Digital Media Law Project, "Defamation," http://www.citmedialaw.org/legal-guide/defamation.

22. Digital Media Law Project, "Cubby v. Compuserve," http://www.citmedialaw.org/threats/cubby-v-compuserve.

23. Digital Media Law Project, "Stratton Oakmont v. Prodigy," http://www.citmedialaw.org/threats/stratton-oakmont-v-prodigy.

24. Digital Media Law Project, "Immunity for Online Publishers under the Communications Decency Act," http://www.citmedialaw.org/legal-guide/immunity-online-publishers-under-communications-decency-act.

25. Digital Media Law Project, "Responding to Strategic Lawsuits against Public Participation (SLAPPs)," http://www.citmedialaw.org/legal-guide/responding-strategic-lawsuits-against-public-participation-slapps.

26. Digital Media Law Project, "Online Activities Not Covered by Section 230," http://www.citmedialaw.org/legal-guide/online-activities-not-covered-section-230.

27. Digital Media Law Project, "Roommates.com—Just How Big a Hole Did the Ninth Circuit Poke in CDA 230?," http://www.citmedialaw.org/blog/2008/roommatescom-just-how-big-hole-did-ninth-circuit-poke-cda-230.

28. Digital Media Law Project, "Online Activities Not Covered." http://www.dmlp.org/legal-guide/online-activities-not-covered-section-230.

29. Digital Media Law Project, "Online Activities Covered by Section 230," http://www.citmedialaw.org/legal-guide/online-activities-covered-section-230.

30. Digital Media Law Project, "Risks Associated with Publication," http://www.citmedialaw.org/legal-guide/risks-associated-publication.

31. There is a growing field of academic work on how counting and algorithms practice politics online in a seemingly apolitical fashion. See D. Beer (2009), Martin and Lynch (2009), and Gillespie (2013).

32. *Consumer Reports* evaluates products through surveys and evaluations: http://www.consumerreports.org/cro/index.htm.

33. Brown and Mikkelson (1997) developed the term *popular epidemiology* to describe how communities experiencing toxic contamination have often organized community research to reveal their shared environmental health problems. In parallel LRC could be described as a platform for "popular sociology," a community-driven study of the habits and culture of leasing and its effects on communities.

34. National Institutes of Health, "General Information on Certificates and the Protections Provided," http://grants.nih.gov/grants/policy/coc/faqs.htm#367.

35. As an ethnographer studying the ExtrAct project, I received a certificate of confidentiality to protect anonymous users of LRC who might be made identifiable by my research.

36. Chapter 9 discusses how ExtrAct tools contribute to community-based participatory research (CBPR). CBPR is an emerging method in health science and sociological research in which communities and researchers work collaboratively to develop research questions and methods that address both pressing academic research questions and advocacy goals (Minkler and Wallerstein 2008). Traditionally CBPR has used community health surveys (Schulz et al. 1998; Clements-Nolle and Bachrach 2008) or environmental monitoring to address communities' environmental and human health issues, such as industrial hog-farming operations (Wing 2002), urban air quality (Morello-Frosch, Pastor, and Sadd 2002; Morello-Frosch et al. 2005), or pesticide exposure (Arcury, Quandt, and McCauley 2000; Quandt, Arcury, and Pell 2001). CBPR researchers have pioneered new media research techniques, such as Photovoice, to enable community engagement (Wang and Burris 1997; Lykes 2001). ExtrAct adds community interconnectivity through digital infrastructures to this field, improving academics' and communities' abilities to collaboratively address and study environmental health threats as the systematic results of shared industrial infrastructures.

37. See Wing (2002) on the challenges faced by researchers whose work threatens socially and economically powerful vested interests. For more on the use of subpoenas to force disclosure of research subjects' identities and the chilling effects on academic research, see P. Fisher (1996), Picou (1996a, 1996b, 1996c), and Traynor (1996).

8. From LRC to WellWatch

1. See the full text of the email from Amy Mall posted on the COG listserv in LRC Appendix.

2. Posted by J Dub on March 18, 2010, within the blog post cited as Fowler (2010).

3. See http://www.landmanreportcard.com/lrc/review/58.

4. See http://www.landmanreportcard.com/lrc/review/53.

5. See http://www.landmanreportcard.com/lrc/review/59.

6. See http://www.landmanreportcard.com/lrc/company/76; http://www.landmanreportcard.com/lrc/review/24.

7. See http://www.landmanreportcard.com/lrc/review/35.

8. See http://www.landmanreportcard.com/lrc/review/45.

9. See http://www.landmanreportcard.com/lrc/review/26.

10. See http://www.landmanreportcard.com/lrc/review/75; http://www.landmanreportcard.com/lrc/review/36.

11. See http://www.landmanreportcard.com/lrc/search?term=Sandy+Freiberg&x=-431&y=-99; http://www.landmanreportcard.com/lrc/search?term=bonnie+foster&x=-431&y=-99.

12. See http://www.landmanreportcard.com/lrc/review/63.

13. *New York Times*, "Drilling Down: Oil and Gas Leases," http://www.nytimes.com/interactive/2011/12/02/us/oil-and-gas-leases.html.

14. *Innovation Trail*, which provides "public media reports on the economy and technology" in upstate New York covered the site twice (Jacobs 2011a, 2011b). *Agroinnovations*, a podcast that covers issues relating to agriculture, interviewed Christina. As of August 2013 the podcast had been downloaded over a thousand times (http://archive.org/details/AgroinnovationsPodcast83TheLandmanReportCard).

15. As discussed in chapter 5, western mapping traditions and the field of geography codeveloped with nation-states and colonialism as tools for orchestrating actions from a distance (Latour 1990; J. Scott 1998; Burnett 2000). Maps are vehicles for dreams that help realize asymmetric power relationships between those able to draw and own maps and those who are mapped (H. Brody 1981; Harley 1988, 1992). Democratized online mapping tools and the emergence of participatory mapping projects such as Google Earth have the potential to contest states' monopoly on maps' meanings and representations (Paglen and Thompson 2006; Thompson 2009; Graham 2010; Paglen 2010).

16. http://www.ushahidi.com/.

17. http://crisismappers.net/.

18. http://www.openstreetmap.org/.

19. https://web.archive.org/web/20100608205043/http://www.boston.com:80/news/local/massachusetts/specials/013009_pothole/.

20. Databases developed, published, and collectively generated by nongovernment, noncorporate groups signify a deep shift in how our collective memory practices are shaped (Bowker 2005). Historically, databases have been vital to the emergence of both states and computers. Databases and computation codeveloped through instruments of bureaucracy like the Census. After their early use for code breaking, computers were speedily taken up for the more mundane aspects of administering nations. The U.S. Census first used a tabulating punch-card machine in 1890 (Truesdell 1965), funded the development of early computers, and received delivery of the first commercial IBM computer in 1951 in order to calculate the Census (Stern 1981; Bashe et al. 1986). Databases are not products of the computer revolution; "if anything the revolution is a product of the drive to database" (Bowker 2005: 109). In the context of distributed information and computation technologies, databases have become a constitutive aspect of daily life. As the Internet culture theorist Lev Manovich noted in 2002: "Database becomes the center of the creative process in the computer age" (227). For theoretical discussions of the social importance of memory devices like databases, see Foucault (1969), Derrida (1998), and Bowker (2005). For more on the relationship between the development of bureaucracies and computers, see Campbell-Kelly and Aspray (1996) and Fuller (2008).

21. The Internet's openness and scalability were neither happenstance nor necessary. It was designed by convention and influenced by the culture of its first major users, academics who sent messages between campuses (Abbate 1999). The original Internet was a peer-to-peer structure, so no user authentication was required when a new computer joined. The machine needed only to be able to use the protocol TCP/IP. Academics who standardized this protocol employed a design ideal called "end to end" that pushed complexity to the system's edges—the devices exchanging information—so that the transit system did the minimum possible work. This transit system is built to receive and send "packets" of a single specific size. Any information—a document, video, or audio file—must be divided up into standardized packets so that they can be managed by a system that cannot differentiate between packet contents but passes them on to the correct "address," called an IP address. Because an IP address is a unique, numerical identifier comprising 32 bits, there is a finite number of IP addresses among which data can be moved. This radically modular structure allows the parts to interoperate without knowledge of their internal state. Internet theorists have argued that its structural openness, scalability, and robustness have made the system astoundingly generative (Zittrain 2006). For more on the history of the Internet, see Krol (1987, 1992), Clark (1988), Campbell-Kelly and Aspray (1996), Leiner et al. (1997), Randall (1997), and Abbate (1999). Wikipedia maintains excellent histories on each aspect of the Internet's development.

22. The World Wide Web (WWW) is probably the most significant software system designed for everyday Internet users. The web was developed in 1990 to use the ability of the Internet to interlink computers so that data could be accessed and read by many different computers, even if they used different software and hardware that would otherwise make it hard to simply transfer files. In 1989, Tim Berners-Lee proposed a system that would allow individuals to call up information from other machines and

interpret it in a shared, machine-readable language. Berners-Lee developed the first web browser, a software program capable of interpreting hypertext, based on its ability to read hypertext markup language (HTML). His browser constructed the hypertext stacks into a human-readable page in real time. This information is not on the user's computer; rather the user's computer uses the Internet to read information from another's computer database or server. Information transfer on the World Wide Web is achieved by another of Berners-Lee's inventions, Hypertext Transfer Protocol, or HTTP. The combination of the Internet's ability to add new computers as long as they have IP addresses and the WWW, in which the browser basically acts as a universal application for calling and interpreting data accessible in networked computers, have generated the web as presently experienced. For histories of the WWW, see Berners-Lee and Fischetti (1999), Cailliau and Gillies (2000), and Herman (2000).

23. http://wikileaks.org/. Scholarship on cyberactivism is growing. See Cleaver (1997), Warf and Grimes (1997), Smith and Smythe (2001), Meikle (2002), Rheingold (2002), Bennett (2003a, 2003b), McCaughey and Ayers (2003), Pickerill (2003), Kahn and Kellner (2004), Shane (2004), and Juris (2005, 2008).

24. Its name a play on GPS, or geographical positioning system, the News Positioning System (NPS) tagged news articles to particular locations on an online map so that a group could collaboratively create a place-based online archive of news articles. Each grassroots group we had met maintained its own paper archive of news. Kari from the Northeast Ohio Gas Accountability Project had carried hers in two huge binders crammed with gas-industry relevant news clippings. We had hoped geo-tagged news articles in NPS could prepopulate WellWatch with data so that the site would be immediately useful for visitors. However, NPS operated until 2012 online. Archives of NPS can be found on the Internet Archive but unfortunately the site's maps are not preserved: https://web.archive.org/web/20111001021604/http://www.newspositioning.com/nps.

It was an effective way of mapping oil and gas news. As of October 1, 2011, 783 articles had been mapped with the site. The ExtrAct team, which included community members, mapped over 543 articles: https://web.archive.org/web/20111001011419/http://www.newspositioning.com/nps/group/1; https://web.archive.org/web/20111001021750/http://www.newspositioning.com/nps/articles?.

These articles were accessible from and mapped on WellWatch. https://web.archive.org/web/20120312132028/http://scrapper.media.mit.edu/wiki/WellWatch.

Articles were organized by location and tags. For example, there were 31 articles on contamination and 92 on natural gas. https://web.archive.org/web/20111001035025/http://www.newspositioning.com/nps/tags.

25. N. Katherine Hayles writes about the differences between human and machine "reading" in *Writing Machines* (2002).

26. ExtrAct benefited from the help of several talented undergraduate research assistants through the MIT co-op program who did the original work of scraping state data.

27. The oil and gas industry's trade association, the American Petroleum Institute, established the API system. Defined as "a unique, permanent, numeric identifier

assigned for identification purposes to a well (hole-in-the-ground) which is drilled for the purpose of finding or producing oil and/or gas or providing related services," the API number was first adopted as the industry standard in 1962 (API 1979). Well locations are important to the gas industry for numerous reasons, including planning further development in drilled areas, reporting, analyzing and predicting national gas yields, and enabling industry data digitization (API 1979). Each number has 12–14 digits. The first two numbers indicate the state in which the well is drilled, the following three indicate the county in which it is located, the next five characters are unique identifiers for the well's county, and the final two numbers are used to separate well bores if multiple wells are bored (in branches) from one well head (API 1979). An additional two characters can be added if a well bore is further subdivided (PPDM 2013). According to Koscinski and Stolle (2011), there are inconsistencies in this system. In 2011, 15–20% of well bores drilled were not recorded in industry databases because multiple bores from one well head were not given unique API numbers. Horizontal drilling that enables companies to drill in multiple directions from one well head has further complicated this numbering system because the well head and well bores may be located in different counties (Koscinski and Stolle 2011). Although all states use the same convention, different individual state agencies give out and maintain their own API number databases (API 1979). Hence there is presently no way to easily look across states online. In 1999, the management of maintaining and updating this system was passed on to the Professional Petroleum Data Management Association.

28. The canonical article that coined the term *semantic web* was written by Tim Berners-Lee (the inventor of the World Wide Web), James Hendler, and Ora Lassila and published in *Scientific American* in 2001.

29. http://www.depreportingservices.state.pa.us/ReportServer/Pages/ReportViewer.aspx?/Oil_Gas/OG_Compliance.

30. The WellWatch design was less concerned with defamation issues raised by terms such as "complaint" because "complaints" on WellWatch are not related to specific individuals as they were for LRC.

31. At the time, we had also changed the name of the project from DrillWell to WellWatch. To prevent confusion I've simply called the project by its latter name, WellWatch, throughout this chapter.

32. For more on the ever-unfinished quality of online software and web databases, see Adrian Mackenzie's discussion of virtuality (2006b).

33. For excellent first-person reporting on how corporate web tracking is both changing what it means to be present and reconfiguring privacy, see Madrigal (2012) and Harwood (2013). For more on how Google's search engine commodifies its unwitting users, see Levy (2010). Vaidhyanathan (2011) further investigates Google's pervasive and infrastructural dominance over information retrieval and representation. Although the quality of their "intelligence" has been questioned, private intelligence firms have been revealed to be using the web, particularly social media, to track activist organizations and disrupt their operations (Gwynne/Austin 1999; Chatterjee 2012; Fisher 2012). These practices were first publicized by the online hacktivist collective Anonymous after it hacked into private firms' email systems and subsequently

published the material on Wikileaks to demonstrate the threats posed by networked digital information systems both to their activist users and to supposed experts in the trafficking of secrets (Chatterjee 2012). The American Civil Liberties Union published a policy paper on the collection of metadata and its threat to personal privacy (Conley 2014).

34. The Five Eyes spying alliance between the United States, Britain, Canada, Australia, and New Zealand, which began in World War II, has been expanded by funding for the War on Terror to become the largest global surveillance and data-mining operation in history (Greenwald and MacAskill 2013).

9. WellWatch

This chapter was adapted from a paper cowritten with the sociologist Len Albright, from Northeastern University's sociology/anthropology department (2014). The paper also appeared in a special issue of the *Journal of Political Ecology* 21 (2014). Professor Albright's review of the WellWatch reports and coauthorship of the paper that this chapter is based on added the perspective of a social scientist who was not involved in the design and development of the site.

All quotes from the WellWatch reports in this chapter have been anonymized and corrected for misspellings, but not corrected for grammar. See WellWatch Appendix—WellWatch complaint number API# 05-045-14674, complaint 0001.

1. See WellWatch Appendix—Parachute CO well not on map.

2. See WellWatch Appendix—WellWatch complaint 37-015-21311, complaint 0001.

3. See WellWatch Appendix—API# 05-045-11375, date of complaint: January 8, 2011.

4. For reports for API# 05-045-19728, see web.archive.org/web/20110314213025/http://scrapper.media.mit.edu/wiki/05-045-19728.

5. See WellWatch Appendix—WellWatch report recovered from Tara Meixsell notes, date of complaint, February 20, 2011.

6. Many WellWatch users lived in rural areas without access to computers beyond the public library. This particular user was elderly. She was assisted in making her report over the phone by the regional WellWatch coordinator, who then digitized her diary of experiences. This case underlines how vital it is to protect evidence of the valuable assistance given by WellWatch coordinators to those without online access (see chapter 7).

7. See WellWatch Appendix—WellWatch complaint 05-045-19108, complaint 0002.

8. See WellWatch Appendix—WellWatch complaint 05-045-19728, complaint 0002.

9. See WellWatch Appendix—WellWatch complaint 05-045-09122, complaint 0002.

10. See WellWatch Appendix—well report recovered from Tara Meixsell, based on Ron Gulla's notes.

11. See web.archive.org/web/20110508201631/http://scrapper.media.mit.edu/wiki/Texas#Complaints; web.archive.org/web/20110508200907/http://scrapper.media.mit.edu/wiki/New_York#Community_Groups; web.archive.org/web/20110508201606/http://scrapper.media.mit.edu/wiki/Pennsylvania#Complaints.

12. See WellWatch Appendix—WellWatch complaint 05-045-14138, complaint 0005.

13. See WellWatch Appendix—WellWatch complaint 05-045-14138, complaint 0006.

14. See WellWatch Appendix—WellWatch complaint, report made by gas-patch employee.

15. See http://publiclab.org/notes/Sara/09-25-2014/hydrogen-sulfide-monitoring-tool-well-received-at-niehs-conference; http://publiclab.org/wiki/wyoming-hydrogen-sulfide-testing-2013-2014.

16. See WellWatch Appendix—WellWatch complaint 05-045-19728, complaint 0001.

17. See WellWatch Appendix—WellWatch complaint, report made by gas-patch employee.

18. See web.archive.org/web/20110508200902/http://scrapper.media.mit.edu/wiki/Community_Resources.

19. Aggressively opposed by the gas industry, the Fracturing Responsibility and Awareness of Chemicals Act is a proposal to define fracking as a federally regulated activity under the Safe Drinking Water Act that would also require gas and oil companies to disclose the chemicals that are used in fracturing fluids (Gibson 2011; Lustgarten 2011b).

20. WellWatch reports can still be viewed online in the WayBackMachine Internet Archive, particularly see http://web.archive.org/web/20110315070603/http://scrapper.media.mit.edu/wiki/Complaints2; and http://web.archive.org/web/20110806000540/http://www.wellwatch.org/wiki/WellWatch. Members of the ExtrAct team also maintained copies of reports, screenshots, code, and other materials.

21. See WellWatch Appendix—WellWatch complaint 05-045-15639, complaint 0003.

22. See WellWatch Appendix—WellWatch complaint 05-045-15639, complaint 0002.

23. See WellWatch Appendix—WellWatch complaint 05-045-10818, complaint 0001.

24. See WellWatch Appendix—WellWatch complaint 37-019-21792, complaint 0001.

25. See WellWatch Appendix—WellWatch complaint recovered from Tara Meixsell notes, date of complaint: February 20, 2011.

26. See WellWatch Appendix—WellWatch complaint recovered from Tara Meixsell notes, date of complaint: February 20, 2011.

27. See WellWatch Appendix—WellWatch complaint 05-045-07374, complaint 0008.

28. See WellWatch Appendix—WellWatch complaint, Pennsylvania landowner report for API number 37-125-22964, recovered from Tara Meixsell.

29. See WellWatch Appendix—WellWatch complaint 05-045-07374, complaint 0008.

30. See WellWatch Appendix—WellWatch complaint recovered from Tara Meixsell notes, date of complaint: February 20, 2011.

31. See WellWatch Appendix—WellWatch complaint recovered from Tara Meixsell notes, date of complaint: February 20, 2011.

32. See WellWatch Appendix—WellWatch complaint 37-019-21792, complaint 0001, recorded by Sara Wylie.

33. See WellWatch Appendix—WellWatch complaint 05-045-07374, complaint 0011, Colorado landowner WellWatch report on frack fluid spill on property recovered.

34. See WellWatch Appendix—WellWatch complaint 05-045-05103, complaint 0002.

35. See WellWatch Appendix, complaints retrieved from Tara Meixsell notes and Ron Gulla notes.

36. See WellWatch Appendix—WellWatch complaint 37-019-21792, complaint 0001.

37. http://www.environmentalhealthproject.org/; http://www.psehealthyenergy.org/.

38. See WellWatch Appendix—WellWatch complaint 05-045-06690, complaint 0001.

39. See WellWatch Appendix—WellWatch complaint 05-045-11375, complaint 0002.

40. See Urbina (2011g).

41. See WellWatch Appendix—WellWatch complaint 05-045-19728, complaint 0009, and 05-045-11375, complaint 0003.

42. See WellWatch Appendix—WellWatch complaint 05-045-11375, complaint 0001.

43. See WellWatch Appendix—WellWatch complaint 05-045-19728, complaint 0005.

44. These were the very oilfields where Joseph Stalin was raised and radicalized to resist some of the worst environmental and social conditions that "capitalism" had yet produced: landscapes of oil and gas extraction.

45. See WellWatch Appendix—WellWatch complaint API# 05-045-14674, complaint 0001.

46. See WellWatch Appendix—WellWatch complaint API# 05-045-07374, complaint 0011.

47. See WellWatch Appendix—WellWatch complaint 05-045-14674, complaint 0001.

48. See WellWatch Appendix—WellWatch complaint, February 20, 2011.

49. See WellWatch Appendix—WellWatch complaint, February 20, 2011.

50. See WellWatch Appendix—WellWatch complaint 37-019-21792, complaint 0001.

51. See WellWatch Appendix—WellWatch complaint 05-045-10463, complaint 0001.

10. The Fossil-Fuel Connection

1. http://www.collinsdictionary.com/submission/6723/fracktivist.

2. "Hydraulic Fracturing for Oil and Gas: Impacts from the Hydraulic Fracturing Water Cycle on Drinking Water Resources," http://www2.epa.gov/hfstudy.

3. http://fracfocus.org/.

4. Additionally, *Public Accountability* has examined how industry funding supports academic research favorable to the natural gas industry at the University of Texas (Connor, Galbraith, and Nelson 2012a), University of Buffalo (Connor, Galbraith, and Nelson 2012b), and MITEI (Connor and Galbraith 2013).

5. Early work on the endocrine disrupting potential of phthalates in 2004 and 2005 included an animal study showing their neurotoxicity (Masuo et al. 2004) and a study showing that such chemicals could be released from medical products and contaminate neonates in intensive care (Green et al. 2005). Subsequent cohort studies have found a significant association with long-term intravenous exposure to phthalates and increased rates of attention deficit disorder (Nutt 2016; Verstraete et al. 2016).

6. "Figure 1: Shale Gas through the Ethane Chain into Manufactured Products," http://energyindepth.org/wp-content/uploads/2012/10/Shale-gas-through-the-ethane-chain.png.

7. https://web.archive.org/web/20140330152604/http://www.globaltrends.com/reports/?doc_id=500539&task=view_details; Tracey Keys, "Corporate Clout 2013: Time for Responsible Capitalism," http://www.slideshare.net/tskeys/corporate-clout-2013-time-for-responsible-capitalism?related=1.

8. Keys, "Corporate Clout 2013."

9. See the letter confirming the use of 2-BE to Amos from the COGCC, hosted online by Earthworks/OGAP at http://www.earthworksaction.org/files/pubs-others/COGCC_Amos_Letter.pdf.

10. The use of nondisclosure agreements limiting the ability of residents who claim that their health and environmental problems are related to fracking has also been documented in Texas, Wyoming, and Arkansas (Efstathiou and Drajem 2013). One case in Pennsylvania that received significant press coverage barred two children (7 and 10 years old at the time) from discussing their family's health issues for the rest of their lives (Goldenberg 2013).

11. See the MSDS for Corexit 9527, Nalco Holding, available at https://inspectapedia.com/oiltanks/Oil_Dispersants_MSDS.php; https://www.msdsonline.com/2010/06/15/msds-information-for-oil-spill-dispersants-corexit-9500-and-corexit-9527-and-safety-info-from-epa-cdc-osha/.

12. MSDS for Corexit 9527.

13. See also a United Kingdom parliamentary memorandum from October 21, 2010, titled "UK Deepwater Drilling Implications of the Gulf of Mexico Oil Spill": "Both of these products were removed from the UK Marine Management Organisation's approved list in 1998, as they proved too toxic in instances where they might end up on rocky shorelines (although existing stocks could be used)" (ECCC 2010: 51).

14. See the MSDS for Corexit 9500, Nalco Holding, at https://web.archive.org/web/20120112045943/http://lmrk.org/corexit_9500_uscueg.539287.pdf.

15. "Whistleblower Witness Statements," http://whistleblower.org/sites/default/files/corexit_report_part2_2014.pdf.

16. https://publiclab.org.

17. http://publiclab.org/wiki/hydrogen-sulfide-sensor.

18. Hydrogeology is the study of water movement through rock.

19. TEDX did two evaluations of the health impacts of chemicals used in Wyoming. The first Wyoming list examined chemicals used within drilling fluid from the Bennett Creek 25-1 well pad by Windsor Energy. The list of chemicals came from ongoing battles that Thomas and the Clark community fought with Windsor Energy over the suspected groundwater contamination from bad drilling practices and a series of waste pits on the pad. The community organized many well-attended public meetings that brought the governor and regulatory agencies to the area. At each meeting Thomas requested a list of the chemicals used in drilling. Her persistence prevailed after an intense meeting when she was handed disks with lists of drilling fluids by an angry Windsor employee who was frustrated by her continual requests. After gathering a four-inch binder filled with MSDS info on the chemicals listed on the disks, the Clark community passed the list on to TEDX for further analysis. The second list was put together after a well blowout in Clark known as the Crosby Blowout, which

released 97 tons of VOCs and 2 tons of BTEX and caused an emergency evacuation of the community (Bleizeffer 2012).

20. The other states were Alabama, Texas, South Dakota, Alaska, Colorado, Kentucky, Louisiana, Michigan, and Tennessee.

21. http://www.iogcc.state.ok.us/hydraulic-fracturing.

22. For EPA program on Environmental Justice, see: https://www.epa.gov/environmentaljustice. For EPA funding on Environmental Justice, see: https://www.epa.gov/environmentaljustice/environmental-justice-small-grants-program. For more on NIEHS work on environmental health disparities and justice, see: https://www.niehs.nih.gov/research/supported/translational/justice/index.cfm. For funding opportunities, see NIEHS, research-to-action environmental justice grants: https://www.niehs.nih.gov/research/supported/translational/peph/prog/rta/index.cfm.

23. "Certificates of Confidentiality," http://grants.nih.gov/grants/policy/coc/.

24. http://public-accountability.org/; http://public-accountability.org/research/littlesis/.

25. "Trash Piles: Map Them! Measure Them! How Big Are They?," http://i.publiclab.org/system/images/photos/000/006/719/original/GM_Forum_20140909_Estimating_Trash_Piles.pdf; "Methods of Estimating the Volume and Weight of Waste Piles through Balloon Mapping," http://publiclab.org/notes/Sara/01-24-2014/methods-of-estimating-the-volume-and-weight-of-waste-piles-through-balloon-mapping.

26. "Estimating Volume and Weight of Petroleum Waste Piles in Southeast Chicago," http://publiclab.org/notes/Holden/03-11-2014/estimating-volume-and-weight-of-petroleum-waste-piles-in-southeast-chicago.

27. Exxon Mobil, "Energy lives here: Electricity with Natural Gas," advertisement campaign, 2014, Exxon Mobil TV Commercial: https://www.youtube.com/watch?v=TSsghIvfzzc.

28. See also Neu et al. (1998); Owen et al. (2000); Berthelot, Cormier, and Magnan (2003); and Conley and Williams (2005).

29. Memorandum of agreement between the EPA and BJ Services Company, Halliburton Energy Services Inc., and Schlumberger Technology Corporation (December 12, 2003), http://www.halliburton.com/public/about_us/pubsdata/hse/pdf/moa_dec12_Final.pdf.

30. 42 U.S.C. § 300h(d). Congress changed the definition of "underground injection" in the Energy Policy Act to exclude "the underground injection of fluids or propping agents (other than diesel fuels) pursuant to hydraulic fracturing operations related to oil, gas, or geothermal production activities."

31. This investigation was spurred by research by the Environmental Working Group (Horwitt 2010) and TEDX on petroleum distillates.

32. See the following pages within the "Vital Signs of the Planet" site: http://climate.nasa.gov/vital-signs/carbon-dioxide/; http://climate.nasa.gov/vital-signs/global-temperature/; http://climate.nasa.gov/vital-signs/arctic-sea-ice/.

33. "Endocrine Disruptors," http://www.niehs.nih.gov/health/topics/agents/endocrine/.

Conclusion

1. For more on embodiment in gender studies, anthropology, and philosophy, see Foucault (1979), Scheper-Hughes and Lock (1987), Csordas (1994), Kleinman and Kleinman (1994), Geurts (2002), Fassin (2007), Lock and Farquhar (2007), and Merleau-Ponty (2008). For work on how bodies are self-consciously shaped through "body work," see Grimlin (2007). For more on the importance of the bodily ways of knowing in the sciences, see Masco (2006), Helmreich (2009b), Roosth (2010), and N. Myers (2015). For more on how biomedicine is challenging and transforming what count as bodies, see Haraway (1997) and Landecker (2010). For works specifically on the body's role in sensing chemical contamination, see Choy (2012b) and Shapiro (2015).

2. New materialism is a branch of feminist research that seeks to move beyond social construction and reassert the agency of "matter" in a nonreductive fashion. It argues that if the social and the biological are coproduced then they both must be indeterminate, that is, there must be ways in which social experiences, cultures, food choices, habits, and histories shape behavior via biological mechanisms, and vice versa. It follows then that both are open to change (Barad 2003; Fausto-Sterling 2005; Coole and Frost 2010; Frost 2011).

3. This is not an exhaustive list of bodily properties; it highlights those relevant to my argument about chemical contamination.

4. Thanks to Dr. Julia Ravell, my former student and this book's editor, for this research, drawn from her course project "When Eating Makes You Ill" on high-fructose corn syrup's illusions of consumer choice and the industry's promotion of chemically addictive foods.

5. Corporations have frequently been critiqued as liquid and fluid in neoliberal capitalism because of the mobility and intangibility of capital (Ho 2009). Rather than rehearse these important arguments, I focus on the material practices of industry and how they make it hard to see industries as agents of social and environmental change, as social movements. Like every other human endeavor, markets and economics are socially constructed and capital is only one metric for counting industrial activity and social health (Mitchell 2011). Capital is a very poor metric for assessing violence or harm because if harm can be translated into "cost," then ceasing violence can be argued to be too costly for companies. We do not accept this kind of logic to manage other forms of violence. It is therefore important for the social sciences and sciences to begin developing other ways of measuring and accounting for industrial actors, similar perhaps to those developed to normalize and discipline citizens and consumers—panoptic surveillance, individuation, progress reports, permanent records, perhaps even psychoanalysis and public service campaigns. This is intended as a serious joke to show that conceptualizing companies as bodies rather than capital opens up entirely new methods for approaching their management.

6. See the strategies of disembodiment above for why this is erroneous.

7. Numerous spills and explosions have been reported from accidents (Gray 2006; Hrin 2012; Finley 2013a; Blake 2014; EPA Region 5 2014; Fontaine 2014; Schafer 2014; Wertz 2014).

REFERENCES

Abbate, Janet. 1999. *Inventing the Internet*. Cambridge, MA: MIT Press.

Abramowitz, Michael, and Steven Mufson. 2007. "Papers Detail Industry's Role in Cheney's Energy." *Washington Post*, July 18.

Adams, C., P. Brown, R. Morello-Frosch, J. Brody, R. Rudel, A. Zota, et al. 2011. "Disentangling the Exposure Experience: The Roles of Community Context and Report-Back of Environmental Exposure Data." *Journal of Health and Social Behavior* 52:180.

Adamson, Joni, Mei Mei Evans, and Rachel Stein, eds. 2002. *The Environmental Justice Reader: Politics, Poetics, and Pedagogy.* Tucson: University of Arizona Press.

Addams, Jane. 1892. "A New Impulse to an Old Gospel." *Forum* 14 (November): 350, 351.

ADEQ (Arkansas Department of Environmental Quality). 2011. "Emissions Inventory and Ambient Air Monitoring of Natural Gas Production in the Fayetteville Shale Region." North Little Rock, AR. November 22. http://www.adeq.state.ar.us/air/default.htm.

Agamben, Giorgio. [1995] 1998. *Homo Sacer: Sovereign Power and Bare Life.* Translated by Daniel Heller-Roazen. Stanford, CA: Stanford University Press.

Agre, Philip E. 1997. "Toward a Critical Technical Practice: Lessons Learned in Trying to Reform AI." In *Social Science, Technical Systems, and Cooperative Work: Beyond the Great Divide,* edited by Geoff Bowker, Les Gasser, Leigh Star, and Bill Turner. Mahwah, NJ: Erlbaum.

Agricola, Georgius. 1546. *De Ortu et Causis.* Book III.

———. 1556. *De Re Metallica.* English trans. H. C. Hoover and L. C. Hoover. 1950. New York: Dover. Published online Project Gutenburg 2011. https://www.gutenberg.org/files/38015/38015-h/38015-h.htm#Footnote_15_15.

Alinsky, Saul. 1971. *Rules for Radicals: A Pragmatic Primer for Realistic Radicals.* New York: Random House.

Allan, R. J., et al. 1991. *Toxic Chemicals in the Great Lakes and Associated Effects.* Vo1. 1. Ottawa, Canada: Department of Fisheries and Oceans.

Allen, Barbara L. 2003. *Uneasy Alchemy: Citizens and Experts in Louisiana's Chemical Corridor Disputes.* Cambridge, MA: MIT Press.

———. 2004. "Shifting Boundary Work: Issues and Tensions in Environmental Health Science in the Case of Grand Bois, Louisiana." *Science as Culture* 13 (4): 429–48.

Altman, Rebecca G. 2008. "Chemical Body Burden and Place-based Struggles for Environmental Health and Justice (a Multi-site Ethnography of Biomonitoring Science)." PhD diss., Brown University.

Alvarez, R. A., S. W. Pacala, J. J. Winebrake, W. L. Chameides, and S. P. Hamburg. 2012. "Greater Focus Needed on Methane Leakage from Natural Gas Infrastructure." *Proceedings of the National Academy of Sciences* 109:6435–40.

Amos, John. 2012. "Bakken Shale-Oil Drilling and Flaring Lights Up the Night Sky." *Skytruth*, March 8. https://www.skytruth.org/2012/03/bakken-shale-oil-drilling-and-flaring/.

Amos, Laura. 2012. "PreHearing Statement of Laura Amos." The Oil and Gas Conservation Commission of the State of Colorado in the Matter of Changes to the Rules and Regulations of the Oil and Gas Conservation Commission Statewide Water Sampling and Monitoring, case no. 1R, docket no. 1211-RM03.

Amos, Laura, and Larry Amos. 2005. "Living in the Gas Fields: One Family's Story." Earthworks, November 22. http://www.earthworksaction.org/voices/detail/laura_amos#.VKcb84061jo.

Anderson, Benedict. 1991. *Imagined Communities: Reflections on the Origin and Spread of Nationalism.* Rev. ed. London: Verso.

Anderson, Debra. 2009. *Split Estate.* Red Rock Pictures LLC. http://www.splitestate.com/.

Anderson, Hil. 2005. "'Fracking' Regulation May Undo Energy Bill." *United Press International,* April 15.

Antonetta, Susanne. 2002. *Body Toxic: An Environmental Memoir.* Berkeley, CA: Counterpoint.

Anway, M. D., C. Leathers, and M. K. Skinner. 2006. "Endocrine Disruptor Vinclozolin Induced Epigenetic Transgenerational Adult Onset Disease." *Endocrinology* 147:5515–23.

Anway, M. D., M. Memon, M. Uzumcu, and M. K. Skinner. 2006. "Transgenerational Effect of the Endocrine Disruptor Vinclozolin on Male Spermatogenesis." *Journal of Andrology* 27:868–79.

AP (Associated Press). 2003. "Bush Tells Bureau to Open Land." *Washington Times,* August 9.

API (American Petroleum Institute). 1979. "The API Well Number and Standard State and County Numeric Codes Including Offshore Waters." API Bulletin D12A. January 1979 (a revision that supersedes April 1968 and December 1970 editions).

Appadurai, Arjun. 1990. "Disjuncture and Difference in the Global Cultural Economy." *Public Culture* 2:1–24.

Appel, Hannah. 2011. "Futures: Oil and the Making of Modularity in Equatorial Guinea." Ph.D. diss., Department of Anthropology, Stanford University.

Appel, Hannah, Arthur Mason, and Michael Watts, eds. 2015. *Subterranean Estates, Life Worlds of Oil and Gas.* Ithaca, NY: Cornell University Press.

Applebome, Peter. 2008. "Gas Drillers in Race for Hearts and Land." *New York Times*, June 29.

Arcury, T., S. Quandt, and L. McCauley. 2000. "Farmworkers and Pesticides: Community-Based Research." *Environmental Health Perspectives* 108:787–92.

Auty, Richard M. 1993. *Sustaining Development in Mineral Economies: The Resource Curse Thesis.* London: Routledge.

Auyero, Javier, and Debora Swistun. 2009. *Flammable: Environmental Suffering in an Argentine Shantytown.* Oxford: Oxford University Press.

Bamberger, M., and R. E. Oswald. 2012. "Impacts of Gas Drilling on Human and Animal Health." *New Solutions* 22 (1): 51–77. doi: http://dx.doi.org/10.2190/NS.22.1.e.

———. 2013. "Science and Politics of Shale Gas Extraction." *New Solutions* 23 (1): 137–66.

Barad, Karen. 2003. "Posthumanist Performativity: Toward an Understanding of How Matter Comes to Matter." *SIGNS: Journal of Women in Culture and Society* 28 (3): 801–31.

Barry, Andrew. 2006. "Technological Zones." *European Journal of Social Theory* 9 (2): 239–53.

Bashe, Charles J., Lyle R. Johnson, John H. Palmer, and Emerson W. Pugh. 1986. *IBM's Early Computers.* Cambridge, MA: MIT Press.

BBC Research and Consulting. 2008. "Northwest Colorado Socioeconomic Analysis and Forecasts." Produced for Associated Governments of Northwest Colorado. April 4.

Beans, Laura. 2013. "Fracking Wastewater Spill Kills Rare Fish in KY, Puts Entire Species at Risk." *EcoWatch*, August 29. http://ecowatch.com/2013/08/29/fracking-wastewater-spill-puts-entire-species-at-risk/.

Beck, Ulrich. 1992. *Risk Society: Towards a New Modernity.* New Delhi: Sage.

Beer, David. 2009. "Power through the Algorithm? Participatory Web Cultures and the Technological Unconscious." *New Media and Society* 11 (6): 985–1002.

Beer, John Joseph. 1959. *The Emergence of the German Dye Industry.* Champaign: University of Illinois Press.

Begos, Kevin. 2012. "Tax Breaks, U.S. Research Play Big Part in Success of Fracking." Associated Press, September 12. http://www.cleveland.com/nation/index.ssf/2012/09/tax_breaks_us_research_play_bi.html.

Bender, Bryan. 2013. "MIT's Ernest Moniz Advised Oil and Gas Investors." *Boston Globe*, March 30. http://www.bostonglobe.com/news/politics/2013/03/30/omama-energy-pick-discloses-industry-ties/DZZakSojOGacmzoqogxHFL/story.html.

Benjamin, Walter. 1999. *The Arcades Project.* Translated by Howard Eiland and Kevin McLaughlin. Cambridge, MA: Harvard University Press.

Benkler, Yochai. 2006. *The Wealth of Networks: How Social Production Transforms Markets and Freedom.* New Haven, CT: Yale University Press.

Bennett, W. Lance. 2003a. "Communicating Global Activism." *Information, Communication and Society* 6 (2): 143–68.

———. 2003b. "New Media Power." In *Contesting Media Power*, edited by Nick Couldry and James Curran. Lanham, MD: Rowman and Littlefield.

Bennis, P., J. Berger, N. Klein, H. Kureishi, T. Nguyen, A. Roy, et al. 2007. *War with No End.* New York: Verso.

Bensaude-Vincent, Bernadette, and Isabelle Stengers. 1996. *A History of Chemistry.* Cambridge, MA: Harvard University Press.

Ben-Yeoshua, S., et al. 2005. *Environmentally Friendly Technologies for Agricultural Produce Quality.* Boca Raton, FL: CRC Press.

Bera, R., and A. Hrybyk. 2013. "iWitness Pollution Map: Crowdsourcing Petrochemical Accident Research." *Scientific Solutions* 23 (3): 521–33.

Berkman Center for Internet and Society. 2010. "Media Laws in the Digital Age: The Rules Have Changed, Have You?" CLE conference, Media Law in a Digital Age. http://www.scribd.com/doc/38347145/Media-Law-in-the-Digital-Age-The-Berkman-Center.

Berners-Lee, Tim, and Mark Fischetti. 1999. *Weaving the Web: The Original Design and Ultimate Destiny of the World Wide Web by Its Inventor.* San Francisco: Harper.

Berners-Lee, Tim, James Hendler, and Ora Lassila. 2001. "The Semantic Web: A New Form of Web Content That Is Meaningful to Computers Will Unleash a Revolution of New Possibilities." *Scientific American*, May 17.

Berry, D. 2008. "The Poverty of Networks." *Theory, Culture and Society* 25 (7–8): 36–37.

Berthelot, Sylvie, Denis Cormier, and Michel Magnan. 2003. "Environmental Disclosure Research: Review and Synthesis." *Journal of Accounting Literature* 22:1–44.

Betts, Kellyn. 2011. "A Study in Balance: How Microbiomes Are Changing the Shape of Environmental Health." *Environmental Health Perspectives* 119 (8): 340–46.

Bholer, Megan. 2008. *Digital Media and Democracy.* Cambridge, MA: MIT Press.

Bialik, C. 2004. "Get the Word Out: Protesters, Delegates Use New Technology to Send Rapid-Fire Cell Phone Messages." *Wall Street Journal Online*, August 31.

Bijker, W. E. 1995. *Of Bicycles, Bakelites, and Bulbs: Toward a Theory of Sociotechnical Change.* Cambridge, MA: MIT Press.

Bijker, W. E., T. P. Hughes, and T. J. Pinch, eds. 1989. *The Social Construction of Technological Systems: New Directions in the Sociology and History of Technology.* Cambridge, MA: MIT Press.

Bisdorf, Robert J., and David V. Fitterman. 2004. "Schlumberger Soundings in the Long Beach-Wilmington Area, County of Los Angeles, California." U.S. Department of the Interior, U.S. Geological Survey. U.S. Government Open-File: Report-1053.

Bishop, Ronald. 2013. "Historical Analysis of Oil and Gas Well Plugging in New York: Is the Regulatory System Working?" *New Solutions* 23 (1): 137–66.

Black, Brian. 2000. *Petrolia: The Landscape of America's First Oil Boom.* Baltimore: Johns Hopkins University Press.

Blake, Mariah. 2014. "Halliburton Fracking Spill Mystery: What Chemicals Polluted an Ohio Waterway?" *Mother Jones*, July 24. http://www.motherjones.com/politics/2014/07/halliburton-ohio-river-spill-fracking.

Bleizeffer, Dustin. 2012. "Crew Stops Flow of Gas at Well Blowout." *WyoFile: People, Places, Policy*, April 25. http://www.wyofile.com/brief/blowout-north-of-douglas-still-uncontrolled/.

Borrell, Brendan. 2010. "The Big Test for Bisphenol A." *Nature* 464:1122–24.
Boudia, Soraya, and Nathalie Jas. 2014. *Powerless Science? Science and Politics in a Toxic World.* New York: Berghahn.
Bowker, Geoffrey C. 1987. "A Well Ordered Reality: Aspects of the Development of Schlumberger, 1920–39." *Social Studies of Science* 17 (4): 611–55.
———. 1994. *Science on the Run.* Cambridge, MA: MIT Press.
———. 2005. *Memory Practices in the Sciences.* Cambridge, MA: MIT Press.
Bowker, Geoffrey C., and Susan Leigh Star. 2000. *Sorting Things Out: Classification and Its Consequences.* Cambridge, MA: MIT Press.
Bozman, Jean S. 1993. "Oil Firm Powers Up Fujitsu Super CPU." *Computerworld* 27: 9:10.
Brand, Stewart. 1988. *The Media Lab: Inventing the Future at MIT.* New York: Penguin.
Brandt, Allen. 2007. *The Cigarette Century: The Rise, Fall, and Deadly Persistence of the Product That Defined America.* New York: Basic Books.
Brandt, A. R., G. A. Heath, E. A. Kort, F. O'Sullivan, G. Petron, S. M. Jordaan, et al. 2014. "Methane Leaks from North American Natural Gas Systems." *Science* 343 (6172): 733–35.
Brandt, E. N. 1997. *Growth Company: Dow Chemical's First Century.* East Lansing: Michigan State University Press.
Brasier, Katherine J., Matthew R. Filteau, Diane K. McLaughlin, Jeffrey Jacquet, Richard C. Stedman, and Timothy W. Kelsy. 2011. "Residents' Perceptions of Community and Environmental Impacts from Development of Natural Gas in the Marcellus Shale: A Comparison of Pennsylvania and New York Cases." *Journal of Rural Social Sciences* 26 (1): 92–106.
Bray, Hiawatha. 2003. "Website Turns Tables on Government Officials." *Boston Globe*, July 4.
Breisford, R. 2014. "Rising Demand, Low-Cost Feed, Spur Ethylene Capacity Growth." *Oil and Gas Journal*, July 7. http://www.ogj.com/articles/print/volume-112/issue-7/special-report-ethylene-report/rising-demand-low-cost-feed-spur-ethylene-capacity-growth.html.
Brindle, James. 2010. "The Iraq War: A History of Wikipedia Changelogs." http://booktwo.org/notebook/wikipedia-historiography/.
Brody, Hugh. 1981. *Maps and Dreams.* New York: Pantheon.
Brody, J. G., R. Morello-Frosch, P. Brown, R. Rudel, R. Altman, M. Frye, et al. 2007. "Is It Safe? New Ethics for Reporting Personal Exposures to Environmental Chemicals." *American Journal of Public Health* 97:1547–54.
Brody, J. G., R. Morello-Frosch, A. Zota, P. Brown, C. Perez, and R. A. Rudel. 2009. "Linking Exposure Assessment Science with Policy Objectives for Environmental Justice and Breast Cancer Advocacy: The Northern California Household Exposure Study." *American Journal of Public Health* 99:S600–S609.
Brown, Joe. 2007. *National Sacrifice Zone: Colorado and the Cost of Energy Independence.* Stars Down to Earth Films.
Brown, Phil. 2007. *Toxic Exposures: Contested Illnesses and the Environmental Health Movement.* New York: Columbia University Press.

Brown, Phil, B. Mayer, S. Zavestoski, T. Luebke, J. Mandelbaum, and S. McCormick. 2003. "The Health Politics of Asthma: Environmental Justice and Collective Illness Experience in the United States." *Social Science and Medicine* 57:453–64.

Brown, Phil, S. McCormick, B. Mayer, S. Zavestoski, R. Morello-Frosch, R. Gasior Altman, and L. Senier. 2006. "'A Lab of Our Own': Environmental Causation of Breast Cancer and Challenges to the Dominant Epidemiological Paradigm." *Science, Technology and Human Values*, September 31, 499–536.

Brown, Phil, and Edwin Mikkelson. 1997. *No Safe Place: Toxic Waste, Leukemia, and Community Action.* Berkeley: University of California Press.

Brown, Phil, Steve Zavestoski, Sabrina McCormick, Brian Mayer, Rachel Morello-Frosch, and Rebecca Gasior. 2004. "Embodied Health Movements: Uncharted Territory in Social Movement Research." *Sociology of Health and Illness* 26:1–31.

Brulle, Robert. 2013. "Institutionalizing Delay: Foundation Funding and the Creation of U.S. Climate Change Counter-Movement Organizations." *Climate Change* 122 (4): 681–94.

Brulle, Robert, and David N. Pellow. 2006. "Environmental Justice: Human Health and Environmental Inequalities." *Annual Review of Public Health* 27:103–24.

Bruns, A. 2008. *Blogs, Wikipedia, Second Life and Beyond: From Production to Produsage.* New York: Peter Lang.

Buchanan, D. 2006. "Greed Wins One over Nature." *Grand Junction Daily Sentinel*, February 12.

Bucher, John R. 2002. "The National Toxicology Program Rodent Bioassay: Designs, Interpretations, and Scientific Contributions." *Annals of New York Academy of Science* 982:198–207.

Bullard, Robert. 1990. *Dumping in Dixie: Race, Class and Environmental Quality.* Boulder, CO: Westview.

Burdeau, Cain, and Janet McConnaughey. 2011. "Louisiana Chemical Plant Explodes, Forces Evacuation." Associated Press, June 15. http://www.masslive.com/news/index.ssf/2011/06/louisiana_chemical_plant_explo.html.

Burnett, Graham. 2000. *Masters of All They Surveyed: Exploration, Geography and a British El Dorado.* Chicago: University of Chicago Press.

Burri, Regula V., and Joseph Dumit. 2007. "Social Studies of Scientific Imaging and Visualization." In *Handbook of Science and Technology Studies,* 3rd ed., edited by Ed Hackett et al. Cambridge, MA: MIT Press.

BW (Business Wire). 2001. "Schlumberger Opens New Generation iCenter to Support Realtime Reservoir Management; Houston iCenter Brings Latest Immersive Technology to North and South America Clients." October 18.

———. 2002a. "Schlumberger DeXa.Net Provides Remote Connectivity for Titanic Expedition and Documentary; Real-Time Data Network Supports Complex Communications Project." August 6.

———. 2002b. "Schlumberger Launches DeXa.Net Secure Private Network; MPLS-Based Connectivity Allows Faster, Simpler, Multi-Application Data Transmission." August 13.

———. 2003. "BG Selects Inside Reality from Schlumberger Information Solutions for Its Visualization Center." January 31.

CAE (Critical Art Ensemble). 1996. "Electronic Civil Disobedience and Other Unpopular Ideas." *Autonomedia* 11. http://www.critical-art.net/books/ecd/ecd2.pdf.

———. 2001. "Digital Resistance: Explorations in Tactical Media." http://www.critical-art.net/books/digital/.

Cailliau, Robert, and James Gillies. 2000. *How the Web Was Born: The Story of the World Wide Web.* Oxford: Oxford University Press.

Calafat, Antonia M., Zsuzsanna Kukleynik, John A. Reidy, Samuel P. Caudill, John Ekong, and Larry L. Needham. 2005. "Urinary Concentrations of Bisphenol A and 4-nonylphenol in a Human Reference Population." *Environmental Health Perspectives* 113 (4): 391–95.

Calafat, Antonia M., X. Ye, L. Y. Wong, John A. Reidy, and Larry L. Needham. 2008. "Exposure of the U.S. Population to Bisphenol A and 4-tertiary-octylphenol: 2003–2004." *Environmental Health Perspectives* 116 (1): 39–44.

Campbell-Kelly, Martin, and William Aspray. 1996. *Computer: A History of the Information Machine.* New York: Basic Books.

Campo, Rafael. 2006. "'Anecdotal Evidence': Why Narratives Matter to Medical Practice." *PLOS Medicine* 3 (10): e423.

Carson, Rachel. 1962. *Silent Spring.* Boston: Houghton Mifflin.

Carton, H., et al. 2006. "Schlumberger Seismic Vessel Geco Searcher Provides Unprecedented Images of the Great Andaman Sumatra Earthquake Megathrust Rupture Plane." American Geophysical Union, Fall meeting. Abstract #U53A-0026.

Cartwright, E. 2013. "Eco-Risk and the Case of Fracking." In *Cultures of Energy: Power, Practices, Technologies,* edited by Sarah Strauss, Stephanie Rupp, and Thomas Love. Walnut Creek, CA: Left Coast Press. 201–12.

Casey, J., D. Savitz, S. Rasmussen, E. Ogburn, J. Pollak, D. Mercer, and B. Schwartz. 2015. "Unconventional Natural Gas Development and Birth Outcomes in Pennsylvania, USA." *Epidemiology* 1.

Casey, Tina. 2014. "Interior Department Blows Off Koch Stranglehold on Onshore Wind Power." *Clean Technica*, August 17. http://cleantechnica.com/2014/08/17/interior-dept-atlantic-offshore-wind-power/.

Casper, Monica. 2003. *Synthetic Planet: Chemical Politics and the Hazards of Modern Life.* New York: Routledge.

Casselman, Ben. 2008. "Corporate News: Oil-Field Services Firms Prosper on Drilling Boom." *Wall Street Journal*, July 23.

Castañeda, Christopher J. 1993. *Regulated Enterprises: Natural Gas Pipelines and Northeastern Markets, 1938–1954.* Columbus: Ohio State University Press.

———. 1999. *Invisible Fuel: Manufactured and Natural Gas in America, 1800–2000.* New York: Twayne.

Castañeda, Christopher J., and Clarence M. Smith. 1996. *Gas Pipelines and the Emergence of America's Regulatory State: A History of Panhandle Eastern Corporation, 1928–1993.* Cambridge: Cambridge University Press.

Castells, Manuel. [1996] 2000. *The Information Age: Economy, Society, and Culture.* Oxford: Blackwell.

Cathles, L. M., L. Brown, M. Taam, and A. Hunter. 2012. "A Commentary on 'The Greenhouse Gas Footprint of Natural Gas in Shale Formations' by R. W. Howarth, R. Santoro, and A. Ingraffea." *Climatic Change* 113:525–35. doi: 10.1007/s10584-011-0333-0.

Caulton, D. R., P. B. Shepson, R. L. Santoro, J. P. Sparks, R. W. Howarth, A. R. Ingraffea, et al. 2014. "Toward a Better Understanding and Quantification of Methane Emissions from Shale Gas Development." *Proceedings of the National Academy of Sciences USA* 111 (17): 6237–42.

Cave, Susan. 2004. "Open Letter to Governor Robert Taft." Ohio Municipal League. May 11.

Cefkin, Melissa, ed. 2008. *Ethnography at Work: New Social Science Research in and for Industry*. New York: Berghahn.

———. 2010. *Ethnography and the Corporate Encounter: Reflections on Research in and of Corporations*. New York: Berghahn.

Chan, Anita. 2004. "Coding Free Software, Coding Free States: Free Software Legislation and the Politics of Code in Peru." *Anthropological Quarterly* 77 (3): 531–45.

Chandler, Alfred D., Jr. 1977. *The Visible Hand: The Managerial Revolution in American Business*. Cambridge, MA: Belknap Press.

———. 2005. *Shaping the Industrial Century: The Remarkable Story of the Evolution of the Modern Chemical and Pharmaceutical Industries*. Cambridge, MA: Harvard University Press.

Chandler, David. 2013. "MIT's Ernest Moniz Nominated Secretary of Energy." *MIT News*, March 3.

Charon, Rita. 2006. *Narrative Medicine: Honoring the Stories of Illness*. New York: Oxford University Press.

Charon, Rita, and Martha Montello, eds. 2002. *Stories Matter: The Role of Narrative in Medical Ethics*. New York: Routledge.

Chatterjee, Pratap. 2012. "WikiLeaks' Stratfor Dump Lifts Lid on Intelligence-Industrial Complex." *Guardian*, February 2012.

Chaves, Ana Lúcia Soares, and Paulo Celso de Melo-Farias. 2006. "Ethylene and Fruit Ripening: from Illumination Gas to the Control of Gene Expression, More Than a Century of Discoveries." *Genetics and Molecular Biology* 29, no. 3: 508–15.

Chesapeake Energy. 2011. "Chesapeake Energy Corporation Provides Full Disclosure of Chemicals Used in the Hydraulic Fracture Well Completion Process through a National Publicly Accessible Registry." 4-Traders, April 11. http://www.4-traders.com/CHESP-ENEGY-12055/news/CHESP-ENEGY-Chesapeake-Energy-Corporation-Provides-Full-Disclosure-of-Chemicals-Used-in-the-Hydrauli-13599373/.

Choy, Tim. 2012a. "Air's Substantiations." In *Lively Capital: Biotechnologies, Ethics, and Governance in Global Markets*, edited by Kaushik Sunder Rajan, 121–52. Durham, NC: Duke University Press.

———. 2012b. *Ecologies of Comparison: An Ethnography of Endangerment in Hong Kong*. Durham, NC: Duke University Press.

City of Fort Worth. 2011. "Natural Gas Air Quality Study (Final Report)." Eastern Research Group and the City of Fort Worth, July 13.

Clark, David D. 1988. "The Design Philosophy of the DARPA Internet Protocols." *Computer Communications Review* 18 (4): 106–14.

Clarke, Adele, and Joan H. Fujimura, eds. 1992. *The Right Tools for the Job: At Work in Twentieth-Century Life Sciences.* Princeton, NJ: Princeton University Press.

Clarren, Rebecca. 2006a. "EPA to Citizens: Frack You." *Salon.com*, May 5.

———. 2006b. "Voices from the Gas Field." *Orion* (November–December).

Cleaver, Harry. 1997. "The Zapatista Effect: The Internet and the Rise of an Alternative Political Fabric." *Journal of International Affairs* 51 (2): 621–40.

Clements-Nolle, K., and A. Bachrach. 2008. "CBPR with a Hidden Population: The Transgender Community Health Project a Decade Later." In *Community-based Research for Health: From Process to Outcomes*, edited by M. Minkler and N. Wallerstein. Hoboken, NJ: Jossey-Bass.

Colborn, Theo. 2004a. "Commentary: Setting Aside Tradition When Dealing with Endocrine Disruptors." *ILAR Journal* 45 (4): 394–400.

———. 2004b. "Endocrine Disruption Overview: Are Males at Risk?" In *Hypospadias and Genital Development*, edited by L. Baskin, 189–201. New York: Kluwer Academic/Plenum.

———. 2004c. "Neurodevelopment and Endocrine Disruption." *Environmental Health Perspectives* 112 (9): 944–49.

———. 2006. "A Case for Revisiting the Safety of Pesticides: A Closer Look at Neurodevelopment." *Environmental Health Perspectives* 114 (1): 10–17.

———. 2007. "Written Testimony of Theo Colborn, PhD, President of TEDX, Paonia, Colorado, before the House Committee on Oversight and Government Reform, Hearing on the Applicability of Federal Requirements to Protect Public Health and the Environment from Oil and Gas Development," October 31. https://s3.amazonaws.com/propublica/assets/natural_gas/colburn_testimony_071025.pdf.

———. 2013. "The Fossil Fuel Connection." *EarthFocus*, October 1. Linktv.org.

———. 2014. "The Overlooked Connection between Human Health and Fossil Gases." Unpublished article. The Endocrine Disruption Exchange, November 14. http://endocrinedisruption.org/assets/media/documents/FFC%20Nov%2014%202014.pdf.

Colborn, Theo, and L. E. Carroll. 2007. "Pesticides, Sexual Development, Reproduction, and Fertility: Current Perspective and Future Direction." *Human and Ecological Risk Assessment* 13 (5): 1078–110.

Colborn, Theo, and Coralie Clement, eds. 1992. *Chemically-Induced Alterations in Sexual and Functional Development: The Wildlife/Human Connection.* Princeton, NJ: Princeton Scientific Publishing.

Colborn, Theo, A. Davidson, S. N. Green, R. A. Hodge, C. L. Jackson, and R. A. Liroff, eds. 1990. *Great Lakes: Great Legacy?* Washington, DC: Conservation Foundation and the Institute for Research on Public Policy.

Colborn, Theo, Dianne Dumanoski, and John Myers. 1997. *Our Stolen Future.* New York: Penguin.

Colborn, Theo, Carol Kwiatowski, Kim Schultz, and Mary Bachran. 2011. "Natural Gas Operations from a Public Health Perspective." *Human and Ecological Risk Assessment* 17 (5): 1039–56. doi: 10.1080/10807039.2011.605662.

Colborn, Theo, K. Schultz, L. Herrick, and C. Kwiatkowski. 2014. "An Exploratory Study of Air Quality Near Natural Gas Operations." *Human and Ecological Risk Assessment* 20 (1): 86–105.

Cole, Luke W., and Sheila R. Foster. 2001. *From the Ground Up: Environmental Racism and the Rise of the Environmental Justice Movement.* New York: New York University Press.

Cole, M., P. Lindeque, C. Halsband, and T. Galloway. 2011. "Microplastics as Contaminants in the Marine Environment: A Review." *Marine Pollution Bulletin* 62 (12): 2588–97.

Coleman, Gabriella. 2004. "The Political Agnosticism of Free and Open Source Software and the Inadvertent Politics of Contrast." *Anthropology Quarterly* 77 (3): 507–19.

———. 2009. "Code Is Speech: Legal Tinkering, Expertise, and Protest among Free and Open Source Software Developers." *Cultural Anthropology* 24 (3): 420–54.

———. 2010. "Ethnographic Approaches to Digital Media." *Annual Review of Anthropology* 39:487–505.

———. 2012. *Coding Freedom: The Ethics and Aesthetics of Hacking.* Princeton, NJ: Princeton University Press.

Colorado School of Mines. 2009. "Potential Gas Committee Reports Unprecedented Increase in Magnitude of U.S. Natural Gas Resource Base," June 18. https://web.archive.org/web/20090621070205/https://www.mines.edu/Potential-Gas-Committee-reports-unprecedented-increase-in-magnitude-of-U.S.-natural-gas-resource-base.

Colson, John. 2010a. "MIT Team's Websites Offer Information about Oil and Gas Industry Activity." *Post Independent*, October 26.

———. 2010b. "Woman Who Lived Near Gas Fields Dies." *Post Independent*, November 16.

Comaroff, Jean, and John Comaroff. 2000. "Millennial Capitalism and the Culture of Neoliberalism." *Public Culture* 3:291–575.

Cone, Marla. 2009. "A New Window into Hormone-Altering Chemicals." *Environmental Health News*, February 9. http://www.ehn.org.

Conley, Chris. 2014. "Metadata: Piecing Together a Privacy Solution." ACLU of Northern California, February 2014. https://www.aclunc.org/sites/default/files/Metadata%20report%20FINAL%202%2021%2014%20cover%20%2B%20inside%20for%20web%20%283%29.pdf.

Conley, John M., and Cynthia A. Williams. 2005. "Engage, Embed, Embellish: Theory versus Practice in the Corporate Social Responsibility Movement." UNC Legal Studies Research Paper No. 05-16.

Connor, Steven. 2005. "The Menagerie of the Senses." Lecture given at the Sixth Synapsis Conference, I cinque sensi (per tacer del sesto), Bertinoro, Italy, September 1. http://stevenconnor.com/menagerie.pdf.

Connor, Steven, and Rob Galbraith. 2013. "Industry Partner or Puppet: How MIT's Influential Study of Fracking Was Authored, Funded and Released by Oil and Gas Industry Insiders." *Public Accountability Initiative* (March).

Connor, Steven, Rob Galbraith, and Ben Nelson. 2012a. "Contaminated Inquiry: How a University of Texas Fracking Study Led by a Gas Industry Insider Spun the Facts and Misled the Public." *Public Accountability Initiative* (July). http://public-accountability.org/wp-content/uploads/ContaminatedInquiry.pdf.

———. 2012b. "The UB Shale Play: Distorting the Facts about Fracking." *Public Accountability Initiative* (May). http://public-accountability.org/wp-content/uploads/UBShalePlay.pdf.

Cooke, Claude E., Jr., Francis E. Dollarhide, Jacques L. Elbel, C. Robert Fast, Robert R. Hannah, Larry J. Harrington, et al. 2010. *Legends of Hydraulic Fracturing.* Society of Petroleum Engineers. CD-ROM.

Coole, Diana, and Samatha Frost, eds. 2010. *New Materialism: Ontology, Agency and Politics.* Durham, NC: Duke University Press.

Coons, Teresa, and Russell Walker. 2005. "Community Health Risk Analysis of Oil and Gas Industry Impacts in Garfield County." Grand Junction, CO: St. Mary's Saccomanno Research Institute and Mesa State College.

Corburn, Jason. 2005. *Street Science: Community Knowledge and Environmental Health Justice.* Cambridge, MA: MIT Press.

Coronil, Fernando. 1997. *The Magical State: Nature, Money, and Modernity in Venezuela.* Chicago: University of Chicago Press.

Crampton, J. W. 2009. "Cartography: Maps 2.0." *Progress in Human Geography* 33:91–100.

Crimaldi, Laura. 2015. "Cape Wind Vows to Move Forward." *Boston Globe*, March 1.

Cronon, William, ed. 1995. *Uncommon Ground: Toward Reinventing Nature.* New York: W. W. Norton.

Crowley, John. 2013. "Connecting Grassroots and Government for Disaster Response." Public Policy Scholar Commons Lab, Wilson Center (October). http://www.wilsoncenter.org/publication-series/commons-lab.

Crutzen, P. J., and E. F. Stoermer. 2000. "The 'Anthropocene.'" *Global Change Newsletter* 41:17–18.

Csikszentmihályi, Chris, et al. 2007. "MIT Center for Future Civic Media: Engineering the Fifth Estate." http://civic.mit.edu/sites/civic.mit.edu/files/C4-Knight-foundation-proposal.pdf.

Csordas, Thomas J. 1994. *Embodiment and Experience: The Existential Ground of Culture and Self.* Cambridge: Cambridge University Press.

da Costa, Beatriz, and Philip Kavita, eds. 2008. *Tactical Biopolitics: Art, Activism, and Technoscience.* Cambridge, MA: MIT Press.

Darrah, Thomas H., Avner Vengosh, Robert B. Jackson, Nathaniel R. Warner, and Robert J. Poreda. 2014. "Noble Gases Identify the Mechanisms of Fugitive Gas Contamination in Drinking-Water Wells Overlying the Marcellus and Barnett Shales." *Proceedings of the National Academy of Sciences* 111 (39): 14076–81. doi: 10.1073/pnas.1322107111.

Daston, Lorraine. 2012. "The Sciences of the Archive." *Osiris* 27 (1): 156–87. http://www.jstor.org/stable/10.1086/667826.

Daston, Lorraine, and Peter Galison. 2007. *Images of Objectivity.* New York: Zone Books.
Davies, R., G. Foulger, A. Bindley, and P. Styles. 2013. "Induced Seismicity and Hydraulic Fracturing for the Recovery of Hydrocarbons." *Marine and Petroleum Geology* 45:171–85.
Davis, Devra. 2003. *When Smoke Ran Like Water: Tales of Environmental Deception and the Battle against Pollution.* New York: Basic Books.
———. 2007. *The Secret History of the War on Cancer.* New York: Basic Books.
Davis, Steven, and Christine Shearer. 2014. "Climate Change: A Crack in the Natural-Gas Bridge." *Nature* 514:436–37. doi: 10.1038/nature13927.
de Certeau, Michel. 1984. *The Practice of Everyday Life.* Berkeley: University of California Press.
Deiber, Ronald J., John G. Palfrey, Rafal Rohozinski, and Jonathan Zittrain. 2010. *Access Controlled.* Cambridge, MA: MIT Press.
Deleuze, Gilles, and Félix Guattari. 1987. *A Thousand Plateaus: Capitalism and Schizophrenia.* Minneapolis: University of Minnesota Press.
Demirjian, Joan. 2009. "Waterline Wait Wears on Gas-Well Neighbors." *Chagrin Valley Times*, February 11. http://www.chagrinvalleytoday.com/communities/chagrinvalleyarchives/article_cd517070-74e4-51ca-9ab9-3079a06ed2a0.html.
———. 2010. "Bainbridge Makes 'Demand' on Gas-Well Lawsuit." *Chagrin Valley Times*, October 27. http://www.neogap.org/neogap/2011/04/10/bainbridge-makes-demand-on-gas-well-lawsuit/.
Dennis, Brady, and Juliet Eilperin. 2017. "EPA Remains Top Target with Trump Administration Proposing 31 Percent Budget Cut." *Washington Post*, May 23.
Derrida, Jacques. 1998. *Archive Fever: A Freudian Impression.* Translated by Eric Prenowitz. Chicago: University of Chicago Press.
Devine, Shanna, and Tom Devine. 2013. "Deadly Dispersants in the Gulf: Are Public Health and Environmental Tragedies the New Norm for Oil Spill Cleanups?" *Government Accountability Project.* http://whistleblower.org/sites/default/files/Corexit_Report_Part1_041913_compressed.pdf.
Diamanti-Kandarakis, E., et al. 2009. "Endocrine-Disrupting Chemicals: An Endocrine Society Scientific Statement." *Endocrine Reviews* 30 (4): 293–342. http://www.endo-society.org/journals/scientificstatements/upload/edc_scientific_statement.pdf.
Di Chiro, Giovanna. 2010. "Polluted Politics? Confronting Toxic Discourse, Sex Panic, and Eco-Normativity." In *Queer Ecologies: Sex, Nature, Politics, Desire*, edited by C. Mortimer-Sandilands and B. Erickson, 199–230. Bloomington: Indiana University Press.
DiSalvo, Carl. 2009. "Design and the Construction of Publics." *Design Issues* 25 (1): 48–63.
DiSalvo, Carl, Kirsten Boehner, Nicholas A. Knouf, and Phoebe Sengers. 2009. "Nourishing the Ground for Sustainable HCI: Considerations from Ecologically Engaged Art." In *Proceedings of the SIGCHI Conference on Human Factors in Computing Systems*, 385–394.

DiSalvo, Carl, Hrönn Brynjarsdóttir, and Phoebe Sengers. 2010. "Mapping the Landscape of Sustainable HCI." In *Proceedings of the SIGCHI Conference on Human Factors in Computing Systems*, 1975–1984.
DiSalvo, Carl, Marti Louw, Julina Coupland, and MaryAnn Steiner. 2009. "Local Issues, Local Uses: Tools for Robotics and Sensing in Community Contexts." In *Proceedings of the ACM Conference on Creativity and Cognition*, 245–54.
DiSalvo, Carl, and Jonathan Lukens. 2009. "Towards a Critical Technological Fluency: The Confluence of Speculative Design and Community Technology Programs." In *Proceedings of the 2009 Digital and Arts and Culture Conference*, 1–5.
Dodds, E. C., and W. Lawson. 1936. "Synthetic Oestrogenic Agents without the Phenanthrene Nucleus." *Nature* 137 (3476): 996.
Dominguez-Bello, M. G., E. K. Costello, M. Contreras, M. Magris, G. Hidalgo, N. Fierer, and R. Knight. 2010. "Delivery Mode Shapes the Acquisition and Structure of the Initial Microbiota across Multiple Body Habitats in Newborns." *Proceedings of the National Academy of Sciences USA* 107 (26): 11971–75.
Doran, David, and Bob Cather. 2013. *Construction Materials Reference Book*. 2nd ed. New York: Routledge.
Dosemagen, Shannon, Sara Wylie, and Jeff Warren. 2011. "Grassroots Mapping: Creating a Participatory Map-Making Process Centered on Discourse." *Journal of Aesthetics and Protest* 8.
Dourish, Paul. 2010. "HCI and Environmental Sustainability: The Politics of Design and the Design of Politics." In *Proceedings of the ACM Conference on Designing Interactive Systems*, 1–10.
Downey, Gary. 1998. *The Machine in Me: An Anthropologist Sits among Computer Engineers*. New York: Routledge.
———. 2007. "Low Cost, Mass Use: American Engineers and the Metrics of Progress." *History and Technology* 23 (3): 289–308.
Downey, Gary, and Juan Rogers. 1995. "On the Politics of Theorizing in a Postmodern Academy." *American Anthropologist* 97 (2): 269–81.
Doyle, Jack. 2004. *Trespass against Us: Dow Chemical and the Toxic Century*. Environment Health Fund. Monroe, ME: Common Courage Press.
Drilling Contractor. 2000. "Alabama Lawsuit Poses Threat to Hydraulic Fracturing across US," January–February, 42–43. www.iadc.org/dcpi/dc-janfeb00/j-coalbed.pdf.
Dugan, Emily. 2010. "Oil Spill Creates Huge Undersea 'Dead Zones.'" *Independent*, May 30.
Dumit, Joseph. 2004. *Picturing Personhood: Brain Scans and Biomedical Identity*. Princeton, NJ: Princeton University Press.
———. 2006. "Illnesses You Have to Fight to Get: Facts as Forces in Uncertain, Emergent Illnesses." *Social Science and Medicine* 62:577–90.
———. 2012. *Drugs for Life*. Durham, NC: Duke University Press.
———. 2014. *Fracking Systems, Earth Dispossessed: STS and Anthropological Research through Game Invention*. Rice University Anthropology Speaker Series. Departments of Anthropology and Science and Technology Studies, March 25.

Dumit, Joseph, and Robbie E. Davis-Floyd, eds. 1998. *Cyborg Babies: From Techno-Sex to Techno-Tots.* New York: Routledge.

ECCC (Energy and Climate Change Committee). 2010. "UK Deepwater Drilling—Implications of the Gulf of Mexico Oil Spill." Second report of Session 2010–11: Volume 1. London: House of Commons, United Kingdom. http://www.publications.parliament.uk/pa/cm201011/cmselect/cmenergy/450/450i.pdf.

Edelstein, Micheal. 2003. *Contaminated Communities: Coping with Residential Toxic Exposure.* 2nd ed. Boulder, CO: Westview, 2003.

EDF (Environmental Defense Fund). 1998. "Toxics Release Inventory Shows Need for Pollution Prevention." Press release, June 18. http://www.edf.org/pressrelease.cfm?contentID=1510.

Edney, M. H. 1990. *Mapping an Empire: The Geographical Construction of British India.* Chicago: University of Chicago Press.

Edwards, Paul N. 2010. *A Vast Machine: Computer Models, Climate Data, and the Politics of Global Warming.* Cambridge, MA: MIT Press.

Efstathiou, Jim, and Mark Drajem. 2013. "Drillers Silence Fracking Claims with Sealed Settlements." *Bloomberg News*, June 6. www.bloomberg.com/news/2013-06-06/drillers-silence-fracking-claims-with-sealed-settlements.html.

Efstathiou, Jim, and Jim Snyder. 2013. "Obama Seen Expanding Natural Gas Exports on Production Records." *Bloomberg News*, May 13. https://www.bloomberg.com/news/articles/2013-05-13/obama-seen-expanding-natural-gas-exports-on-production-records.

Eglash, Ron. 1999. *African Fractals: Modern Computing and Indigenous Design.* New Brunswick, NJ: Rutgers University Press.

Eglash, Ron, Audrey Bennett, Casey O'Donnell, Sybillyn Jennings, and Margaret Cintorino. 2006. "Culturally Situated Design Tools: Ethnocomputing from Field Site to Classroom." *American Anthropologist* 108 (2): 347–62.

Ehn, P., and Morten Kyng. 1991. "Cardboard Computers: Mocking-It-up or Hands-on the Future." In *Design at Work,* edited by Joan Greenbaum and Morten Kyng, 169–196. Hillsdale, NJ: Laurence Erlbaum.

EIA (Energy Information Administration). 2009. "Top 100 Oil and Gas Fields." https://web.archive.org/web/20100827024714/https://www.eia.gov/pub/oil_gas/natural_gas/data_publications/crude_oil_natural_gas_reserves/current/pdf/top100fields.pdf.

———. 2013a. "EIA Drilling Productivity Report for Center on Global Energy Policy, Columbia University." http://www.eia.gov/pressroom/presentations/sieminski_10292013_drilling.pdf.

———. 2013b. "Rankings: Natural Gas Marketed Production, 2015." http://www.eia.gov/state/rankings/#/series/47.

———. 2015. "Natural Gas Weekly Update." http://www.eia.gov/naturalgas/weekly/.

Energy and Commerce Committee Democrats. 2011. "Waxman, Markey, and DeGette Investigation Finds Continued Use of Diesel in Hydraulic Fracturing Fluids," January 31.

Energy Policy Act. 2005. "Hydraulic Fracturing." Public Law 109–58. August 8, sec. 322 H.

Enteen, Jillana. 2006. "Spatial Conceptions of URLs: Tamil Eelam Networks on the World Wide Web." *New Media Society* 8:229.
EPA (Environmental Protection Agency). 2002. "EPA Statement Regarding Endocrine-Disruptor Low Dose Hypothesis," March 26. http://www.epa.gov/endo/pubs/edmvs/lowdosepolicy.pdf.
———. 2003. "How Are the Toxic Release Inventory Data Used? Government, Business, Academic, and Citizen Uses." EPA-260-R-002–004.
———. 2004. "Evaluation of Impacts to Underground Sources of Drinking Water by Hydraulic Fracturing of Coalbed Methane Reservoirs." EPA 816-R-04–003. http://water.epa.gov/type/groundwater/uic/class2/hydraulicfracturing/wells_coalbedmethanestudy.cfm.
———. 2010. "Analysis of Eight Oil Spill Dispersants Using In Vitro Tests for Endocrine and Other Biological Activity."
———. 2016. "Hydraulic Fracturing for Oil and Gas: Impacts from the Hydraulic Fracturing Water Cycle on Drinking Water Resources in the United States (Final Report)." Washington, DC: EPA/600/R-16/236F.
EPA Region 5. 2014. "Pollution/Situation Report: Statoil Eisenbarth Well Response—Removal Polrep," June 29. https://web.archive.org/web/20140809010406/http://theoec.org/sites/default/files/Eisenbarth%20well%20pad%20fire.pdf.
EPA Region 8. 2009. "Pavillion Area Groundwater Investigation SI—ARR, URS Operating Services, Inc. Contract No. EP-W-05-050." Revision, August 1.
———. 2011. "Draft Investigation of Ground Water Contamination Near Pavillion, Wyoming," December 8. http://www2.epa.gov/region8/draft-investigation-ground-water-contamination-near-pavillion-wyoming.
EPN (Environmental Protection Network). 2017. "Analysis of Trump Administration Proposals for FY2018 Budget for the Environmental Protection Agency," March 22. http://www.4cleanair.org/sites/default/files/Documents/EPA_Budget_Analysis_EPN_3-22-2017.pdf.
Epstein, Steven. 1996. *Impure Science: AIDS, Activism and the Politics of Knowledge.* Berkeley: University of California Press.
Erbentraut, Joseph. 2015. "How One Community Is Kicking the Koch Brothers' Harmful Black Dust Out of Their Neighborhood." *Huffington Post*, February 27. http://www.huffingtonpost.com/2015/02/27/chicago-petcoke-koch-brothers_n_6755040.html.
Eriksen, M., L. C. M. Lebreton, H. S. Carson, M. Thiel, C. J. Moore, et al. 2014. "Plastic Pollution in the World's Oceans: More Than 5 Trillion Plastic Pieces Weighing over 250,000 Tons Afloat at Sea." *PLOS One* 9 (12): e111913. doi: 10.1371/journal.pone.0111913.
ES (Endocrine Society). 2011. "Society Endorses EDC Bill to Strengthen NTP and NIEHS Programs." *Endocrine Insider*, July 13.
Esswein, Eric J., Michael Breitenstein, John Snawder, Max Kiefer, and W. Karl Sieber. 2013. "Occupational Exposures to Respirable Crystalline Silica during Hydraulic Fracturing." *Journal of Occupational Environmental Hygiene* 10 (7): 347–56. doi: 10.1080/15459624.2013.788352.

Esswein, Eric J., John Snawder, Bradley King, Michael Breitenstein, Marissa Alexander-Scott, and Max Kiefer. 2014. "Evaluation of Some Potential Chemical Exposure Risks during Flowback Operations in Unconventional Oil and Gas Extraction: Preliminary Results." *Journal of Occupational and Environmental Hygiene* 11 (10): D174–D184. doi: 10.1080/15459624.2014.93396.

Ester, Leslie. 2005. *Synthetic Worlds: Nature, Art and the Chemical Industry.* Harmondsworth, UK: Reaktion.

EV *World.* 2005. "Cape Wind Suffers Another Blow," March 19.

EWG (Environmental Working Group). 2005. "Body Burden: The Pollution in Newborns, a Benchmark Investigation of Industrial Chemicals, Pollutants and Pesticides in Umbilical Cord Blood," July 14. http://www.ewg.org/research/body-burden-pollution-newborns.

———. 2008. "Colorado's Chemical Injection," June 10. https://web.archive.org/web/20130301024304/http://www.ewg.org/research/colorados-chemical-injection.

———. 2009. "Free Pass for Oil and Gas: Environmental Protections Rolled Back as Western Drilling Surges: Oil and Gas Industry Exemptions."

Faber, Daniel. 2008. *Capitalizing on Environmental Justice: The Polluter-Industrial Complex in the Age of Globalization.* Lanham, MD: Rowman and Littlefield.

———. 2011. "The Sociology of Environmental Justice: Merging Research and Action." In *Sociologists in Action,* edited by Kathleen Odell Korgen, Jonathan White, and Shelley White. Thousand Oaks, CA: Sage.

Farman, Jason. 2010. "Mapping the Digital Empire: Google Earth and the Process of Postmodern Cartography." *New Media and Society* 12 (6): 869–88.

Farney, Jonathan. 2013. "Obama's Energy Plan Takes Aim at Coal Plants." Associated Press, June 25. http://www.startribune.com/obama-s-climate-plan-takes-aim-at-coal-plants/213093021/.

Farrell, Patrick. 2005. "Methamphetamine Fuels the West's Oil and Gas Boom." *High Country News,* October 3.

Fassin, Didier. 2007. *When Bodies Remember: Experiences and Politics of* AIDS *in South Africa.* California Series in Public Anthropology. Berkeley: University of California Press.

Fausto-Sterling, Anne. 2005. "The Bare Bones of Sex: Part 1—Sex and Gender." *SIGNS: Journal of Women in Culture and Society* 30 (2): 1491–528.

Feller, Joseph, Brian Fitzgerald, Scott Hissam, and Karim Lakhani, eds. 2005. *Perspectives on Free and Open Source Software.* Cambridge, MA: MIT Press.

Fenichell, Stephan. 1996. *Plastics: The Making of a Synthetic Century.* New York: HarperCollins.

Ferguson, James. 1994. *The Anti-Politics Machine: "Development," Depoliticization, and Bureaucratic Power in Lesotho.* Minneapolis: University of Minnesota Press.

———. 2005. "Seeing Like an Oil Company: Space, Security, and Global Capital in Neoliberal Africa." *American Anthropologist* 107 (3): 377–82.

———. 2006. "New Spatializations of Order and Disorder in Neoliberal Africa." In *Global Shadows: Essays on Africa in the Neoliberal World Order.* Durham, NC: Duke University Press.

Feriancek, Jeanine. 2008. "Minerals and Mining Law." *Findlaw*, March 26. http://corporate.findlaw.com/business-operations/minerals-amp-mining-law.html.
Fig, David. 2005. "Manufacturing Amnesia: Corporate Social Responsibility in South Africa." *International Affairs* 81 (3): 599–617.
Finley, Bruce. 2013a. "Big Fracking Fluid Spill Near Windsor Is Cleaned Up, Company Says." *Denver Post*, February 14. http://www.denverpost.com/ci_22593942/big-fracking-fluid-spill-near-windsor-is-cleaned.
———. 2013b. "Seismic Surveying Rattles Colorado Homeowners." *Denver Post*, March 16.
Fischer, Douglas. 2013. "'Dark Money' Funds Climate Change Denial Effort." *Scientific American*, December 23. http://www.scientificamerican.com/article/dark-money-funds-climate-change-denial-effort/.
Fischer, Michael M. J. 1999a. "Emergent Forms of Life: Anthropologies of Late or Postmodernities." *Annual Review in Anthropology* 28:455–78.
———. 1999b. "Worlding Cyberspace: Towards a Crucial Ethnography in Time, Space, Theory." In *Critical Anthropology Now: Unexpected Context, Shifting Constituencies, Changing Agendas*, edited by George Marcus, 245–304. Santa Fe, NM: SAR Press.
———. 2003. *Emergent Forms of Life and the Anthropological Voice.* Durham, NC: Duke University Press.
———. 2009. *Anthropological Futures.* Durham, NC: Duke University Press.
———. 2010. "Dr. Folkman's Decalogue and Network Analysis." In *A Reader in Medical Anthropology*, edited by B. Good et al. Hoboken, NJ: Wiley-Blackwell.
Fisher, Max. 2012. "Stratfor Is a Joke and So Is Wikileaks for Taking It Seriously." *Atlantic*, February 27.
Fisher, P. M. 1996. "Science and Subpoenas: When Do the Courts Become Instruments of Manipulation?" *Law Contemporary Problems* 59:159–67.
Fitzgerald, Deborah. 1993. "Farmers Deskilled: Hybrid Corn and Farmers' Work." *Technology and Culture* 34:324–43.
———. 2003. *Every Farm a Factory: The Industrial Ideal in American Agriculture.* New Haven, CT: Yale University Press.
Fitzgerald, Sandy. 2012. "EPA: Ethylene Cracker Plant Could Cause Major Pollution." *Newsmax.com*, March 27. http://www.newsmax.com/US/pollution-epa-pennsylvania/2012/03/27/id/433951/.
Fontaine, Tom. 2014. "Truck Crash Causes Fracking Water, Diesel Spill into Chartiers Creek." *Pittsburgh Tribune*, April 21.
Forsythe, Diana. 2001. *Studying Those Who Study Us: An Anthropologist in the World of Artificial Intelligence.* Stanford, CA: Stanford University Press.
Fortun, Kim. 2001. *Advocacy after Bhopal: Environmentalism, Disaster, New Global Orders.* Chicago: University of Chicago Press.
———. 2004. "From Bhopal to the Informating of Environmentalism: Risk Communication in Historical Perspective." *Osiris* 19:283–96.
———. 2009. "Biopolitics and the Informating of Environmentalism." February 9. http://figuringoutmethods.wikispaces.com/file/view/KFLivelyCapitalFeb09.pdf.

———. 2011. "Biopolitics and the Informating of Environmentalism." In *Lively Capital: Biotechnologies, Ethics, and Governance in Global Markets*, edited by Kaushik Sunder Rajan. Durham, NC: Duke University Press.

———. 2012. "Ethnography in Late Industrialism." *Cultural Anthropology* 27 (3): 446–64.

Fortun, Kim, Erik Bigras, Brandon Costelloe-Kuehn, Allison Kenner, Tahereh Saheb, Jerome Crowder, et al. 2013. "Asthma, Culture, and Cultural Analysis." In *Heterogeneity in Asthma: Translational Profiling and Phenotyping*, edited by Allan Brasier, 321–32. New York, NY: Springer.

Fortun, Kim, and Michael Fortun. 2005. "Scientific Imaginaries and Ethical Plateaus in Contemporary U.S. Toxicology." *American Anthropologist* 107 (1): 43–54.

Foster, Andrea, Peter Fairley, and Rick Mullin. 1998. "Scorecard Hits Home: Web Site Confirms Internet's Reach." *Chemical Week*, June 3.

Foucault, Michel. 1969. *The Archaeology of Knowledge*. Translated by A. M. Sheridan Smith. London: Routledge, 2002.

———. 1979. *Discipline and Punish: The Birth of the Prison*. New York: Penguin.

———. 2007. *Security, Territory, Population: Lectures at the Collège de France 1977–1978*. Translated by Graham Burchell. Basingstoke, UK: Palgrave Macmillan.

Fowler, Tom. 2010. "Help, There's a Landman at my Door." *Fuel Fix Blog*, March 18. http://fuelfix.com/energywatch/2010/03/18/help-theres-a-landman-at-my-door/.

Fox, Josh. 2010. *Gas Land*. HBO Documentary Film and International WOW Productions.

———. 2013. *Gasland Part II*. HBO Documentary Film and International WOW Productions.

Frankowski, Eric. 2008. "Industry Secrets and a Nurse's Story." *High Country News*, July 28. http://www.hcn.org/wotr/gas-industry-secrets-and-a-nurses-story.

Frazier, Reid R. 2013. "Shale Gas Fuels Gulf's Chemical Industry." *Allegheny Front, Environmental Radio*, October 18. http://www.alleghenyfront.org/story/shale-gas-fuels-gulfs-chemical-industry.

———. 2016. "When a Fracking Boom Goes Bust." *Inside Energy*, March 28. http://insideenergy.org/2016/03/28/when-a-fracking-boom-goes-bust/.

Frey, David. 2006. "Something in the Air?" *Aspen Daily News*, May 3.

Frickel, Scott, and Michelle Edwards. 2014. "Untangling Ignorance in Environmental Risk Assessment." In *Powerless Science? Science and Politics in a Toxic World*, edited by Soraya Boudia and Nathalie Jas. New York: Berghahn.

Frickel, Scott, Sara Gibbon, Jeff Howard, Joanna Kempner, Gwen Ottinger, and David Hess. 2010. "Undone Science: Charting Social Movement and Civil Society Challenges to Research-Agenda-Setting." *Science, Technology and Human Values* 35:444–73.

Friedenberger, Amy. 2011. "Pitt Professor Resigns over Beliefs about Marcellus Shale." *Pitt News*, April 10.

Fritz, S., I. McCallum, C. Schill, C. Perger, R. Grillmayer, F. Achard, et al. 2009. "Geo-Wiki.Org: The Use of Crowd-Sourcing to Improve Global Land Cover." *Remote Sensing* 1 (3): 345–54.

Frolik, Cory. 2006. "Halliburton Spill Results in Acid Cloud: More Than 220 People Evacuated to Mall." *Daily Times* (Farmington, NM), June 7.

Frost, Samantha. 2011. "The Implications of the New Materialisms for Feminist Epistemology." In *Feminist Epistemology and Philosophy of Science: Power in Knowledge*, edited by H. E. Grasswick, 69–83. New York, NY: Springer.

Fry, Richard, and Rakesh Kochhar. 2014. "America's Wealth Gap between Middle-Income and Upper-Income Families Is Widest on Record." Pew Research Center, December 17. http://www.pewresearch.org/fact-tank/2014/12/17/wealth-gap-upper-middle-income/.

FS (Forest Service). 2005. "Environmental Assessment Spaulding Peak Natural Gas Exploration and Development Area Wide Plan Grand Valley Ranger District Grand Mesa, Uncompahgre and Gunnison National Forests Delta County, Colorado, United States." Department of Agriculture and Forest Service (October).

Fuller, Matthew. 2008. *Software Studies: A Lexicon.* Cambridge, MA: MIT Press.

Fung, Archon, and D. O'Rourke. 2000. "Reinventing Environmental Regulation from the Grassroots Up: Explaining and Expanding the Success of the Toxics Release Inventory." *Environmental Management* 25:115–27.

Galison, Peter. 1997. *Image and Logic: A Material Culture of Microphysics.* Chicago: University of Chicago Press.

Galloway, A. R. 2004. *Protocol: How Control Exists after Decentralization.* Cambridge, MA: MIT Press.

GAO (General Accounting Office). 2003. "Energy Task Force: Process Used to Develop the National Energy Policy." Report to Congressional Requesters, GAO-03-894, August.

Garcia, David, and Geert Lovink. 1997. "The ABCs of Tactical Media." Nettime. http://www.ljudmila.org/nettime/zkp4/74.htm.

GCM (Global Community Monitor). 2011. "GASSED! Citizen Investigation of Toxic Air Pollution from Natural Gas Development," July 21.

Gellman, B., and L. Poitras. 2013. "U.S., British Intelligence Mining Data from Nine U.S. Internet Companies in Broad Secret Program." *Washington Post*, June 6.

George, Dev. 2000. "Schlumberger Introduces Q-Marine and Q-Borehole Seismic Technology at Annual SEG Conference in Calgary." *Oil and Gas Online*, August 8. http://www.oilandgasonline.com/doc/schlumberger-introduces-q-marine-q-borehole-s-0001.

Gerlach, Luther P., and Virginia H. Hine. 1970. *People, Power, Change: Movements of Social Transformation.* Indianapolis: Bobbs-Merrill.

Geurts, Kathryn Linn. 2002. *Culture and the Senses: Embodiment, Identity, and Well-Being in an African Community.* Berkeley: University of California Press.

Gibson, Connor. 2011. "Big Oil Front Group Poured Millions into Lobby against Fracking Regulation." *Polluter Watch Blog*, February 18. http://www.polluterwatch.com/blog/big-oil-front-group-poured-millions-lobby-against-fracking-regulation.

Gillespie, Tarleton. 2013. "The Relevance of Algorithms." In *Media Technologies*, edited by Tarleton Gillespie, Pablo Boczkowski, and Kirsten Foot. Cambridge, MA: MIT Press.

Goldenberg, Suzanne. 2013. "Children Given Lifelong Ban on Talking about Fracking." *Guardian*, August 5.

Goldstein, Josh, and Juliana Rotich. 2008. "Digitally Networked Technology in Kenya's Crisis." Internet and Democracy Case Study Series, September 29. Harvard, MA: Berkman Center.

Graham, M. 2010. "Neogeography and the Palimpsests of Place: Web 2.0 and the Construction of a Virtual Earth." *Tijdschrift voor Economische en Sociale Geografie* 101 (4): 422–36.

Grand Junction Daily Sentinel. 2006. "Landowners in the Dark on 'Split-Estate' Auctions," January 18.

Graves, John. 2012. *Fracking: America's Alternative Energy Revolution*. Ventura, CA: Safe Harbor.

Gray, D. 2006. "Water from Gas Production Spills into Mamm Creek." *Glenwood Springs Post Independent*, February 8.

Green, R., R. Hauser, A. M. Calafat, J. Weuve, T. Schettler, S. Ringer, K. Huttner, and H. Hu. 2005. "Use of Di(2-ethylhexyl) phthalate-containing Medical Products and Urinary Levels of Mono(2-ethylhexyl) Phthalate in Neonatal Intensive Care Unit Infants." *Environmental Health Perspectives* 113:1222–25.

Greenpeace. 2009. Letter to Jack Gerard, President of American Petroleum Institute, August 12. https://www.desmogblog.com/sites/beta.desmogblog.com/files/GP%20API%20letter%20August%202009-1.pdf.

Greenwald, G., and E. MacAskill. 2013. "NSA Prism Program Taps into User Data of Apple, Google and Others." *Guardian*, June 6.

Grefe, Edward, and Marty Linsky. 1995. *The New Corporate Activism: Harnessing the Power of Grassroots Tactics for Your Organization*. New York: McGraw-Hill.

Grimlin, Debra. 2007. "What Is 'Body Work'? A Review of the Literature." *Sociology Compass* 1 (1): 353–70.

Grossman, E., L. Vandenberg, K. Thayer, and L. Birnbaum. 2015. "Theodora (Theo) Colborn: 1927–2014." *Environmental Health Perspectives* 123:A54. http://dx.doi.org/10.1289/ehp.1509743.

Grow, B., J. Schneyer, and A. Driver. 2012. "Special Report: The Casualties of Chesapeake's 'Land Grab' across America." Reuters, October 2.

Guillette, Louis J., Jr., and D. Andrew Crain. 2000. *Environmental Endocrine Disruptors: An Evolutionary Perspective*. New York: Taylor and Francis.

GWPC (Ground Water Protection Council). 2009. "Modern Shale Gas Development in the United States," April. https://energy.gov/sites/prod/files/2013/03/f0/ShaleGasPrimer_Online_4-2009.pdf.

Gwynne, Sam. 1999. "Spies Like Us." *Time Magazine*, January 25.

Hadden, Susan. 1994. "Citizen Participation in Environmental Policy Making." In *Learning from Disaster: Risk Management after Bhopal*, edited by Sheila Jasanoff, 179–191. Philadelphia: University of Pennsylvania Press.

Halavais, A. 2000. "National Borders on the World Wide Web." *New Media and Society* 2 (1): 7–28.

Halber, Deborah. 1999. “Software to Aid Research on Earth's Crust.” MIT News Office, November 3. http://newsoffice.mit.edu/1999/gift-1103.

———. 2006. “Acoustic Data May Reveal Hidden Gas, Oil Supplies.” MIT News Office, December 21.

Hamilton, Douglas. 1998. *Fooling with Nature.* PBS program, June 2.

Hanel, Joe. 2008a. “Chemical Exposure Leads to Liver, Heart, Lung Failure.” *Durango* (CO) *Herald,* July 17.

———. 2008b. “Nurse Sick after Aiding Gas Worker: Driller Refuses to Disclose Chemicals That Would Have Helped Treatment.” *Durango* (CO) *Herald,* July 17.

———. 2008c. “S. Ute Land Was Site of Frac Fluid Spill.” *Durango* (CO) *Herald,* August 1.

Haraway, Donna. 1988. “Situated Knowledges: The Science Question in Feminism and the Privilege of Partial Perspectives.” *Feminist Studies* 14 (3): 575–99.

———. 1991. *Simians, Cyborgs, and Women: The Reinvention of Nature.* New York: Routledge.

———. 1997. *Modest_Witness@Second_Millennium.* New York: Routledge.

———. 2003. *The Companion Species Manifesto: Dogs, People, and Significant Otherness.* Chicago: Prickly Paradigm.

Harding, Susan. 2000. *The Book of Jerry Falwell.* Princeton, NJ: Princeton University Press.

———, ed. 2004. *The Feminist Standpoint Theory Reader: Intellectual and Political Controversies.* New York: Routledge.

Hargrove, Brantley. 2014. “Earthquakes and the Texas Miracle.” *D Magazine* (May). http://www.dmagazine.com/publications/d-magazine/2014/may/earthquakes-and-the-texas-miracle?ref=mostcommented.

Harley, J. B. 1988. “Maps, Knowledge, and Power.” In *The Iconography of Landscape: Essays on the Symbolic Representation, Design and Use of Past Environments.* New York: Cambridge University Press. 277–312.

———. 1992. “Deconstructing the Map.” In *Writing Worlds: Discourse, Text and Metaphor in the Representation of Landscape,* edited by Trevor J. Barnes and James S. Duncan, 231–47. London: Routledge.

Hartmann, T. 2002. *Unequal Protection: The Rise of Corporate Dominance and the Theft of Human Rights.* San Francisco: Berrett-Koehler.

Harwood, Matthew. 2013. “My Life in Circles: Why Metadata Is Incredibly Intimate.” American Civil Liberties Union Blog, July 29. https://www.aclu.org/blog/technology-and-liberty-national-security/my-life-circles-why-metadata-incredibly-intimate.

Hayles, N. Katherine. 2002. *Writing Machines.* Cambridge, MA: MIT Press.

Heiman, Jeremy. 2004. “Well Comes under Suspicion in West Divide Creek Gas Seep.” (Glenwood Springs, CO) *Post Independent,* April 18.

Heller, Peter. 2012. “The Fight over Fracking in Colorado's North Fork Valley.” *Bloomberg Businessweek,* July 12. https://www.bloomberg.com/news/articles/2012-07-12/the-fight-over-fracking-in-colorados-north-fork-valley.

Helmreich, Stefan. 1998. *Silicon Second Nature: Culturing Artificial Life in a Digital World.* Berkeley: University of California Press.

———. 2009a. *Alien Ocean: Anthropological Voyages in Microbial Seas.* Berkeley: University of California Press.

———. 2009b. "Intimate Sensing." In *Simulation and Its Discontents,* edited by Sherry Turkle, 129–150. Cambridge: MIT Press.

Herman, Andrew. 2000. *The World Wide Web and Contemporary Cultural Theory: Magic, Metaphor, Power.* New York: Routledge.

Hess, David. 1999. "Social Reporting: A Reflexive Law Approach to Corporate Social Responsiveness." *Journal of Corporation Law* 25 (1): 41–84.

———. 2001. "Regulating Corporate Social Performance: A New Look at Corporate Social Accounting, Auditing and Reporting." *Business Ethics Quarterly* 11 (2): 307–30.

———. 2004. "Guest Editorial: Health, the Environment and Social Movements." *Science as Culture* 13 (4): 421–27.

———. 2007a. *Alternative Pathways in Science and Industry: Activism, Innovation, and the Environment in an Era of Globalization.* Cambridge, MA: MIT Press.

———. 2007b. "Social Reporting and New Governance Regulation: The Prospects of Achieving Corporate Accountability through Transparency." *Business Ethics Quarterly* 17 (3): 453–76.

———. 2008. "The Three Pillars of Corporate Social Reporting as New Governance Regulation: Disclosure, Dialogue and Development." *Business Ethics Quarterly* 18 (4): 447–82.

———. 2015. "Undone Science, Industrial Innovation, and Social Movements." In *Routledge International Handbook of Ignorance Studies,* edited by Matthias Gross and Linsey McGoey. New York: Routledge.

Hightower, Jim. 1978. *Hard Tomatoes, Hard Times: A Report of the Agribusiness Accountability Project on the Failure of America's Land Grant College Complex.* Rochester, VT: Schenkman.

Hill, E. L. 2012. "Unconventional Natural Gas Development and Infant Health: Evidence from Pennsylvania." Charles Dyson School of Applied Economics and Management at Cornell University.

Hindman, Matthew. 2009. *The Myth of Digital Democracy.* Princeton, NJ: Princeton University Press.

Hirsch, Tad, and John Henry. 2005. "TXTmob: Text Messaging for Protest Swarms." Proceedings of the Conference on Human Factors in Computing Systems, Portland, OR.

Ho, Karen. 2009. *Liquidated: An Ethnography of Wall Street.* Durham, NC: Duke University Press.

Hodges, Tony. 2004. *Angola: Anatomy of an Oil State.* Oxford: James Currey.

Hofrichter, Richard. 1993. *Toxic Struggles: The Theory and Practice of Environmental Justice.* Gabriola Island, BC: New Society Publishers.

Holmes, Douglas, and George Marcus. 2008. "Collaboration Today and the Classic Scene of Fieldwork Encounter." *Collaborative Anthropologies* (1): 81–101. Lincoln: University of Nebraska Press.

Horn, Steve. 2011. "Smeared but Still Fighting, Cornell's Tony Ingraffea Debunks Gas Industry Myths." *DeSmogBlog.com*, December 2. http://www.desmogblog.com/smeared-still-fighting-cornell-s-tony-ingraffea-debunks-gas-industry-myths.

Horwitt, Dusty. 2008. "Coming Up Dry." Environmental Working Group, July 17. http://www.ewg.org/research/coming-dry.

———. 2010. "Drilling around the Law." Environmental Working Group, January 20. http://www.ewg.org/drillingaroundthelaw.

———. 2011. "Cracks in the Façade: 25 Years Ago, EPA Linked Fracking to Water Contamination." Environmental Working Group, August 3. http://www.ewg.org/research/cracks-facade.

Houen, Alex, ed. 2014. *Critical Terrorism Studies: States of War since 9/11: Terrorism, Sovereignty and the War on Terror.* New York: Taylor and Francis.

House Committee on Energy and Commerce. 2013. Statement of Dr. Ernest J. Moniz, U.S. Department of Energy, June 13. https://energy.gov/articles/secretary-ernest-j-moniz-s-written-testimony-climate-change-house-committee-energy-and.

Howarth, Robert W. 2014. "A Bridge to Nowhere: Methane Emissions and the Greenhouse Gas Footprint of Natural Gas." *Energy Science and Engineering* 2 (2): 47–60.

Howarth, Robert W., Renee Santoro, and Anthony Ingraffea. 2011. "Methane and the Greenhouse Gas Footprint of Natural Gas from Shale Formations." *Climatic Change*, March 13.

Howarth, Robert W., D. Shindell, R. Santoro, A. Ingraffea, N. Phillips, and A. Townsend-Small. 2012. "Methane Emissions from Natural Gas Systems." Background paper prepared for the National Climate Assessment, February 25. Reference number 2011–0003. http://www.eeb.cornell.edu/howarth/publications/Howarth_et_al_2012_National_Climate_Assessment.pdf.

Howdeshell, K. L., P. H. Peterman, B. M. Judy, J. A. Taylor, C. E. Orazio, R. L. Ruhlen, et al. 2003. "Bisphenol A Is Released from Polycarbonate Animal Cages into Water at Room Temperature." *Environmental Health Perspectives* 111:1180–87.

Howe, Jeff. 2008. *Crowdsourcing: Why the Power of the Crowd Is Driving the Future of Business.* New York: Random House.

Hrin, Eric. 2012. "Acid Spill Reported in Leroy Co." (Scranton, PA) *Times-Tribune*, November 12.

Humphreys, Macartan, Jeffrey Sachs, and Joseph Stiglitz, eds. 2007. *Escaping the Resource Curse.* New York: Columbia University Press.

Hunn, David. 2016. "Woe in the Oilfield: 213 Companies Have Now Declared Bankruptcy." *Fuel Fix*, October 25. http://fuelfix.com/blog/2016/10/25/213-oil-companies-have-declared-bankruptcy-how-hard-it-is-to-be-an-oilfield-service-company/.

Hurdle, Jon. 2015. "FracFocus Upgrades Availability of Data on Oil and Gas Operations." State Impact, National Public Radio, May 8. https://stateimpact.npr.org/pennsylvania/2015/05/08/fracfocus-upgrades-availability-of-data-on-oil-gas-operations/.

IHS. 2011. "The Economic and Employment Contributions of Shale Gas in the United States." Prepared for America's Natural Gas Alliance, December. http://anga.us/media/content/F7D1750E-9C1E-E786-674372E5D5E98A40/files/shale-gas-economic-impact-dec-2011.pdf.

———. 2014. "Energy 50, The Definitive Annual Ranking of the World's Largest Listed Energy Firms." https://www.ihs.com/pdf/IHS-Energy-50-Final-2014_209412110913052332.pdf.

Innovation Quarterly. 2008. "A Day in the Life of . . . Olivier Peyret." http://www.eirma.org/eiq/015/pages/eiq-2008-015-0002.html.

IOGCC (Interstate Oil and Gas Compact Commission). 2005. "Mature Region, Youthful Potential: Oil and Natural Gas Resources in the Appalachian and Illinois Basins." Report by the Appalachian and Illinois Basin Directors of the Interstate Oil and Gas Compact Commission, September.

Jackson, R. B., A. Vengosh, T. Darrah, et al. 2013. "Increased Stray Gas Abundance in a Subset of Drinking Water Wells Near Marcellus Shale Gas Extraction." *PNAS: Proceedings of the National Academy of Sciences* 110 (28): 11250–55. doi: 10.1073/pnas.1221635110.

Jacobs, Emma. 2011a. "The Front Lines of Fracking Get Personal." *Innovation Trail*, August 15. http://innovationtrail.org/post/front-lines-fracking-get-personal.

———. 2011b. "How to Monitor the Gas Industry Online." *Innovation Trail*, August 23. http://innovationtrail.org/post/how-monitor-gas-industry-online.

Jacobson, Rebecca. 2015. "Fracking Brine Leak in North Dakota Reaches Missouri River, Prompts State Democrats to Call for More Regulation." *PBS Newshour*, January 26. http://www.pbs.org/newshour/rundown/fracking-brine-leak-north-dakota-reaches-missouri-river-prompts-state-democrats-call-regulation/.

Jalbert, Kirk, Abby Kinchy, and Simona Perry. 2014. "Civil Society Research and Marcellus Shale Natural Gas Development: Results of a Survey of Volunteer Water Monitoring Organizations." *Journal of Environmental Studies and Sciences* 4 (1): 2190–6483.

Jasanoff, Sheila, and Sang-Hyun Kim. 2009. "Containing the Atom: Sociotechnical Imaginaries and Nuclear Regulation in the U.S. and South Korea." *Minerva* 47 (2): 119–46.

———. 2013. "Sociotechnical Imaginaries and National Energy Policies." *Science as Culture* 22 (2): 189–96. doi: 10.1080/09505431.2013.786990.

Jenkins, Henry. 2006. *Convergence Culture*. New York: New York University Press.

Johnson, Richard, and Timothy Gower. 2008. *The Sugar Fix: The High-Fructose Fallout That Is Making You Fat and Sick*. New York: Rodale.

Johnston, Laura. 2009. "Ohio Laws Governing Gas Drilling among Most Lenient in Nation, Experts Say." (Cleveland, OH) *Plain Dealer*, July 5. http://blog.cleveland.com/metro/2009/07/ohio_panel_that_hears_gas_well.html.

Judson, Richard S., Matthew T. Martin, David M. Reif, Keith A. Houck, Thomas B. Knudsen, Daniel M. Rotroff, et al. 2010. "Analysis of Eight Oil Spill Dispersants Using Rapid, In Vitro Tests for Endocrine and Other Biological Activity." *Environmental Science and Technology* 44 (15): 5979–85.

Junkins, Casey. 2013. "Landowner Discovers Explosives Company Testing for Natural Gas Deposits." *Intelligencer/Wheeling News-Register*, September 1.
Juris, Jeffrey S. 2005. "The New Digital Media and Activist Networking within Anti-Corporate Globalization Movements." *Annals of the American Academy of Political and Social Science* 597:189–208.
———. 2008. *Networking Futures: The Movements against Corporate Capitalism*. Durham, NC: Duke University Press.
Kahn, Richard, and Douglas Kellner. 2004. "New Media and Internet Activism: From the 'Battle of Seattle' to Blogging." *New Media and Society* 6:88.
Kaiser, David, ed. 2010. *Becoming MIT: Moments of Decision*. Cambridge, MA: MIT Press.
———. 2011. "The Search for Clean Cash." *Nature* 472:30–31.
Kalfayan, Leonard. 2008. *Production Enhancement with Acid Stimulation*. 2nd ed. Tulsa, OK: PennWell.
Kamalick, Joe. 2009. "Frack Attack on US Natgas." *ICIS Chemical Business* (June): 22–28.
Karion, A., C. Sweeney, G. Petron, G. Frost, R. M. Hardesty, J. Kofler, et al. 2013. "Methane Emissions Estimate from Airborne Measurements over a Western United States Natural Gas Field." *Geophysical Research Letters* 40:4393.
Karkkainen, Bradley C. 2001. "Information as Environmental Regulation: TRI and Performance Benchmarking, Precursor to a New Paradigm?" *Georgetown Law Journal* 89:257–370.
Kassotis, Christopher D., Donald E. Tillitt, J. Wade Davis, Annette M. Hormann, and Susan C. Nagel. 2014. "Estrogen and Androgen Receptor Activities of Hydraulic Fracturing Chemicals and Surface and Ground Water in a Drilling-Dense Region." *Endocrinology* 155 (3): 897–907.
Kawai, K., S. Murakami, E. Senba, T. Yamanaka, Y. Fujiwara, C. Arimura, T. Nozaki, M. Takii, and C. Kubo. 2007. "Changes in Estrogen Receptors α and β Expression in the Brain of Mice Exposed Prenatally to Bisphenol A." *Regulatory Toxicology and Pharmacology*, 47 (2): 166–70. http://dx.doi.org/10.1016/j.yrtph.2006.04.002.
Keck, F., and A. Lakoff, eds. 2013. "Sentinel Devices." *Limn*, issue 3. http://limn.it/issue/03/.
Kelso, Matt. 2014. "Updated PA Data and Trends." FracTracker Alliance, December 8. http://www.fractracker.org/2014/12/updated-pa-data-trends/.
Kelty, Chris. 2008. *Two Bits: The Cultural Significance of Free Software and the Internet*. Durham, NC: Duke University Press.
Kemp, John. 2014. "The Real Shale Revolution." Reuters, July 14. http://www.reuters.com/article/2014/07/15/shale-usa-kemp-idUSL6N0PP3FR20140715.
Kenworthy, Tom. 2010. "Onshore Oil and Gas Drilling Done Right: Obama Administration Moves to Protect Lands in American West." Center for American Progress, May 25.
Kerry, John. 2009. The Endocrine Disruption Prevention Act. 111th Congress, 1st Session, S. 2828.

———. 2011. The Endocrine-Disrupting Chemicals Exposure Elimination Act of 2011. 112th Congress, 1st Session.
Khatib, F., F. DiMaio, Foldit Contenders Group, Foldit Void Crushers Group, S. Cooper, et al. 2011. "Crystal Structure of a Monomeric Retroviral Protease Solved by Protein Folding Game Players." *Nature Structural and Molecular Biology* 18:1175–77.
Kilburn, Kaye. 1995. "Hydrogen Sulfide and Reduced-Sulfur Gases Adversely Affect Neurophysiological Functions." *Toxicology and Industrial Health* 11 (2): 192–93.
———. 2003. "Effects of Hydrogen Sulfide on Neurobehavioral Function." *Southern Medical Journal* 96 (7): 639–46.
Kilburn, Kaye, J. D. Thrasher, and M. R. Gray. 2010. "Low-Level Hydrogen Sulfide and Central Nervous System Dysfunction." *Toxicology and Industrial Health* 26 (7): 387–405.
Kim, W. Y. 2013. "Induced Seismicity Associated with Fluid Injection into a Deep Well in Youngstown, Ohio." *Journal of Geophysical Research: Solid Earth* 118:3506–18. doi: 10.1002/jgrb.50247.
Kirby, David. 2013. "Corexit: An Oil Spill Solution Worse Than the Problem?" Takepart.com, April 22. https://web.archive.org/web/*/http://www.takepart.com/article/2013/04/17/corexit-deepwater-horizon-oil-spill.
Klein, Naomi. 2000. *No Logo: Taking Aim at the Brand Bullies.* Toronto: Knopf.
Klein, R., M. Kellermeier, D. Touraud, E. Müller, and W. Kunz. 2013. "Choline Alkylsulfates—New Promising Green Surfactants." *Journal of Colloid and Interface Science* 392:274–80.
Kleinman, Arthur, and Joan Kleinman. 1994. "How Bodies Remember: Social Memory and Bodily Experience of Criticism, Resistance, and Delegitimation Following China's Cultural Revolution." *New Literary History* 25 (3): 707–23.
Kliewer, Gene. 2014. "Seismic Vessel Survey Is Expanded to Include Additional Vessel Types." *Offshore Magazine*, March 10. http://www.offshore-mag.com/articles/print/volume-74/issue-3/geology-geophysics1/seismic-vessel-survey-is-expanded-to-include-additional-vessel-types.html.
Klopott, Freeman. 2013. "New York Assembly Approves Two-Year Moratorium on Fracking." *Bloomberg News*, March 6. https://www.bloomberg.com/news/articles/2013-03-06/new-york-assembly-approves-two-year-moratorium-on-fracking.
Kloppenburg, Jack, Jr. 1988. *First the Seed: The Political Economy of Plant Biotechnology, 1492–2000.* Cambridge: Cambridge University Press.
Knorr-Cetina, Karin. 1971. *The Manufacture of Knowledge: An Essay on the Constructivist and Contextual Nature of Science.* Oxford: Pergamon.
Knorr-Cetina, Karin, Hermann Strasser, and Hans Georg Zilian. 1975. *Determinants and Controls of Scientific Development.* Dordrecht, Netherlands: Reidel.
Knorr-Cetina, Karin, and Richard Whitley, eds. 1980. "The Social Process of Scientific Investigation." In *Sociology of Sciences Yearbook*, vol. 4. Dordrecht, Netherlands: Reidel.
Koehler, K. E., R. C. Voigt, S. Thomas, et al. 2003. "When Disaster Strikes: Rethinking Caging Materials." *Lab Animal* (NY) 32:24–27.

Kohler, Robert E. 1994. *Lords of the Fly: Drosophila Genetics and the Experimental Life.* Chicago: University of Chicago Press.
———. 2002. *Landscapes and Labscapes: Exploring the Lab-Field Border in Biology.* Chicago: University of Chicago Press.
Konschnik, K., M. Holden, and A. Shasteen. 2013. "Legal Fractures in Chemical Disclosure Laws: Why the Voluntary Chemical Disclosure Registry FracFocus Fails as a Regulatory Compliance Tool." Harvard Law School, Environmental Law Program, Policy Initiative Study, April 23.
Korfmacher, Katrina Smith, Walter A. Jones, Samantha L. Malone, and Leon F. Vinci. 2013. "Public Health and High Volume Hydraulic Fracturing." *New Solutions* 23 (1): 13–31.
Koscinski, P., and J. Stolle. 2011. "History, Industry Use, and Importance of the API Number to the Oil and Gas Producing States." Chesapeake Energy Corporation.
Kosnik, Renee Lewis. 2007. "The Oil and Gas Industry's Exclusions and Exemptions to Major Environmental Statutes." Oil and Gas Accountability Project, October.
Krimsky, Sheldon. 2000. *Hormonal Chaos.* Baltimore: Johns Hopkins University Press.
———. 2005. "The Weight of Scientific Evidence in Policy and Law." *American Journal of Public Health* 95 (1): S129–S136.
Krol, Ed. 1987. *Hitchhiker's Guide to the Internet.* Washington, DC: National Science Foundation.
———. 1992. *Whole Internet User's Guide and Catalog.* Sebastopol, CA: O'Reilly and Associates.
Kroll-Smith, Steve, ed. 2000. *Illness and the Environment: A Reader in Contested Medicine.* New York: New York University Press.
Kuklick, Henrika, and Robert E. Kohler, eds. 1996. *Science in the Field. Osiris*, vol. 11. Chicago: University of Chicago Press.
Kuns, Laura. 2006. Complaint investigation form. Lake County General Health District, ID# 20060499, July 24. Received from Kari Matsko.
LaGrone, C. C., S. A. Baumgartner, and R. A. Woodroof, Jr. 1985. "Chemical Evolution of a High Temperature Fracturing Fluid." *Society of Petroleum Engineers Journal* 25 (5): 623–28.
Lakoff, George, and Mark Johnson. 2003. *Metaphors We Live By.* Chicago: University of Chicago Press.
Landecker, Hannah. 2010. *Culturing Life: How Cells Became Technologies.* Cambridge, MA: Harvard University Press.
———. 2011. "Food as Exposure: Nutritional Epigenetics and the New Metabolism." *BioSocieties* 6:167–94.
Langston, Nancy. 2010. *Toxic Bodies: Endocrine Disruptors and the Lessons of History.* New Haven, CT: Yale University Press.
Latour, Bruno. 1987. *Science in Action: How to Follow Scientists and Engineers through Society.* Cambridge, MA: Harvard University Press.
———. 1988. *The Pasteurization of France.* Cambridge, MA: Harvard University Press.
———. 1990. "Drawing Things Together." In *Representation in Scientific Practice*, edited by Michael Lynch and Steve Woolgar. Cambridge, MA: MIT Press.

———. 2004. "Why Has Critique Run Out of Steam? From Matters of Fact to Matters of Concern." *Critical Inquiry* 30 (2): 225–48.

Latour, Bruno, and Steve Woolgar. 1979. *Laboratory Life: The Construction of Scientific Facts.* Princeton, NJ: Princeton University Press.

Lavandera, Ed. 2008. "Urban Drilling Bonanza Pits Neighbor against Neighbor." CNN, August 20.

Lax, Simon, Daniel P. Smith, Jarrad Hampton-Marcell, et al. 2014. "Longitudinal Analysis of Microbial Interaction between Humans and the Indoor Environment." *Science* 345 (6200): 1048–52.

LEAF (Legal Environmental Assistance Foundation). 1997. *Legal Environmental Assistance Foundation Inc. v. United States Environmental Protection Agency.* 118 F.3d 1467 (11th Circ. 1997).

———. 2001. *Legal Environmental Assistance Foundation Inc., Petitioner, v. United States Environmental Protection Agency, Respondent.* 276 F.3d 1253 (11th Cir. 2001).

Leiner, Barry M., et al. 1997. "The Past and Future History of the Internet." *Communications of the* ACM 40 (2).

Lele, Shreeniwas. 2006. Personal letter to the University Hospital Health Systems. Received from Kari Matsko.

Lerner, Steve. 2005. *Diamond: A Struggle for Environmental Justice in Louisiana's Chemical Corridor.* Cambridge, MA: MIT Press.

———. 2010. *Sacrifice Zones: The Front Lines of Toxic Chemical Exposure in the United States.* Cambridge, MA: MIT Press.

Lessig, Lawrence. 1999. *Code: And Other Laws of Cyberspace.* New York: Basic Books.

———. 2006. *Code Version 2.0.* New York: Basic Books.

Levy, Steven. 2010. "Exclusive: How Google's Algorithm Rules the Web." *Wired*, February 22. https://www.wired.com/2010/02/ff_google_algorithm/.

Liboiron, Max. 2012. "Redefining Pollution: Plastics in the Wild." Diss., New York University, Media, Culture, and Communication. http://www.academia.edu/3102377/Redefining_Pollution_Plastics_in_the_wild.

Lock, Margaret, and Judith Farquhar. 2007. *Beyond the Body Proper: Reading the Anthropology of Material Life.* Durham, NC: Duke University Press.

Lofholm, Nancy. 2006. "Watershed Drilling Plan Riles Palisade." *Denver Post*, July 17.

———. 2007. "Oil and Gas Wellness Checkup." *Denver Post*, May 6.

Lomax, Simon. 2012. "Key Concessions You'll Never Hear About in New TEDX Air Report." *Energy in Depth*, November 26. http://energyindepth.org/mtn-states/key-concessions-youll-never-hear-about-in-new-tedx-air-report/.

Lovink, Geert. 2002. *Dark Fiber: Tracking Critical Internet Culture.* Cambridge, MA: MIT Press.

———. 2008. *Zero Comments: Blogging and Critical Internet Culture.* New York: Routledge.

Lury, Celia. 1996. *Consumer Culture.* Oxford: Polity Press.

———. 2004. *Brands: The Logos of the Global Economy.* London: Routledge.

Lustgarten, Abrahm. 2008a. "Buried Secrets: Is Natural Gas Drilling Endangering U.S. Water Supplies?" ProPublica, November 13. http://www.propublica.org/article/buried-secrets-is-natural-gas-drilling-endangering-us-water-supplies-1113.

———. 2008b. "Drill for Natural Gas, Pollute Water." *Scientific American*, November 17.

———. 2010. "Chemicals Meant to Break Up BP Oil Spill Present New Environmental Concerns." ProPublica, April 30. http://www.propublica.org/article/bp-gulf-oil-spill-dispersants-0430.

———. 2011a. "Fracking Cracks the Public Consciousness in 2011." ProPublica, December 29.

______. 2011b. "Opponents to Fracking Disclosure Take Big Money from Industry." ProPublica, January 14. http://www.propublica.org/article/opponents-to-fracking-disclosure-take-big-money-from-industry#naturalgascaucus.

———. 2013. "EPA's Abandoned Wyoming Fracking Study One Retreat of Many." ProPublica, July 3.

———. 2014. "New York State Bans Fracking after Years of Delays and Debate, Gov. Andrew Cuomo Decides Risks Outweigh Rewards." ProPublica, December 17.

Lykes, M. B. 2001. "Creative Arts and Photography in Participatory Action Research in Guatemala." In *Handbook of Action Research: Participative Inquiry and Practice*, edited by P. W. Reason and H. Bradbury, 363–371. Thousand Oaks, CA: Sage.

Lynch, David, and Alan Bjerga. 2013. "Taxpayers Turn US Farmers into Fat Cats with Subsidies." *Bloomberg News*, September 9. https://www.bloomberg.com/news/articles/2013-09-09/farmers-boost-revenue-sowing-subsidies-for-crop-insurance.

Lynch, Michael, and Steve Woolgar. 1990. *Representation in Scientific Practice*. Cambridge, MA: MIT Press.

Macey, Gregg, Ruth Breech, Mark Chernaik, Caroline Cox, Denny Larson, Deb Thomas, and David O. Carpenter. 2014. "Air Concentrations of Volatile Compounds Near Oil and Gas Production: A Community-Based Exploratory Study." *Environmental Health* 13 (82). http://www.ehjournal.net/content/13/1/82.

Mackenzie, Adrian. 2006a. *Cutting Code: Software and Sociality*. New York: Peter Lang.

———. 2006b. "Java: The Practical Virtuality of Internet Programming." *New Media and Society* 8:441. http://nms.sagepub.com/content/8/3/441.

———. 2012. "More Parts Than Elements: How Databases Multiply." *Environment and Planning D: Society and Space* 30 (2): 335–50.

MacKenzie, Donald. 1990. *Inventing Accuracy: A Historical Sociology of Nuclear Missile Guidance*. Cambridge, MA: MIT Press.

Madrigal, Alexis. 2012. "I'm Being Followed: How Google—and 104 Other Companies—Are Tracking Me on the Web." *Atlantic*, February 29.

Mäkinen, Maarit, and Mary Wangu Kuira. 2008. "Social Media and Postelection Crisis in Kenya." *International Journal of Press/Politics* 13 (3): 328–35.

Maksoud, Judy. 2005. "Exploration Picks Up in Eastern Canada's Deepwater Basins." *Geology and Geophysics* (October).

Malewitz, Raymond. 2014. *The Practice of Misuse: Rugged Consumerism in Contemporary American Culture.* Stanford, CA: Stanford University Press.
Manama. 2010a. "Halliburton's Q2 Zooms Despite Spill Turmoil." *Oil and Gas News,* July 27.
———. 2010b. "Halliburton Q3 Profit Surges on Drilling." *Oil and Gas News,* October 25.
Manovich, Lev. 2002. *The Language of New Media.* Cambridge, MA: MIT Press.
Marcus, George E. 1995a. "Ethnography in/of the World System: The Emergence of Multi-Sited Ethnography." *Annual Review of Anthropology* 24:95–117.
———, ed. 1995b. *Technoscientific Imaginaries: Conversations, Profiles, and Memoirs.* Chicago: University of Chicago Press.
———, ed. 1998. *Corporate Futures: The Diffusion of the Culturally Sensitive Corporate Form.* Chicago: University of Chicago Press.
———, ed. 2000a. *Para-Sites: A Casebook against Cynical Reason.* Chicago: University of Chicago Press.
———. 2000b. "The Twistings of Geography and Anthropology in Winds of Millennial Transition." In *Cultural Turns/Geographical Turns: Perspectives on Cultural Geography,* edited by I. Cook, D. Crouch, S. Naylor, and J. R. Ryan, 13–25. New York: Prentice Hall.
Marcus, George E., and Michael M. J. Fischer. 1996. *Anthropology as Cultural Critique.* 2nd ed. Chicago: University of Chicago Press.
Marris, Emma. 2010. "Birds Flock Online." *Nature.* doi: 10.1038/news.2010.395. http://www.nature.com/news/2010/100810/full/news.2010.395.html.
Martin, Aryn, and Michael Lynch. 2009. "Counting Things and People: The Practices and Politics of Counting." *Social Problems* 56 (2): 243–66.
Martin, Emily. 1987. *The Woman in the Body.* Boston: Beacon.
———. 1994. *Flexible Bodies.* Boston: Beacon.
Masco, Joseph. 2006. *The Nuclear Borderlands.* Princeton, NJ: Princeton University Press.
Massaro, Joe. 2012. "Professor Ingraffea: The Next Monster under the Bed." *Energy in Depth,* September 26. http://energyindepth.org/marcellus/professor-ingraffea-the-next-monster-under-the-bed/.
Masuo, Y., M. Ishido, M. Morita, S. Oka. 2004. "Effects of Neonatal Treatment with 6-hydroxydopamine and Endocrine Disruptors on Motor Activity and Gene Expression in Rats." *Neural Plasticity* 11:59–76.
Mathes, Adam. 2004. "Folksonomies—Cooperative Classification and Communication through Shared Metadata." Ph.D. diss., Graduate School of Library and Information Science, University of Illinois at Urbana-Champaign.
Matz, Jacob, and Daniel Renfrew. 2014. "Selling 'Fracking': Energy in Depth and the Marcellus Shale." *Environmental Communication.* doi: 10.1080/17524032.2014.929157.
Maule, Alexis L., Colleen M. Makey, Eugene B. Benson, Isaac J. Burrows, and Madeleine K. Scammell. 2013. "Disclosure of Hydraulic Fracturing Fluid Chemical Additives: Analysis of Regulations." *New Solutions* 23 (1): 167–87.
Mayeux, Debra. 2011. "The Playground's Neighbor Is a Gas Well." *Talon* 19 (9): 1.

McCaughey, Martha, and Michael D. Ayers, eds. 2003. *Cyberactivism: Online Activism in Theory and Practice.* New York: Routledge.

McJeon, H., J. Edmonds, N. Bauer, et al. 2014. "Limited Impact on Decadal-Scale Climate Change from Increased Use of Natural Gas." *Nature* 514:482–85. doi: 10.1038/nature13837.

McKenzie, L., R. Guo, R. Z. Witter, D. A. Savitz, L. S. Newman, et al. 2014. "Birth Outcomes and Maternal Residential Proximity to Natural Gas Development in Rural Colorado." *Environmental Health Perspectives* 122 (4): 412–17.

McKenzie, L., R. Witter, L. Newman, and J. Adgate. 2012. "Human Health Risk Assessment of Air Emissions from Development of Unconventional Natural Gas Resources." *Journal of the Science of the Whole Environment* May 1 (424): 79–87.

McKibbin, M. 2004. "Wells Caused Rare Tumor, Woman Says." *Grand Junction Sentinel*, December 22.

McKinley, Ryan. 2003. "Open Government Information Awareness." Master's thesis, Department of Architecture, Program in Media Arts and Sciences, Massachusetts Institute of Technology.

McNeill, John Robert. 2000. *Something New under the Sun: An Environmental History of the Twentieth-Century World.* New York: W. W. Norton.

Meikle, Graham. 2002. *Future Active: Media Activism and the Internet.* London: Pluto.

———. 2008. "Whacking Bush: Tactical Media as Play." In *Digital Media and Democracy: Tactics in Hard Times,* edited by Megan Boler, 367–382. Cambridge, MA: MIT Press.

Meikle, Jeffery I. 1995. *American Plastics: A Cultural History.* New Brunswick, NJ: Rutgers University Press.

Meixsell, Tara. 2010. *Collateral Damage: A Chronicle of Lives Devastated by Gas and Oil Development and the Valiant Grassroots Fight to Effect Political and Legislative Change over the Impacts of the Gas and Oil Industry in the United States.* Lexington, KY: CreateSpace Independent Publishing.

Merleau-Ponty, Maurice. 2008. *Phenomenology of Perception.* London: Routledge.

Messaoud, Malik A., Mohammed Z. Boulegroun, Aziza Gribi, Rachid Kasmi, et al. 2005. "New Dimensions in Land Seismic Technology." *Oilfield Review* (autumn): 42–53.

Milbank, Dana, and Justin Blum. 2005. "Document Says Oil Chiefs Met with Cheney Task Force." *Washington Post*, November 16.

Miller, Daniel. 1998. *A Theory of Shopping.* Ithaca, NY: Cornell University Press.

Miller, Ellen. 2006. "City Hits Dry Hole in Drilling-Lease Bid; Grand Junction Tries to Buy Control of Watershed." *Rocky Mountain News*, February 10.

Miller, S. M., S. C. Wofsy, A. M. Michalak, E. A. Kort, et al. 2013. "Anthropogenic Emissions of Methane in the United States." *Proceedings of the National Academy of Sciences USA* 110 (50): 20018–22.

Milman, Oliver. 2014. "Full Scale of Plastic in the World's Oceans Revealed for First Time." *Guardian*, December 10. http://www.theguardian.com/environment/2014/dec/10/full-scale-plastic-worlds-oceans-revealed-first-time-pollution.

Mindell, David. 2002. *Between Human and Machine: Feedback, Control, and Computing before Cybernetics*. Baltimore: Johns Hopkins University Press.
———. 2008. *Digital Apollo: Human and Machine in the First Six Lunar Landings*. Cambridge, MA: MIT Press.
Minkler, Meredith, and Nina Wallerstein. 2008. *Community-Based Participatory Research for Health: From Process to Outcomes*. San Francisco: Jossey-Bass.
Mitchell, Timothy. 1991. *Colonizing Egypt*. Berkeley: University of California Press.
———. 2011. *Carbon Democracy: Political Power in the Age of Oil*. New York: Verso.
MITEI (MIT Energy Initiative). 2009a. "Founding Member Program," March 2. https://web.archive.org/web/20090302153000/http://web.mit.edu/mitei/support/founding.html.
———. 2009b. "Sustaining Member Program," March 2. https://web.archive.org/web/20090302153005/http://web.mit.edu/mitei/support/sustaining.html.
———. 2010. "The Future of Natural Gas: An Interdisciplinary MIT Study." Interim Report. http://web.mit.edu/ceepr/www/publications/Natural_Gas_Study.pdf.
———. 2011. "The Future of Natural Gas: An Interdisciplinary MIT Study." http://energy.mit.edu/publication/future-natural-gas/; https://web.archive.org/web/20111218170612/http://web.mit.edu/ceepr/www/publications/Natural_Gas_Study.pdf.
———. 2012. "About MITEI," August 20. https://web.archive.org/web/20120820104758/http://web.mit.edu/mitei/about/index.html.
———. 2014a. "MITEI Hydrocarbon Products and Processing." http://mitei.mit.edu/research/innovations/hydrocarbon-products-and-processing.
———. 2014b. "MITEI Members." http://mitei.mit.edu/about/members.
MIT LabCAST. 2009. "#36 extrACT." MIT Media Lab, February 25. http://labcast.media.mit.edu/?p=69.
MIT News. 1999. "MIT Receives $5 Million in Software from Schlumberger." http://web.mit.edu/newsoffice/1999/schlumberger.html.
———. 2007. "BP-MIT Partnership to Focus on Energy Conversion Technologies," September 25. http://newsoffice.mit.edu/2007/bp-energy-0925.html.
———. 2010. "MITEI-Led Study Offers Comprehensive Look at the Future of Natural Gas," June 25. http://newsoffice.mit.edu/2010/gas-report-0625.
MIT Press Release. 2010. "MIT Releases Major Report: The Future of Natural Gas," June 25. https://web.archive.org/web/20130922134515/http://web.mit.edu/press/2010/natural-gas.html.
Mobaldi, Steve. 2007. "Oil and Gas Exemptions from Environmental Health Protections." Hearing before the Committee on Oversight and Government Reform, House of Representatives, 110th Congress, 1st session, October 31. Serial No. 110–98.
Mol, Annemarie. 2002. *The Body Multiple: Ontology in Medical Practice*. Durham, NC: Duke University Press.
Moniz, Ernest. 2013. Testimony of Secretary Ernest Moniz, U.S. Department of Energy. House Committee on Energy and Commerce, Subcommittee on Energy and Power, June 13.
Montgomery, Carl T., and Michael B. Smith. 2010. "Hydraulic Fracturing: History of an Enduring Technology." *Journal of Petroleum Technology*, December.

Moore, C. W., B. Zielinska, G. Petron, and R. B. Jackson. 2014. "Air Impacts of Increased Natural Gas Acquisition, Processing and Use: A Critical Review." *Environmental Science Technology* 48 (15): 8349–59.
Morello-Frosch, R. A., M. Pastor, and J. Sadd. 2002. "Integrating Environmental Justice and the Precautionary Principle in Research and Policy-Making: The Case of Ambient Air Toxics Exposures and Health Risks among School Children in Los Angeles." *Annals of the American Academy of Political and Social Science* 584:47–68.
Morello-Frosch, R., M. Pastor, J. Sadd, C. Porras, and M. Prichard. 2005. "Citizens, Science and Data Judo: Leveraging Community-Based Participatory Research to Build a Regional Collaborative for Environmental Justice in Southern California." In *Methods for Community-Based Participatory Research for Health*, edited by Barbara Israel, Eugenia Eng, Amy Shultz, and Edith Parker. Hoboken, NJ: Jossey-Bass.
Moscou, Jim. 2008. "Oil and Gas Exploration: Is 'Fracking' Safe?" *Newsweek*, August 19.
Motta, Michael J. 2015. "Powering and Puzzling: Offshore Wind Energy Policy Innovation, Implementation, and Learning in Massachusetts." Diss., School of Public Policy and Urban Affairs, Northeastern University.
Moynihan, Colin. 2008. "City Subpoenas Creator of Text Messaging Code." *New York Times*, March 30.
Mufson, Steven. 2013. "Ernest Moniz, MIT Physicist, Nominated as Energy Secretary." *Washington Post*, March 4.
Mullarkey, Peter, Srinagesh Gavirneni, Grant Butler, and Douglas Morrice McCombs. 2007. "Schlumberger Uses Simulation in Bidding and Executing Land Seismic Surveys." *Interfaces* 37 (2): 120–32.
Munro, Robert, et al. 2013. *Aerial Damage Assessment for Hurricane Sandy.* Proceedings of the 10th International ISCRAM Conference. Baden-Baden, Germany.
Murphy, Michelle. 2004. "Uncertain Exposures and the Privilege of Imperception: Activist Scientists and Race at the U.S. Environmental Protection Agency." *Osiris*, 2nd ser., 19: 266–82.
———. 2006. *Sick Building Syndrome and the Politics of Uncertainty: Environmental Politics, Technoscience and Women Workers.* Durham, NC: Duke University Press.
Myers, J. P., F. S. vom Saal, B. T. Akingbemi, K. Arizono, et al. 2009. "Why Public Health Agencies Cannot Depend on Good Laboratory Practices as a Criterion for Selecting Data: The Case of Bisphenol A." *Environmental Health Perspectives* 117 (3): 309–15.
Myers, Natasha. 2015. *Rendering Life Molecular.* Durham, NC: Duke University Press.
Myhre, G., D. F.-M. Shindell, W. Bréon, J. Collins, et al. 2013. "Anthropogenic and Natural Radiative Forcing." In *Climate Change 2013: The Physical Science Basis. Contribution of Working Group I to the Fifth Assessment Report of the Intergovernmental Panel on Climate Change*, edited by T. F. Stocker, D. Qin, G.-K. Plattner, M. Tignor, S. K. Allen, J. Boschung, A. Nauels, Y. Xia, V. Bex, and P. M. Midgley. New York: Cambridge University Press.
Nace, T. 2003. *Gangs of America: The Rise of Corporate Power and the Disabling of Democracy.* San Francisco: Berrett-Koehler.

NACEPT (National Advisory Council for Environmental Policy and Technology). 2016. "Environmental Protection belongs to the Public: A Vision for Citizen Science at EPA," December. EPA-219-R-16-001.

Nader, Ralph, and Wesley J. Smith. 1998. *No Contest: Corporate Lawyers and the Perversion of Justice in America.* New York: Random House.

Nash, June. 1993. *We Eat the Mines and the Mines Eat Us: Dependency and Exploitation in Bolivian Tin Mines.* New York: Columbia University Press.

Nash, Linda. 2004. "The Fruits of Ill-Health: Pesticides and Workers' Bodies in Post–World War II California." *Osiris,* 2nd ser., 19:203–19.

———. 2006. *Inescapable Ecologies: A History of Environment, Disease and Knowledge.* Berkeley: University of California Press.

National Institute of Environmental Health Sciences. 2010. "Endocrine Disruptors." https://web.archive.org/web/20110320180525/http://www.niehs.nih.gov/research/resources/library/research/env_health/endocrine.cfm.

Negroponte, Nicholas P. 1995. *Being Digital.* New York: Knopf.

Neslin, David. 2009. "Understanding the COGCC Rulemaking." *IOGCC Mid-Year Summit: Stepping Lightly: Balancing Energy and the Environment,* May 11–13.

Neu, D., et al. 1998. "Managing Public Impressions: Environmental Disclosures in Annual Reports." *Accounting Organizations and Society* 23:265–82.

New York Times. 2001. "Alberta Energy Says It Is Buying Ballard Petroleum," January 19. http://www.nytimes.com/2001/01/19/business/company-news-alberta-energy-says-it-is-buying-ballard-petroleum.html.

———. 2009. "The Halliburton Loophole." Opinion, November 2. http://www.nytimes.com/2009/11/03/opinion/03tue3.html.

Ng, Marie, et al. 2014. "Global, Regional, and National Prevalence of Overweight and Obesity in Children and Adults during 1980–2013: A Systematic Analysis for the Global Burden of Disease Study 2013." *Lancet* 384 (9945): 766–81.

NHANES (National Health and Nutrition Examination Survey). 2009. "Fourth National Report on Human Exposure to Environmental Chemicals." Department of Health and Human Services, Centers for Disease Control and Prevention. http://www.cdc.gov/exposurereport/pdf/FourthReport.pdf.

———. 2013. "Fourth National Report on Human Exposure to Environmental Chemicals: Updated Tables." Department of Health and Human Services, Centers for Disease Control and Prevention, March. https://www.cdc.gov/exposurereport/pdf/FourthReport_UpdatedTables_Mar2013.pdf.

Niederer, Sabine, and José van Dijck. 2010. "Wikipedia as a Sociotechnical System: Wisdom of the Crowd or Technicity of Content?" *New Media and Society* 12:1368–87.

Nijhuis, Michelle. 2006. "How Halliburton Technology Is Wrecking the Rockies." *On Earth Magazine* (summer).

Nixon, Rob. 2011. *Slow Violence and the Environmentalism of the Poor.* Cambridge, MA: Harvard University Press.

Noble, David F. 1977. *America by Design: Science, Technology and the Rise of Corporate Capitalism.* Oxford: Oxford University Press.

NRDC (Natural Resources Defense Council). 2002a. "Cheney Energy Task Force." http://www.nrdc.org/air/energy/taskforce/searchinx.asp.

———. 2002b. "Hydraulic Fracturing of Coalbed Methane Wells: A Threat to Drinking Water." Comments on Senate Bill 1766, Section 604, January. https://web.archive.org/web/20061007101803/https://www.earthworksaction.org/pubs/200201_NRDC_HydrFrac_CBM.pdf.

Nutt, Amy Ellis. 2016. "Banned Chemical Still Used in Hospital IVs Is Linked to Attention Deficit Disorder." *Washington Post*, April 6.

O'Brien, Erin. 2009. "Drill, Baby, Drill." *Cleveland Scene*, September 30. http://www.clevescene.com/cleveland/drill-baby-drill/Content?oid=1659409.

ODNR (Ohio Department of Natural Resources). 2007. Personal letter to Kari Matsko, December 10. Received from Kari Matsko.

Office of Congressman Moran. 2009. "Kerry, Moran Respond to Phenomenon of Intersex Fish: Joint Bill Would Enable Gov't Action on Chemicals Found Dangerous to the Environment/Public Health." Press release, December 3.

Offshore. 2005. "Worldwide Seismic Vessel Survey." 65 (3): 68.

OGA (Ohio General Assembly). 2004. House Bill Number 278.

OGAP (Oil and Gas Accountability Project). 2006. "Disclosure of Chemicals Used in All Phases of Oil and Gas Development." Letter to Gary Baughman, Director, Hazardous Materials and Waste Management Division; Steve Gunderson, Director, Water Quality Control Division, CO; Margie Perkins, Director, Air Pollution Control Division, CO; Brian Macke, Director, Oil and Gas Conservation Commission; and Colorado Department of Natural Resources, CO, June 14.

———. 2007a. "Colorado Surface Owner Protection Act. National Precedent Set by Landowner Protection Bill Signed into Law by Colorado Governor Ritter Effective September 1," May 29.

———. 2007b. "New Mexico Passes Precedent-Setting Landowner Protection Act," June 27.

Ong, Aihwa, and Stephen J. Collier, eds. 2005. *Global Assemblages: Technology, Politics and Ethics as Anthropological Problems*. Oxford: Basil Blackwell.

Oreskes, Naomi, and Erik M. Conway. 2010. *Merchants of Doubt: How a Handful of Scientists Obscured the Truth on Issues from Tobacco Smoke to Global Warming*. New York: Bloomsbury.

OSC (Oil Spill Commission). 2011. "Deep Water: The Gulf Oil Disaster and the Future of Offshore Drilling." Report to the President, National Commission BP Deepwater Horizon Oil Spill and Offshore Drilling.

Ottinger, Gwen. 2010. "Buckets of Resistance: Standards and the Effectiveness of Citizen Science." *Science, Technology, and Human Values* 35 (2): 244–70.

Ottinger, Gwen, and Benjamin R. Cohen, eds. 2011. *Technoscience and Environmental Justice: Expert Cultures in a Grassroots Movement*. Cambridge, MA: MIT Press.

Oudshoorn, Nelly. 1994. *Making the Natural Body: An Archeology of Sex Hormones*. New York: Routledge.

Oudshoorn, Nelly, and T. J. Pinch, eds. 2003. *How Users Matter: The Co-construction of Users and Technology*. Cambridge, MA: MIT Press.

OVE (Ohio Valley Energy). 2007. Personal letter to Mark Scoville, August 20. http://www.landmanreportcard.com/lrc/review/21.
Owen, David L., et al. 2000. "The New Social Audits: Accountability, Managerial Capture or the Agenda of Social Champions?" *European Accounting Review* 9:81–87.
Page, K. A., et al. 2013. "Effects of Fructose vs. Glucose on Regional Cerebral Blood Flow in Brain Regions Involved with Appetite and Reward Pathways." *Journal of the American Medical Association* 309 (1): 63–70.
Paglen, Trevor. 2010. *Blank Spots on the Map: The Dark Geography of the Pentagon's Secret World.* New York: Penguin.
Paglen, Trevor, and A. C. Thompson. 2006. *Torture Taxi: On the Trail of the CIA's Rendition Flights.* Brooklyn, NY: Melville House.
Palanza, P., L. Giolosa, F. S. vom Saal, and S. Parmigiani. 2008. "Effects of Developmental Exposure to Bisphenol A on Brain and Behavior in Mice." *Environmental Research* 108 (2): 150–57.
Palfrey, John, and Catherine Bracy. 2011. "Review of the MIT Center for Future Civic Media." Knight Foundation. http://www.knightfoundation.org/media/uploads/publication_pdfs/Center-for-Future-Civic-Media-Assessment-Report-0621v1.pdf.
Palfrey, John, Jonathan Zittrain, Ron Deibert, and Rafal Rohozinski. 2008. *Access Denied: The Practice and Policy of Global Internet Filtering.* Cambridge, MA: MIT Press.
Papworth, Stuart. 2009. "Stepping Up Land Seismic." *E&P*, March.
Pauly, Philip. 1987. *Controlling Life: Jacques Loeb and the Engineering Ideal in Biology.* Oxford: Oxford University Press.
Pellow Naguib, David, and Robert J. Brulle. 2005. *Power, Justice and the Environment: A Critical Appraisal of the Environmental Justice Movement.* Cambridge, MA: MIT Press.
Peluso, Nancy, and Michael Watts. 2001. *Violent Environments.* Ithaca, NY: Cornell University Press.
Pennigroth, Stephen M., Matthew M. Yarrow, Abner X. Figueroa, Rebecca J. Bowen, and Soraya Delgado. 2013. "Community-Based Risk Assessment of Water Contamination from High-Volume Horizontal Hydraulic Fracturing." *New Solutions* 23 (1): 137–66.
Pennsylvania Department of Environmental Protection. 2000. "Pennsylvania's Plan for Addressing Problem Abandoned Wells and Orphaned Wells." Bureau of Oil and Gas Management. Doc# 550-0800-001, April 10. http://www.elibrary.dep.state.pa.us/dsweb/Get/Version-48262/.
Perin, Monica. 2002. "Oil Technology Spawns 'Titanic' Documentary." *Houston Business Journal*, August 11. http://www.bizjournals.com/houston/stories/2002/08/12/story8.html?page=all.
Perry, Simone. 2011. "Energy Consequences and Conflicts across the Global Countryside: North American Agricultural Perspectives." *Forum on Public Policy* 2:1–23.
———. 2012. "Development, Land Use, and Collective Trauma: The Marcellus Shale Gas Boom in Rural Pennsylvania." *Culture, Agriculture, Food, and Environment* 34 (1): 81–92.

———. 2013. "Using Ethnography to Monitor the Community Health Implications of Onshore Unconventional Oil and Gas Developments: Examples from Pennsylvania's Marcellus Shale." *New Solutions* 23 (1): 33–53.

Pétron, G., G. Frost, B. Miller, A. Hirsch, et al. 2012. "Hydrocarbon Emissions Characterization in the Colorado Front Range: A Pilot Study." *Journal of Geophysical Research: Atmospheres* 117: D4. doi: 10.1029/2011JD016360.

Pétron, G., A. Karion, C. Sweeney, B. R. Miller, et al. 2014. "A New Look at Methane and Non-Methane Hydrocarbon Emissions from Oil and Natural Gas Operations in the Colorado Denver-Julesburg Basin." *Journal of Geophysical Research: Atmospheres* 119 (11): 6836–52.

Phillips, Nathan, Robert Ackley, Eric R. Crosson, Adrian Down, Lucy R. Hutyra, Max Brondfield, et al. 2013. "Mapping Urban Pipeline Leaks: Methane Leaks across Boston." *Environmental Pollution* 173:1–4.

Pickerill, Jenny. 2003. *Cyberprotest: Environmental Activism On-line.* Manchester, UK: Manchester University Press.

Picou, J. S. 1996a. "Compelled Disclosure of Scholarly Research: Some Comments on 'High Stakes Litigation.'" *Law and Contemporary Problems* 59:149–57.

———. 1996b. "Sociology and Compelled Disclosure: Protecting Respondent Confidentiality." *Sociological Spectrum* 16:209–37.

———. 1996c. "Toxins in the Environment, Damage to the Community: Sociology and the Toxic Tort." In *Witnessing for Sociology: Sociologists in Court*, edited by P. J. Jenkins and S. Kroll-Smith, 212–24. Westport, CT: Praeger.

Piketty, Thomas. 2014. *Capital in the Twenty-First Century.* Translated by Arthur Goldhammer. Cambridge, MA: Harvard University Press.

Plumer, Brad. 2014. "Remarks of President Barack Obama—as Prepared for Delivery of State of the Union Address, Tuesday, January 28, 2014." *Washington Post.*

Pollan, Michael. 2001. *The Botany of Desire: A Plant's-eye View of the World.* New York: Random House.

PPDM (Professional Petroleum Data Management Agency). 2013. "The API Number Standard: Illustrations of API Number Assignment," May 31.

Pring, George W., and Penelope Canan. 1996. *SLAPPs: Getting Sued for Speaking Out.* Philadelphia: Temple University Press.

Proctor, R., and L. Schiebinger. 2008. *Agnotology: The Making and Unmaking of Ignorance.* Stanford, CA: Stanford University Press.

Quandt, S., T. Arcury, and A. Pell. 2001. "Something for Everyone? A Community and Academic Partnership to Address Farmworker Pesticide Exposure in North Carolina." *Environmental Health Perspectives* 109 (suppl. 3): 435–41.

Rabinow, Paul. 1996. *Making PCR.* Chicago: University of Chicago Press.

Rajak, Dinah. 2014. "Corporate Memory: Historical Revisionism, Legitimation and the Invention of Tradition in a Multinational Mining Company." *PoLAR: Political and Legal Anthropology Review* 37 (2): 259–80.

Ramirez-Andreotta, M. D., M. L. Brusseau, J. F. Artiola, R. M. Maier, and A. J. Gandolfi. 2015. "Building a Co-created Citizen Science Program with Gardeners

Neighboring a Superfund Site: The Gardenroots Case Study." *International Public Health Journal* 7 (1).

Rampton, Sheldon, and John Stauber. 2000. *Trust Us, We're Experts: How Industry Manipulates Science and Gambles with Your Future.* New York: Tarcher.

Randall, Neil. 1997. *The Soul of the Internet: Net Gods, Netizens and the Wiring of the World.* London: International Thomson Computer Press.

Rappaport, S. M., and M. T. Smith. 2010. "Epidemiology: Environment and Disease Risks." *Science* 330 (6003): 460–61.

Ratto, M., S. Wylie, and K. Jalbert. 2014. "Introduction to the Special Forum on Critical Making as Research Program." *Information Society* 30 (2): 85–95.

Readings, Bill. 1996. *The University in Ruins.* Cambridge, MA: Harvard University Press.

Reed, Stanley. 2008. "The Stealth Oil Giant." *Business Week*, January 14.

Reid, S., N. Greene, and A. Nogee. 2009. "Re: Cape Wind Offshore Wind Energy Project and AERU Regulations." Letter to Secretary of Interior, Ken Salazar, from the Union of Concerned Scientists, Natural Resources Defense Council, and Conservation Law Foundation, March 9.

Reuters. 2012. "Amid Fracking Boom, Unlicensed Middlemen Often Closing the Deals—Reuters Investigates," October 1. https://web.archive.org/web/20121228001552/http://www.reuters.com/video/2012/10/01/amid-fracking-boom-unlicensed-middlemen?videoId=238108258.

Rheinberger, Hans-Jörg. 1997. *Toward a History of Epistemic Things: Synthesizing Proteins in the Test Tube.* Stanford, CA: Stanford University Press.

Rheingold, Howard. 2002. *Smart Mobs: The Next Social Revolution.* London: Perseus.

Rimassa, Shawn M., Paul Howard, Bruce MacKay, Kristel Blow, and Noel Coffman. 2011. "Case Study: Evaluation of an Oxidative Biocide during and after a Hydraulic Fracturing Job in the Marcellus Shale." Society of Petroleum Engineers International Symposium on Oilfield Chemistry, April 11–13.

Roberts, Celia. 2007. *Messengers of Sex: Hormones, Biomedicine and Feminism.* Cambridge: Cambridge University Press.

Rodney, Walter. 1972. *How Europe Underdeveloped Africa.* London: Bogle-L'Ouverture.

Roosth, Sophia. 2010. "Crafting Life: A Sensory Ethnography of Fabricated Biologies." Ph.D. diss., MIT.

Rosner, David, and Gerald Markowitz. 2002. *Deceit and Denial: The Deadly Politics of Industrial Pollution.* Berkeley: University of California Press.

Ross, Michael. 2006. "A Closer Look at Oil, Diamonds, and Civil War." *Annual Review of Political Science* 9:265–300.

Roubini, Sonia, and Jason Tashea. 2014. "After Sunflower Movement, Taiwan's Gov Uses Open Source to Open the Government." *Tech President*, November 5. http://techpresident.com/news/wegov/25339/sunflower-movement-gov-taiwan-open-government.

Rowland, D., and J. Siegel. 2015. "Ohio's Oil-and-Gas Industry Donations, Ruling Tied?" *Columbus Dispatch*, February 22. http://www.dispatch.com/content/stories/local/2015/02/22/industry-donations-ruling-tied.html.

Ruhlen, R. L., J. A. Taylor, J. Mao, J. Kirkpatrick, W. V. Welshons, and F. S. vom Saal. 2011. "Choice of Animal Feed Can Alter Fetal Steroid Levels and Mask Developmental Effects of Endocrine Disrupting Chemicals." *Journal of Developmental Origins of Health and Disease* 2:36–48.

Rusek, Joan Cooper. 2009. "English Drive Homes Now Worthless, Banks Say." (Cleveland, OH) *West Geauga Sun*, May 28.

Russell, Edmund. 2001. *War and Nature: Fighting Humans and Insects with Chemicals from World War I to Silent Spring.* Cambridge: Cambridge University Press.

Saberi, Pouné. 2013. "Navigating Medical Issues in Shale Territory." *New Solutions* 23 (1): 209–21.

Saez, Emmanuel, and Gabriel Zucman. 2014. "Wealth Inequality in the United States Since 1913: Evidence from Capitalized Income Tax Data." Working Paper 20625. *National Bureau of Economic Research*, October. http://www.nber.org/papers/w20625.

Salazar, Ken. 2010. "Statement of Ken Salazar, Secretary of the Interior, before the Senate Committee on Environment and Public Works Regarding Offshore Oil and Gas Development," May 18.

Santiago, Myrna I. 2006. *The Ecology of Oil: Environment, Labor and the Mexican Revolution 1900–1938.* Cambridge: Cambridge University Press.

Santo, Alysia. 2011. "The Landman Cometh: Innovation Trail and Other New York Outlets Help Readers Prepare for Fracking Prospectors." *Columbia Journalism Review*, December 1. http://www.cjr.org/the_news_frontier/the_landman_cometh.php?page=all.

Savage, Charlie. 2008. "Sex, Drug Use and Graft Cited in Interior Department." *New York Times*, September 11.

Sawyer, Suzana. 2004. *Crude Chronicles: Indigenous Politics, Multinational Oil and Neoliberalism in Ecuador.* Durham, NC: Duke University Press.

Saxton, Dvera. 2014. "Strawberry Fields as Extreme Environments: The Ecobiopolitics of Farmworker Health." *Medical Anthropology: Cross-Cultural Studies in Health and Illness*. doi: 10.1080/01459740.2014.959167.

Schafer, Jackie. 2014. "Three Tanker Trucks Crash in Canton Township, Spilling Fracking Water, Diesel Fuel, Henderson Avenue Closed While Hazmat, DEP Teams Respond." *WTAE* (Pittsburgh Action News), April 21.

Scheper-Hughes, Nancy, and Margaret Lock. 1987. "The Mindful Body: A Prolegomenon to Future Work in Medical Anthropology." *Medical Anthropology Quarterly* 1:6–41.

Scheyder, Ernest. 2016. "Plains Crash: In North Dakota's Oil Patch, a Humbling Comedown." Reuters, May 18. http://www.reuters.com/investigates/special-report/usa-northdakota-bust/.

Schlumberger. 1998. "Schlumberger 1998 Third Quarter Earnings." *Business Wire*, October 22. http://www.thefreelibrary.com/Schlumberger+1998+Third+Quarter+Earnings.-a053111602.

———. 2015. "Schlumberger Announces Full-Year and Fourth Quarter 2014 Results." Schlumberger Investor Center. http://investorcenter.slb.com/phoenix.zhtml?c=97513&p=irol-newsArticle_print&ID=2008146.

Schneyer, Joshua, and Brian Grow. 2011. "Energy Giant Hid behind Shells in 'Land Grab.'" Reuters, December 28.

Schor, Elana. 2010. "Ingredients of Controversial Dispersants Used on Gulf Spill Are Secrets No More." *New York Times*, June 9.

Schulz, A., E. Parker, B. Israel, A. Becker, B. Maciak, and R. Hollis. 1998. "Conducting a Participatory Community-Based Survey for a Community Health Intervention on Detroit's East Side." *Journal of Public Health Management and Practice* 4 (2): 10–24.

Scott, Dayna Nadine. 2009. "'Gender-Benders': Sex and Law in the Constitution of Polluted Bodies." *Feminist Legal Studies* 17:241–65.

———. 2010. "Body Polluted: Questions of Scale, Gender, and Remedy." *Loyola of Los Angeles Law Review* 44:121–56.

Scott, James C. 1985. *Weapons of the Weak: Everyday Forms of Peasant Resistance.* New Haven, CT: Yale University Press.

———. 1998. *Seeing Like a State: How Certain Schemes to Improve the Human Condition Have Failed.* New Haven, CT: Yale University Press.

Seelye, Katharine. 2013. "Koch Brother Wages 12-Year Fight over Wind Farm." *New York Times*, October 22.

SEHN (Science and Environmental Health Network). 2000. "The Precautionary Principle" (January).

Sellers, Christopher C. 1997. *Hazards of the Job: From Industrial Disease to Environmental Health Science.* Chapel Hill: University of North Carolina Press.

Serres, Michel. 1982. *The Parasite*. Translated, with notes, by Lawrence R. Schehr. Baltimore: Johns Hopkins University Press.

Shabrawi, Ayman, et al. 2005. "How Single-Sensor Seismic Improved Image of Kuwait's Minagish Field." *First Break* 23:63–69.

Shane, Peter M., ed. 2004. *Democracy Online: The Prospects for Political Renewal through the Internet.* New York: Routledge.

Shapin, Steven, and Simon Schaffer. 1985. *Leviathan and the Air Pump.* Princeton, NJ: Princeton University Press.

Shapiro, Nick. 2015. "Attuning to the Chemosphere: Domestic Formaldehyde, Bodily Reasoning, and the Chemical Sublime." *Cultural Anthropology* 3 (30): 368–93.

Shell, Hanna. 2012. *Hide and Seek: Camouflage, Photography and the Media of Reconnaissance.* Cambridge, MA: MIT Press.

Shellenberger, Michael. 2011. "Interview with Dan Steward, Former Mitchell Energy Vice President." *Breakthrough Institute*, December 11. http://thebreakthrough.org/archive/interview_with_dan_steward_for.

Shellenberger, Michael, Ted Nordhaus, Alex Trembath, and Jesse Jenkins. 2012. "Where the Shale Gas Revolution Came From." *Breakthrough Institute*, May 23. http://thebreakthrough.org/index.php/programs/energy-and-climate/where-the-shale-gas-revolution-came-from/.

Sheppard, Kate. 2009. "Astroturf Wars Continue as More Info Comes to Light on 'Energy Citizen' Rallies." *Grist*, August 17. http://grist.org/article/2009-08-17-astroturf-wars-continue-api-energy-citizen-rallies/.

Shirky, Clay. 2008. *Here Comes Everybody.* New York: Penguin.
Siegler, Kirk. 2007. "Oil Company Isolates Workers to Fight Drug Abuse." National Public Radio, March 26. http://www.npr.org/templates/story/story.php?storyId=9138938.
Silverstein, Ken. 2013. "Energy Secretary Moniz Signals LNG Exports Will Soon Get Moving." *Forbes*, June 15. www.forbes.com/sites/kensilverstein/2013/06/15/energy-secretary-moniz-signals-lng-exports-will-soon-get-moving/.
Skone, Timothy. 2012. "Role of Alternative Energy Sources: Natural Gas Power Technology Assessment." National Energy Technology Laboratory, U.S. Department of Energy, June 30. http://www.netl.doe.gov/energy-analyses/pubs/NGTechAssess.pdf.
Skrtic, Lana. 2006. "Hydrogen Sulfide, Oil and Gas, and People's Health." Master's thesis, Energy and Resources Group, University of California, Berkeley.
Slater, Donald. 1996. *Consumer Culture and Modernity.* Oxford: Polity Press.
Smith, Grant. 2014. "U.S. Overtakes Saudi Arabia as World's Biggest Oil Producer." *Bloomberg News*, July 4.
Smith, Peter J., and Elizabeth Smythe. 2001. "Globalization, Citizenship, and Technology." In *Culture and Politics in the Information Age*, edited by F. Weber, 183–206. London: Routledge.
Snedegar, Jean. 2008. "Oil and Gas Drillers Pushing Landowners for Fast Deals." *Morning West Virginia*, TV broadcast.
Soja, Edward W. 1989. *Postmodern Geographies: The Reassertion of Space in Critical Social Theory?* London: Verso.
Song, Lisa. 2011. "MIT Web Tools Help Small Landowners Navigate Gas Leasing Frenzy." Reuters, May 1.
Soraghan, Mike. 2013a. "FracFocus Has 'Serious Flaws,' Harvard Study Says." EnergyWire, April 23. www.eenews.net/stories/1059979931.
———. 2013b. "One-Fifth of FracFocus Reports in Colo., Pa. Were Late in 2012." EnergyWire, June 7. www.eenews.net/stories/1059982441/print.
Soto, A. M., H. Justicia, J. W. Wray, and C. Sonnenschein. 1991. "P-Nonyl-Phenol: An Estrogenic Xenobiotic Released from 'Modified' Polystyrene." *Environmental Health Perspectives* 92:167–73.
Spaulding, S. 2006a. "City Outbid on Mesa Land." *Grand Junction Daily Sentinel*, February 10.
———. 2006b. "Landowners Can't Keep Up with Fast, High Bids." *Grand Junction Daily Sentinel*, February 10.
———. 2006c. "Water Runs Deep in Drilling Debates; Palisade Votes No on Drilling in Watershed." *Grand Junction Daily Sentinel*, January 11, 1A.
Stanhope, Kimber, et al. 2009. "Consuming Fructose-Sweetened, Not Glucose-Sweetened, Beverages Increases Visceral Adiposity and Lipids and Decreases Insulin Sensitivity in Overweight/Obese Humans." *Journal of Clinical Investigation* 119 (5): 1322–34.
Stauffer, Nancy. 2006. "MIT Math Model Could Aid Natural Gas Production." MIT Energy Initiative, November 21. http://mitei.mit.edu/news/mit-math-model-could-aid-natural-gas-production.

Steinberg, Theodore. 2002. *Down to Earth: Nature's Role in American History.* Oxford: Oxford University Press.
———. 2010. "Can Capitalism Save the Planet? On the Origins of Green Liberalism." *Radical History Review* 107:10–24.
Steingraber, Sandra. 1998. *Living Downstream: A Scientist's Personal Investigation of Cancer and the Environment.* New York: Vintage.
Steinzor, Nadia, Wilma Subra, and Lisa Sumi. 2012. "Gas Patch Roulette: How Shale Gas Development Risks Public Health in Pennsylvania." Earthworks/OGAP, October 18. http://www.earthworksaction.org/library/detail/gas_patch_roulette_full_report.
———. 2013. "Investigating Links between Shale Gas Development and Health Impacts through a Community Survey Project in Pennsylvania." *New Solutions* 23 (1): 55–83.
Stengle, Jamie. 2011. "Chemical Mixing Sparks Massive Texas Plant Fire." Associated Press, October 4. http://www.chem.info/news/2011/10/chemical-mixing-sparks-massive-texas-plant-fire.
Stern, Nancy. 1981. *From ENIAC to UNIVAC: An Appraisal of the Eckert-Mauchly Computers.* Clifton, NJ: Digital Press.
Stewart, Thomas E. 2005. "Testimony Regarding Ohio Energy Policy." Ohio Senate Committee on Energy and Public Utilities, May 3.
———. 2007. "Testimony." 127th General Assembly, Ohio House of Representatives Alternative Energy Committee, June 20.
Stilgoe, John. 1985. *Metropolitan Corridor: Railroads and the American Scene.* New Haven, CT: Yale University Press.
Stine, Jeffrey K., and Joel A. Tarr. 1998. "At the Intersection of Histories: Technology and the Environment." *Technology and Culture* 39 (4): 601–40.
Stoll, Steven. 1998. *The Fruits of Natural Advantage: Making the Industrial Countryside in California.* Berkeley: University of California Press.
Strauss, Sarah, Stephanie Rupp, and Thomas Love, eds. 2013. *Cultures of Energy: Power, Practices, Technologies.* Walnut Creek, CA: Left Coast Press.
Stromberg, Joseph. 2013. "Radioactive Wastewater from Fracking Is Found in a Pennsylvania Stream." *Smithsonian.com*, October 2. http://www.smithsonianmag.com/science-nature/radioactive-wastewater-from-fracking-is-found-in-a-pennsylvania-stream-351641/?no-ist.
Subra, Wilma. 2014. "Sasol North America, Inc.—Two New Very Large Projects in an Environmental Justice Area of Calcasieu Parish Resulting in Increased Air Pollution and Increased Risk of Facility Accidents." Subra Company, December 22.
Suchman, Lucy. 1987. *Plans and Situated Actions: The Problem of Human-Machine Communication.* New York: Cambridge University Press.
———. 2007. "Anthropology as 'Brand': Reflections on Corporate Anthropology." Lancaster University Paper, Colloquium on Interdisciplinarity and Society. Oxford University, February 24.
Suchman, Lucy, and Libby Bishop. 2000. "Problematizing 'Innovation' as a Critical Project." *Technology Analysis and Strategic Management* 12 (3): 327–33.

Suchman, Lucy, Jeanette Blomberg, Julian Orr, and Randall Trigg. 1999. "Reconstructing Technologies as Social Practice." *American Behavioral Scientist* 43 (3): 392–408.
Sui, Daniel Z. 2008. "The Wikification of GIS and Its Consequences: Or Angelina Jolie's New Tattoo and the Future of GIS." *Computers, Environment and Urban Systems* 32:1–5.
Sumi, Lisa. 2005. "Oil and Gas at Your Door? A Landowner's Guide to Oil and Gas Development." Washington, DC: Earthworks/OGAP.
———. 2006. "Summary of Recent Incidents Involving the Release of Chemicals." Research Director, OGAP, June 28.
———. 2008. "Shale Gas: Focus on the Marcellus Shale." Earthworks/OGAP, May.
———. 2012. "Enforcement Report-COGCC: Inadequate Enforcement Means Current Colorado Oil and Gas Development Is Irresponsible." Earthworks, March 20. http://www.earthworksaction.org/library/detail/enforcement_report_cogcc.
Surowiecki, James. 2004. *The Wisdom of Crowds*. New York: Little, Brown.
Swanson, Ana. 2014. "America's Oil Boom Is Visible from Space." *Washington Post*, October 20. http://www.washingtonpost.com/blogs/wonkblog/wp/2014/10/20/americas-oil-boom-is-visible-from-space/.
Swinstead, Nick. 1999. "A Better Way to Work." *Oilfield Review* 11 (3).
Szott, Randall. 2006. "What Is Neogeography Anyway?" *Platial News and Neogeography*.
Taussig, Michael. 1986. *Shamanism, Colonialism and the Wild Men: A Study in Terror and Healing*. Chicago: University of Chicago Press.
———. 1993. *Mimesis and Alterity: A Particular History of the Senses*. New York: Routledge.
TEDX (The Endocrine Disruption Exchange). 2005. "End of Year Report."
———. 2006. "Chemicals Used in Natural Gas Extraction," April 25.
———. 2007. "Number of Chemicals Detected in Reserve Pits for 6 Wells in New Mexico That Appear on National Toxic Chemicals Lists." Amended document, November 15.
———. 2008. "Number of Chemicals Detected in Reserve Pits for 6 Wells in New Mexico that Appear on National Toxic Chemicals Lists." Amended document, November.
———. 2009a. "Chemicals Used in Natural Gas Development in Colorado," April 20.
———. 2009b. "Products and Chemicals Used in Fracturing," February 16.
Thompson, Nato, ed. 2009. *Experimental Geography: Radical Approaches to Landscape, Cartography, and Urbanism*. Brooklyn, NY: Melville House.
Thompson, Nato, and Gregory Sholette. 2004. *The Interventionists: Users' Manual for the Creative Disruption of Everyday Life*. North Adams, MA: Mass MoCA Publications.
Thornton, Joseph. 2001. "Evolution of Vertebrate Steroid Receptors from an Ancestral Estrogen Receptor by Ligand Exploitation and Serial Genome Expansions." *Proceedings of the National Academy of Sciences* 98 (10): 5671–76.

Timmer, J. 2010. "Galaxy Zoo Shows How Well Crowdsourced Citizen Science Works." *Ars Technica*, October 26.
Tollefson, Jeff. 2012. "Air Sampling Reveals High Emissions from Gas Fields." *Nature News*, February 7. www.nature.com/news/air-sampling-reveals-high-emissions-from-gas-field-1.9982.
———. 2013. "Methane Leaks Erode Green Credentials of Natural Gas." *Nature* 493 (January 2). doi: 10.1038/493012a.
Travis, Anthony. 1993. *The Rainbow Makers: The Origins of the Synthetic Dyestuffs Industry in Western Europe*. Cranbury, NJ: Associated University Press.
Traweek, Sharon. 1988. *Beamtimes and Lifetimes: The World of High Energy Physicists*. Cambridge, MA: Harvard University Press.
Traynor, M. 1996. "Countering the Excessive Subpoena for Scholarly Research." *Law and Contemporary Problems* 59:119–48.
Truesdell, Leon E. 1965. *The Development of Punch Card Tabulation in the Bureau of the Census, 1890–1940: With Outlines of Actual Tabulation Programs*. Washington, DC: U.S. Government Printing Office.
Tsing, Anna. 2004. *Friction: An Ethnography of Global Connection*. Princeton, NJ: Princeton University Press.
Tufte, E. R. 1997. *Visual and Statistical Thinking: Displays of Evidence for Decision Making*. Cheshire, CT: Graphics Press.
Turkle, Sherry. 2009. *Simulation and Its Discontents*. Cambridge, MA: MIT Press.
Tuz, Susan. 2004. "Schlumberger Doll Research Plans to Move." *News Times*, January 16. http://www.newstimes.com/news/article/Schlumberger-Doll-Research-plans-to-move-42471.php.
Tyl, R. W., C. Myers, M. Marr, C. S. Sloan, N. Castillo, M. M. Veselica, et al. 2008a. "Two-Generation Reproductive Toxicity Study of Dietary Bisphenol A (BPA) in CD-1 (Swiss) Mice." *Toxicological Sciences* 104:362–84.
———. 2008b. "Two-Generation Reproductive Toxicity Evaluation of Dietary 17β-estradiol (E2; CAS No. 50–28–2) in CD-1 (Swiss) Mice." *Toxicological Sciences* 102 (2): 392–412.
Urbina, Ian. 2010. "U.S. Said to Allow Drilling without Needed Permits." *New York Times*, May 13.
———. 2011a. "EPA Calls for More Testing of Pennsylvania Rivers." *New York Times*, March 8.
———. 2011b. "EPA Struggles to Regulate Natural Gas Industry." *New York Times*, March 4.
———. 2011c. "Gas Drillers Recycle Wastewater, but Risks Remain." *New York Times*, March 2.
———. 2011d. "Natural Gas and Polluted Air." Video report in "Regulation Lax as Gas-Wells' Tainted Water Hits Rivers." *New York Times*, February. http://www.nytimes.com/2011/02/27/us/27gas.html?_r=1&hp.
———. 2011e. "Millions of Gallons of Hazardous Chemicals Injected into Wells, Report Says: Oil and Gas Companies Put Toxic Chemicals into Wells in a Drilling

Process Known as Hydraulic Fracturing, According to a Congressional Study." *New York Times*, April 17.

———. 2011f. "Pennsylvania Calls for More Water Tests: Regulators Want Waste Treatment Plants and Drinking Water Facilities to Look for Radioactive Pollutants and Other Contaminants Resulting from the Growth of Natural Gas Drilling." *New York Times*, April 8.

———. 2011g. "Regulation Lax as Gas Wells' Tainted Water Hits Rivers." *New York Times*, February 27.

Urbina, Ian, and Jo McGinty. 2011. "Learning Too Late of the Perils in Gas Well Leases." *New York Times*, December 1. http://www.nytimes.com/2011/12/02/us/drilling-down-fighting-over-oil-and-gas-well-leases.html?pagewanted=3&_r=2&hp.

Vaidhyanathan, Siva. 2011. *The Googlization of Everything and Why We Should Worry.* Berkeley: University of California Press.

Vandenberg, L. N., T. Colborn, T. B. Hayes, J. J. Heindel, et al. 2012. "Hormones and Endocrine-Disrupting Chemicals: Low-Dose Effects and Non-Monotonic Dose Responses." *Endocrine Reviews* 33 (3): 378–455.

Van Sant, Gus, dir. 2012. *Promised Land.* Focus Features. Film.

Verran, Helen, and Michael Christie. 2007. "Using/Designing Digital Technologies of Representation in Aboriginal Australian Knowledge Practices." *Human Technology* 3 (2): 214–27.

Verstraete, S., I. Vanhorebeek, A. Covaci, F. Güiza, G. Malarvannan, P. Joren, G. Van den Berghe. 2016. "Circulating Phthalates during Critical Illness in Children Are Associated with Long-Term Attention Deficit: A Study of a Development and a Validation Cohort." *Intensive Care Medicine* 42:379–92.

Viner, Brian. 2013. "Why the World Isn't Running out of Oil." *Telegraph*, February 19. http://www.telegraph.co.uk/news/earth/energy/oil/9867659/Why-the-world-isnt-running-out-of-oil.html.

Vitalis, Robert. 2007. *America's Kingdom: Myth-making on the Saudi Oil Frontier.* Stanford, CA: Stanford University Press.

Vogel, Sarah A. 2008a. "Battles over Bisphenol A." DefendingScience.org, May 16.

———. 2008b. "From 'the Dose Makes the Poison' to 'the Timing Makes the Poison': Conceptualizing Risk in the Synthetic Age." *Environmental History* 13 (4): 667–73.

———. 2009. "The Politics of Plastics: The Making and Unmaking of Bisphenol A 'Safety.'" *American Journal of Public Health* 99:S559–S566.

———. 2012. *Is It Safe? BPA and the Struggle to Define the Safety of Chemicals.* Berkeley: University of California Press.

vom Saal, Frederick C., and Claude Hughes. 2005. "An Extensive New Literature Concerning Low-Dose Effects of Bisphenol A Shows the Need for a New Risk Assessment." *Environmental Health Perspectives* 113 (8).

vom Saal, Frederick C., C. A. Richter, J. Mao, and W. V. Welshons. 2005. "Commercial Animal Feed: Variability in Estrogenic Activity and Effects on Body

Weight in Mice." *Birth Defects Research Part A: Clinical and Molecular Tetralogy* 73:474–75.

Wade, Louise C. 1967. "The Heritage from Chicago's Early Settlement Houses." *Journal of the Illinois State Historical Society* 60 (4): 411–41.

Wagman, David. 2006. "Tight Gas Is Hitting Its Stride." *Oil and Gas Investor*, March 12.

Walker, M. L., C. E. Shuchart, Y. G. Yaritz, and L. R. Norman. 1995. "Effects of Oxygen on Fracturing Fluids." *Society of Petroleum Engineers*, January 1. doi: 10.2118/28978-MS.

Walley, Chris. 2009. "Deindustrializing Chicago: A Daughter's Story." In *The Insecure American*, edited by Hugh Gusterson and Catherine Besteman, 113–39. Berkeley: University of California Press, 2009.

———. 2013. *Exit Zero: Family and Class in Post-Industrial Chicago.* Chicago: University of Chicago Press.

Wang, C., and M. Burris. 1997. "Photovoice: Concept, Methodology and Use for Participatory Needs Assessment." *Health Education and Behavior* 24 (3): 369–87.

Wang, Marian. 2010. "Experts: Gulf Workers' Levels of Chemical Exposure May Be 'Perfectly Legal, but Not Safe.'" ProPublica, June 11.

Wardrip-Fruin, N., and N. Monfort, eds. 2003. *The New Media Reader.* Cambridge, MA: MIT Press.

Warf, Barney, and John Grimes. 1997. "Counterhegemonic Discourses and the Internet." *Geographical Review* 87:259–74.

Warner, N. R., C. A. Christie, R. B. Jackson, and A. Vengosh. 2013. "Impacts of Shale Gas Wastewater Disposal on Water Quality in Western Pennsylvania." *Environmental Science and Technology* 47 (20): 11849–57.

Watts, Michael. 2003. "Economies of Violence: More Oil, More Blood." *Economic and Political Weekly* 38 (48): 5089–99.

———. 2005. "Righteous Oil? Human Rights, the Oil Complex and Corporate Social Responsibility." *Annual Review of Environment and Resources* 30:9.1–9.35 (373–407).

Watts, Michael, and Ed Kashi. 2008. *Curse of the Black Gold: 50 Years of Oil in the Niger Delta.* New York: PowerHouse.

Waxman, Henry. 2007. "Oil and Gas Development. Exemptions from Health and Environmental Protections Hearing before the Committee on Oversight and Government Reform." House Committee on Oversight and Government Reform.

Webb, Dennis. 2004. "Commission: EnCana Fine 'Sizable.'" *Glenwood Springs Post Independent*, August 16.

———. 2014. "Researcher Tries to Reboot Website for Drilling Concerns." *Grand Junction Sentinel*, April 9. http://www.gjsentinel.com/special_sections/articles/researcher-tries-to-reboot-website-for-drilling-co.

Webb, Ellen, Sheila Bushkin-Bedient, Amanda Cheng, Christopher D. Kassotis, Victoria Balise, and Susan C. Nagel. 2014. "Developmental and Reproductive Effects of Chemicals Associated with Unconventional Oil and Natural Gas Operations." *Reviews on Environmental Health* 29 (4). doi: 10.1515/reveh-2014-0057.

Weber, Max. 1968. *Economy and Society: An Outline of Interpretive Sociology.* New York: Bedminster Press.

Weinhold, B. 2012. "The Future of Fracking: New Rules Target Air Emissions for Cleaner Natural Gas Production." *Environmental Health Perspectives* 120 (7): a272–a279.

Welsh, Jane. 2008. "Memorandum Concerning Seismic Testing in Highway Rights of Way." Chenango County Farm Bureau, NY.

Wertz, Joe. 2014. "Fracking Site Operator Faces Contempt Complaint after Acid Spill." National Public Radio, August 14. http://stateimpact.npr.org/oklahoma/2014/08/14/fracking-site-operator-faces-contempt-complaint-after-acid-spill/.

Wethe, David. 2009. "Schlumberger Presses for Shale-Gas Openness as Regulation Looms." *Bloomberg News*, September 29.

WGAL.com. 2010. "Halliburton Identifies Hydraulic Fracturing Chemicals: Website Lists Some of Commonly Used Chemicals," November 15.

White, Richard. 1991. *It's Your Misfortune and None of My Own: A New History of the American West.* Norman: University of Oklahoma Press.

———. 1995. *The Organic Machine: The Remaking of the Columbia River.* New York: Hill and Wang.

Wiist, William H. 2011. "Citizens United, Public Health, and Democracy: The Supreme Court Ruling, Its Implications, and Proposed Action." *American Journal of Public Health* 101 (7): 1172–79.

Wilber, Tom. 2012. *Under the Surface: Fracking, Fortunes and the Fate of the Marcellus Shale.* Ithaca, NY: Cornell University Press.

Wild, C. 2005. "Complementing the Genome with an 'Exposome': The Outstanding Challenge of Environmental Exposure Measurement in Molecular Epidemiology." *Cancer Epidemiology Biomarkers and Prevention* 14 (8): 1847–50.

Willis, David. 2008. "Gas Boom." BBC *World News America*, April 30. http://www.youtube.com/watch?v=y5iSPFbj6Zc.

Willow, A., and S. Wylie. 2014. "Politics, Ecology and the New Anthropology of Energy: Exploring the Emerging Frontiers of Hydraulic Fracking." *Journal of Political Ecology* 21:222–36.

Wills, Jonathan. 2000. "Muddied Waters: A Survey of Offshore Oilfield Drilling Wastes and Disposal Techniques to Reduce the Ecological Impact of Sea Dumping." *Sakhalin Environment Watch*, May 25.

Wills, Thomas. 2006. "In Depth: The Smell of Malathion in the Evening . . ." *North Fork Merchant Herald*, July 18–August 21, 3–5.

Wilson, Weston. 2004. Personal letter to Representative DeGette, Allard, Campbell, October 8.

Wing, Steve. 2002. "Social Responsibility and Research Ethics in Community-Driven Studies of Industrialized Hog Production." *Environmental Health Perspectives* 110:437–44.

Winichakul, Thongchai. 1994. *Siam Mapped: A History of the Geo-Body of a Nation.* Honolulu: University of Hawaii Press.

Winner, Langdon. 1980. "Do Artifacts Have Politics?" *Daedalus* 109 (1): 121–36.

Winston, Patrick Henry, and Karen A. Prendergast. 1986. *The AI Business: Commercial Uses of Artificial Intelligence.* Cambridge, MA: MIT Press.
Worster, Donald. 1990. "Toward an Agroecological Perspective in History." *Journal of American History* 76(4): 1087–106.
———. 1992. *Under Western Skies: Nature and History in the American West.* New York: Oxford University Press.
WSERC (Western Slope Environmental Resource Council). 2005. "Comments: Environmental Assessment Spaulding Peak Natural Gas Exploration and Development Area Wide Plan," November 25.
Wylie, Sara. 2011a. "Corporate Bodies and Chemical Bonds: An STS Analysis of Natural Gas Development in the U.S." Ph.D. diss., Department of History, Anthropology, and Science, Technology and Society Program, MIT.
———. 2011b. "Hormone Mimics and Their Promise of Significant Otherness." *Science as Culture*, 1–28.
Wylie, Sara, and L. Albright. 2014. "WellWatch: Reflections on Designing Digital Media for Multi-Sited Para-Ethnography." *Journal of Political Ecology* 21:320–48.
Wylie, Sara, K. Jalbert, S. Dosemagen, and M. Ratto. 2014. "Institutions for Civic Technoscience: How Critical Making Is Transforming Environmental Research." *Information Society* 30 (2): 116–26.
Wylie, Sara, and Deborah Thomas. 2014. "New Tools for Detecting and Communicating Environmental Exposures and Risks Associated with Oil and Gas Extraction." Partnerships for Environmental Public Health, National Institute of Environmental Health Science, Annual Meeting, September 22.
Wylie, Sara, Kim Schultz, Deborah Thomas, Chris Kassotis, and Susan Nagel. 2016. "Inspiring Collaboration: The Legacy of Theo Colborn's Trans-disciplinary Research on Fracking." *NEW SOLUTIONS: A Journal of Environmental and Occupational Health Policy* 26 (3): 360–88.
Yang, Q. 2010. "Gain Weight by 'Going Diet'? Artificial Sweeteners and the Neurobiology of Sugar Cravings: Neuroscience." *Yale Journal of Biology and Medicine* 83 (2): 101–8.
Yergin, Daniel. 1991. *The Prize: The Epic Quest for Oil, Money and Power.* New York: Free Press.
Yoquinto, Luke. 2014. "The Robots of Resistance." *Big Roundtable*, October 9.
Zalik, Anna. 2004. "The Niger Delta: Petro-Violence and Partnership Development." *Review of African Political Economy* 101 (4): 401–24.
———. 2008. "Oil Sovereignties: Ecology and Nationality in the Nigerian Delta and the Mexican Gulf." In *Extractive Economies and Conflicts in the Global South,* edited by K. Omeje, 181–198. Ashgate, VT: Burlington.
———. 2009. "Zones of Exclusion: Offshore Extraction, the Contestation of Space and Physical Displacement in the Nigerian Delta and the Mexican Gulf." *Antipode* 41 (3): 557–82.
Zimmer, Michael. 2009. "Renvois of the Past, Present and Future: Hyperlinks and the Structuring of Knowledge from the Encyclopédie to Web 2.0." *New Media and Society* 11:95–113.

Zittrain, Jonathan. 2006. "The Generative Internet." *Harvard Law Review* 119 (7): 1974–2040.

———. 2008. *The Future of the Internet and How to Stop It.* New Haven, CT: Yale University Press.

Zuckerman, Ethan. 2008. "Digital Activists Find Ways to Help Kenya." http://www.ethanzuckerman.com/blog/2008/01/09/digital-activists-find-ways-to-help-kenya.

Zuckerman, Gregory. 2013. *The Frackers: The Outrageous Inside Story of the New Billionaire Wildcatters.* New York: Penguin.

INDEX

Page numbers followed by *f* indicate illustrations.